NUCLEIC ACIDS HYBRIDIZATION

Nucleic Acids Hybridization
Modern Applications

Edited by

ANTON BUZDIN

Shemyakin-Ovchinnikov,
Institute of Bioorganic Chemistry,
Moscow, Russia

and

SERGEY LUKYANOV

Shemyakin-Ovchinnikov,
Institute of Bioorganic Chemistry,
Moscow, Russia

A C.I.P. Catalogue record for this book is available from the Library of Congress.

ISBN 978-1-4020-6039-7 (HB)
ISBN 978-1-4020-6040-3 (e-book)

Published by Springer,
P.O. Box 17, 3300 AA Dordrecht, The Netherlands.

www.springer.com

Printed on acid-free paper

Cover idea is the courtesy of Dr. Lilia M. Ganova-Raeva

All Rights Reserved
© 2007 Springer
No part of this work may be reproduced, stored in a retrieval system, or transmitted in any form or by any means, electronic, mechanical, photocopying, microfilming, recording or otherwise, without written permission from the Publisher, with the exception of any material supplied specifically for the purpose of being entered and executed on a computer system, for exclusive use by the purchaser of the work.

CONTENTS

Contributors xi

Acknowledgments xv

Preface xvii

1. Nucleic Acids Hybridization: Potentials and Limitations 1
   Anton A. Buzdin
   1. Introduction 2
   2. Cloning the Differences: Subtractive Hybridization 4
      2.1 Birth of a Method 4
      2.2 PCR-assisted Subtractive Hybridization 8
      2.3 First Worldwide Success: Representational
          Differences Analysis 10
      2.4 Further Improvements: Suppression Subtractive
          Hybridization, Polymerase Chain Reaction Suppression
          Effect, and Normalization of cDNA Libraries 13
      2.5 Covalently Hybridized Subtraction, Primer Extension
          Enrichment Reaction, and Other Promising Approaches
          in Subtractive Hybridization 18
   3. Finding Common DNA: Coincidence Cloning 20
   4. Hybridization in Solution for the Recovery of
      Genomic Polymorphisms 22
   5. Conclusions 24
   References 25

2. Selective Suppression of Polymerase Chain Reaction
   and Its Most Popular Applications 29
   *Sergey A. Lukyanov, Konstantin A. Lukyanov, Nadezhda G. Gurskaya,
   Ekaterina A. Bogdanova, Anton A. Buzdin*
   1. Introduction 30
   2. Selective PCR Suppression 30
   3. Preparation of ITR-Containing DNA Samples 32
   4. Strategy of Employment of SSP 33
   5. SSP-based Procedures 33
      5.1 Construction of cDNA Libraries from a Small Amount
          of Biological Material 35
      5.2 Detection of Differentially Expressed Genes 36
      5.3 Search for 5′- and 3′-Terminal Fragments of cDNA 40
      5.4 Search for Promoter Sites (Chromosome Walking) 40
      5.5 *In Vitro* Cloning 41
      5.6 Multiplex PCR 42
   6. Detailed Protocols 43
      6.1 Materials 43
      6.2 Methods 45
   References 48

3. Suppression Subtractive Hybridization 53
   Sergey A. Lukyanov, Denis Rebrikov, Anton A. Buzdin
   1. Introduction 54
   2. The Principle of Suppression Subtractive Hybridization 55
   3. The Principle of Mirror-Oriented Selection 57
   4. Technical Comments and Considerations 59
      4.1 Normalization Step and the Efficiency of SSH 59
      4.2 Differential Screening of the Subtracted Libraries 61
      4.3 Starting RNA Materials 61
      4.4 Size of cDNA Fragments 62
   5. Detailed Protocols: Materials 63
      5.1 Oligonucleotides 63
      5.2 Buffers and Enzymes 63
   6. Detailed Protocols: Methods 65
      6.1 Preparation of Subtracted cDNA or Genomic
          DNA Library 65
      6.2 Mirror Orientation Selection 70
      6.3 Cloning of Subtracted cDNAs 74
      6.4 Differential Screening of the Subtracted cDNA Library . 74
   7. Notes 77
   References 83

Contents

4. Stem-Loop Oligonucleotides as Hybridization Probes
   and Their Practical Use in Molecular Biology and Biomedicine 85
   Anton A. Buzdin, Sergey A. Lukyanov
   1. Introduction 86
   2. Molecular Beacons 87
   3. Stem-Loop DNA Probes on Microarrays 92
   4. Conclusions 93
   References 94

5. Normalization of cDNA Libraries 97
   Alex S. Shcheglov, Pavel A. Zhulidov, Ekaterina A. Bogdanova, Dmitry A. Shagin
   1. Introduction 98
      1.1 Normalized cDNA Libraries: What are they needed for? 98
      1.2 Evaluation of cDNA Library Normalization Efficacy 100
      1.3 Basic Approaches to Generate Normalized
           cDNA Libraries 100
   2. Methods of cDNA Normalization 102
      2.1 cDNA Normalization by Means of Hydroxyapatite
           Column Chromatography 102
      2.2 Generation of Normalized Full-Length-Enriched cDNA
           Libraries Using DNA Immobilization on a Solid Support 104
      2.3 Normalization of Full-Length cDNA with the
           use of Biotinylated RNA 105
      2.4 Normalization of Fragmented cDNA by Means
           of Selective Amplification 107
      2.5 cDNA Normalization Using Frequent-Cutter
           Restriction Enzymes 109
      2.6 Normalization of Full-Length-Enriched cDNA
           with Duplex-Specific Nuclease (DSN Normalization) 109
   3. cDNA Preparation for DSN Normalization 111
      3.1 RNA Requirements 111
      3.2 cDNA Synthesis 112
      3.3 cDNA Purification 113
   4. DSN Normalization Protocol 113
      4.1 Materials 113
      4.2 cDNA Precipitation 114
      4.3 Hybridization 115
      4.4 DSN Treatment 116
      4.5 First Amplification of Normalized cDNA 117
      4.6 Preliminary Analysis of the Normalization Results 119
      4.7 Second Amplification of Normalized cDNA 121
   References 122

6. Primer Extension Enrichment Reaction (PEER)
   and Other Methods for Difference Screening 125
   Lilia M. Ganova-Raeva
   1. Introduction 127
   2. Primer Extension Enrichment Reaction 128
      2.1 Method Outline 128
      2.2 Discussion 132
      2.3 PEER Protocol 135
   3. Other Subtraction and Hybridization Based Methods
      for Difference Screening 137
      3.1 Differential Screening 137
      3.2 Subtractive Hybridization 139
      3.3 Subtractive Cloning 140
      3.4 Differential Display 140
      3.5 AFLP, SAGE/CAGE, GSTs, and DARFA 142
      3.6 Representational Differences Analysis 143
      3.7 SPAD–RDA 146
      3.8 Enzymatic Degradation Subtractions (EDS, LCS,
          DSC, NSC, UDG/USA, and CODE) 146
      3.9 Suppression Subtraction Hybridization 150
      3.10 Selective Amplification Via Biotin and
           Restriction-Mediated Enrichment 153
      3.11 DNA Enrichment by Allele-Specific
           Hybridization 154
      3.12 Methods Combining the use of SSH
           and Microarrays 155
      3.13 Conclusions 156
   References 157

7. Subtractive Hybridization with Covalently
   Modified Oligonucleotides 167
   Shi-Lung Lin, Donald Chang, Joseph D. Miller, Shao-Yao Ying
   1. Introduction 168
   2. Subtractive Hybridization Methods 169
   3. Covalent Modification 171
   4. Subtractive Hybridization with Covalently Modified
      Subtracters 176
   5. Applications 179
   6. Protocols 180
      6.1 Preparation of Subtracter and Tester DNA Libraries 180
      6.2 Covalent Modification of Subtracter DNAs 180
      6.3 Subtractive Hybridization and CHS–PCR Amplification ... 181

Contents

	6.4 Covalent Binding Efficiency and Subtractive Stringency of CHS	183
	6.5 Identification of Genomic Deletion Using CHS	184
	References	185

8. Coincidence Cloning: Robust Technique for Isolation of Common Sequences 187
   Anton A. Buzdin
   1. Introduction 188
   2. Cloning Selection of Heteroduplexes 191
   3. Physical Separation of Hybrids 193
   4. PCR-only-based Approaches 193
   5. Future Prospects 202
   6. Protocols 202
      6.1 Cloning Similarities in Genomic DNAs 202
      6.2 Cloning and Presice Mapping of Transcribed Repetitive Elements 204
      6.3 Finding Methylated or Unmethylated CpGs in Large Genomic Contigs 207
   References 209

9. DNA Hybridization in Solution for Mutation Detection 211
   Anton A. Buzdin
   1. Introduction 212
   2. Chemical Approaches 215
   3. Enzymatic Approaches 217
      3.1 Nuclease-Based Mutation Scanning 218
      3.2 Allele-Specific PCR-Based Approaches 229
      3.3 Other Enzymatic Approaches for Mutation Scanning 230
   4. Physical Approaches 233
   5. Bioinformatical Approaches 235
   References 236

10. Current Attempts to Improve the Specificity of Nucleic Acids Hybridization 241
    Anton A. Buzdin
    1. Introduction 242
    2. Improving Hybridization Kinetics 244
       2.1 Simplification of Hybridizing Mixtures 246
       2.2 Chemical Modifications 248
    3. Improving Selection of Perfectly Matched Hybrids 249

    4. Protocols 255
       4.1 Targeted Genomic Difference Analysis 255
       4.2 Using Competitor DNA to Decrease
           the Background of Genomic Repeats 258
       4.3 Mispaired DNA Rejection 260
    References 262

11. Concepts on Microarray Design for Genome
    and Transcriptome Analyses 265
    Helder I. Nakaya, Eduardo M. Reis, Sergio Verjovski-Almeida
    1. Building a Microarray Chip 266
       1.1 Spotted DNA Microarrays 268
       1.2 *In situ* Synthesis 271
    2. Selecting the Probes 277
       2.1 Gene-Oriented Arrays 278
       2.2 Epigenomic Microarrays 279
       2.3 Tiling Arrays 279
    3. Specific Question, Specific Chip 280
       3.1 Transcriptional Profiling 280
       3.2 Comparative Genome Hybridization 282
       3.3 Alternative Splicing 284
       3.4 Transcriptome Annotation 285
       3.5 Small MicroRNA Profiling 289
       3.6 Methylation Pattern 289
       3.7 ChIP-Chip 293
       3.8 Genotyping 295
       3.9 Intronic Transcription 297
    4. Conclusions 299
    References 300

Index 309

CONTRIBUTORS

Ekaterina A. Bogdanova
Shemyakin-Ovchinnikov Institute of Bioorganic Chemistry
Moscow 117997, Miklukho-Maklaya 16/10, Russia

Anton A. Buzdin
Shemyakin-Ovchinnikov Institute of Bioorganic Chemistry
Moscow 117997, Miklukho-Maklaya 16/10, Russia

Nadezhda G. Gurskaya
Shemyakin-Ovchinnikov Institute of Bioorganic Chemistry
Moscow 117997, Miklukho-Maklaya 16/10, Russia

Konstantin A. Lukyanov
Shemyakin-Ovchinnikov Institute of Bioorganic Chemistry
Moscow 117997, Miklukho-Maklaya 16/10, Russia

Sergey A. Lukyanov
Shemyakin-Ovchinnikov Institute of Bioorganic Chemistry
Moscow 117997, Miklukho-Maklaya 16/10, Russia

Denis Rebrikov
Shemyakin-Ovchinnikov Institute of Bioorganic Chemistry
Moscow 117997, Miklukho-Maklaya 16/10, Russia

Dmitry A. Shagin
Shemyakin-Ovchinnikov Institute of Bioorganic Chemistry
Moscow 117997, Miklukho-Maklaya 16/10, Russia

Alex S. Shcheglov
Shemyakin-Ovchinnikov Institute of Bioorganic Chemistry
Moscow 117997, Miklukho-Maklaya 16/10, Russia

Pavel A. Zhulidov
Shemyakin-Ovchinnikov Institute of Bioorganic Chemistry
Moscow 117997, Miklukho-Maklaya 16/10, Russia

Donald Chang
Department of Cell and Neurobiology, Keck School of Medicine,
University of Southern California,
1333 San Pablo Street, BMT-403,
Los Angeles, CA 90033, USA

Shi-Lung Lin
Department of Cell and Neurobiology, Keck School of Medicine,
University of Southern California,
1333 San Pablo Street, BMT-403,
Los Angeles, CA 90033, USA

Joseph Miller
Department of Cell and Neurobiology, Keck School of Medicine,
University of Southern California,
1333 San Pablo Street, BMT-403,
Los Angeles, CA 90033, USA

Shao-Yao Ying
Department of Cell and Neurobiology, Keck School of Medicine,
University of Southern California,
1333 San Pablo Street, BMT-403,
Los Angeles, CA 90033, USA

Lilia M. Ganova-Raeva
Centers for Disease Control and Prevention
Division of Viral Hepatitis
1600 Clifton Rd. NE, MS A-33
Atlanta, Georgia 30329, USA

Contributors

Helder I. Nakaya
*Departamento de Bioquimica, Instituto de Quimica,
Universidade de São Paulo,
05508-900 São Paulo, SP, Brazil*

Eduardo M. Reis
*Departamento de Bioquimica, Instituto de Quimica,
Universidade de São Paulo,
05508-900 São Paulo, SP, Brazil*

Sergio Verjovski-Almeida
*Departamento de Bioquimica, Instituto de Quimica,
Universidade de São Paulo,
05508-900 São Paulo, SP, Brazil*

ACKNOWLEDGMENTS

Anton A. Buzdin:
Many thanks to Professor Eugene D. Sverdlov for his fruitful discussion, innovative ideas, and overall support of this project. Thanks to my colleagues, friends, and family members for their help, patience, and understanding. A. Buzdin was funded by the Molecular and Cellular Biology Program of the Presidium of the Russian Academy of Sciences, by the personal grant from the President of the Russian Federation, and by Russian Foundation for Basic Research grants Nos. 05-04-48682-a, 2006.20034.

Sergey A. Lukyanov, Alex Shcheglov, Pavel Zhulidov, Ekaterina Bogdanova, Dmitry Shagin:
Work on this book was supported by the Molecular and Cellular Biology Program of the Russian Academy of Sciences and Evorogen JSC (Moscow, Russia).

Lilia M. Ganova-Raeva:
Special thanks to Dr. Y. Khudyakov who contributed most to the PEER backbone idea and has been relentlessly resourceful, helpful, and patient throughout the development of the method. Thanks to Dr. H. Fields in whose lab the PEER testing was initiated. Thanks to Dr. X. Zhang for great help with the library screenings and to Dr. F. Cao for introducing better enzymes in the protocol.

Shi-Lung Lin, Donald Chang, Joseph D. Miller, Shao-Yao Ying:
This study was supported by NIH/NCI grant CA-85722. Rhw CHS technology is the property of University of Southern California and protected by US patent numbers, 5,928,872 and 6,130,040.

Helder I. Nakaya, Eduardo M. Reis, Sergio Verjovski-Almeida:
The work in the authors' laboratory was supported by grants and fellowships from Fundação de Amparo à Pesquisa do Estado de São Paulo (FAPESP), and Conselho Nacional do Desenvolvimento Científico e Tecnológico (CNPq), Brazil.

PREFACE

Watson–Crick hybridization of complementary sequences in nucleic acids is one of the most important fundamental processes necessary for molecular recognition *in vivo*, as well as for nucleic acid identification and isolation *in vitro*. This book is devoted to a large family of *in vitro* DNA hybridization-based experimental techniques. A wide spectrum of experimental tasks covered by these approaches includes finding differential sequences in both genomic DNAs and mRNAs, genome walking, multiplex PCR, cDNA library construction starting from minute amount of total RNA, rapid amplification of cDNA 5′- and 3′-ends, effective smoothing of the concentrations of rare and abundant transcripts in cDNA libraries, recovery of promoter active repeats and differentially methylated genomic DNA, identification of common sequences in genomic or cDNA sources, new gene mapping, finding evolutionary conserved DNA and both single-nucleotide and extended mutation discovery, or large-scale monitoring. Several approaches, such as microarray hybridization, have become extremely popular tools for specialists in biochemistry and biomedicine, whereas the potential of many other advantageous techniques seems to be underestimated now.

Analysis of differential gene expression requires application of global approaches that represent a leading tool in postgenomic studies and include transcriptome and proteome analysis, as well as methods allowing population-wide sequence and functional polymorphism analysis. Central to these new technologies are DNA chips designed for quantitative and qualitative uses. Although they are very useful and widely distributed, many popular DNA microarray techniques share a number of shortcomings:

1. The analysis is limited by a number of cDNAs/synthetic oligonucleotides applied on the chip. This number is usually significantly lower than the total

gene quantity of the organisms under study. It creates, therefore, a problem, that many genes escape such an analysis.
2. General transcriptome-wide chip techniques in their actual state hardly distinguish between different gene splice forms.
3. The expression of genes transcribed at low levels cannot be detected by using standard microarray approaches.
4. cDNA-based chips do not differentiate between many gene family members and/or between many transcripts containing repetitive DNA.
5. Microarray chips lack many natural RNAi, cDNAs, or synthetic oligonucleotides and, therefore, cannot be used for the comprehensive study of gene expression regulation at the level of RNA interference by small interfering RNAs.

However, most of these concerns can be effectively addressed by using specific variants of microchip technology, thus making microarrays a truly universal technique (see Chapter 11). Probably, the most important disadvantage of closed systems such as microarrays is that they require preliminary genomic sequence information in order to identify differentially expressed transcripts.

Open systems have the flexibility of identifying uncatalogued sequences. Related experimental techniques, based on DNA hybridization in solution, may be advantageous for many applications, starting from representative cDNA library construction for expressed sequence tag (EST) sequencing, to the identification of evolutionary conserved sequences, differentially expressed genes, or genomic deletions. Unique characteristics of many such techniques make them powerful competitors for well-known approaches that are appreciated worldwide like microarray and competitive genomic hybridizations. Nucleic acid hybridization in solution has few general advantages over hybridization with solid carrier-immobilized nucleic acids: faster hybridization kinetics, better discrimination of proper hybrids, and their availability for further PCR amplification and cloning. Among such in-solution hybridization methods, subtractive hybridization is undoubtedly the most popular technique.

Many techniques have a low efficiency of identifying rare genes that are differentially expressed. This problem is exacerbated when the change in expression level of rare transcripts is small. Since genes expressed at low levels also play a role in establishing differentiated phenotypes, their identification is essential for a complete mechanistic understanding of cellular changes. The major advantage of subtractive hybridization lies in the ability to identify differentially transcribed genes, irrespective of the level of expression, in the absence of sequence information. In addition to preparation of differential cDNA libraries, subtractive hybridization is also extremely useful for identification of genomic DNA differences.

Coincidence cloning, on the contrary, makes it possible to identify sequences which are common for all samples under comparison; cDNA normalization, which is used for smoothing of rare and frequent transcript content in cDNA libraries, may be extremely useful for representative EST library construction. Moreover, several techniques deal with the large-scale DNA polymorphism recovery, including identification of single nucleotide polymorphisms.

Preface

The international team of the authors of this book has tried both to elucidate the current state of the art in hybridization techniques and to help the readers in choosing an appropriate method for performing an experiment in the most efficient way. Enclosed experimental protocols along with both comprehensive and detailed method descriptions make this truly universal book useful to all those interested in the modern life science methodologies.

CHAPTER 1

NUCLEIC ACIDS HYBRIDIZATION: POTENTIALS AND LIMITATIONS

ANTON A. BUZDIN

Shemyakin-Ovchinnikov Institute of Bioorganic Chemistry, Russian Academy of Sciences, 16/10 Miklukho-Maklaya, 117997 Moscow, Russia
Phone: +(7495) 3306329; Fax: +(7495) 3306538; E-mail: anton@humgen.siobc.ras.ru

Abstract: Several nucleic acids hybridization-based approaches, such as microarray, competitive genomic, and Southern or Northern blot hybridization, have become popular tools for specialists in biochemistry and in biomedicine, and are now in routine use. However, the potential of in-solution nucleic acids hybridization-based experimental techniques seems to be underestimated now. Examples are subtractive hybridization (SH), which allows one to efficiently find differences in genomic DNAs or in cDNA samples; coincidence cloning (CC), which, on the contrary, makes it possible to identify sequences that are present in all the samples under comparison; cDNA normalization, which is used for the smoothing of rare and frequent transcript content in cDNA libraries; and TILLING approach, which has demonstrated its great potential for the reverse genetics studies. Finally, several techniques are aimed at the large-scale recovery of DNA polymorphisms, including single nucleotide polymorphisms (SNPs). This book will focus on the above-mentioned and other recent developments in the area of nucleic acids hybridization, including attempts to improve its specificity. In this introductory chapter, I have tried to briefly characterize the current state of the art in in-solution nucleic acids hybridization techniques, and to define their major principles and applications. The advantages and shortcomings of these techniques will be discussed here.

Keywords: Nucleic acids hybridization, cDNA library construction, EST, differentially expressed genes, differential transcripts, differential sequence, microarray, competitive genomic hybridization, subtractive hybridization, coincidence cloning, rare transcript, frequent transcript, normalization, cDNA normalization, cDNA subtraction, genomic subtraction, polymorphism recovery, mutation, single nucleotide polymorphism, SNP, hybridization kinetics, hybridization rate, tracer, tester, driver, genome size, genome complexity, representational differences analysis (RDA), subcloning, restriction fragment length polymorphisms recovery, suppression subtractive hybridization, SSH, genomic polymorphism, Sanger sequencing, mispaired nucleotides, mutant strand, wild-type strand, mutant allele, wild-type allele, rapid amplification of cDNA ends, RACE, differentially methylated.

A. Buzdin and S. Lukyanov (eds.), Nucleic Acids Hybridization, 1–28.
© 2007 Springer.

Abbreviations: BAC, bacterial artificial chromosome; CC, coincidence cloning; cDNA, complementary DNA; CHS, covalently hybridized subtraction; dNTP, deoxyribonucleotidetriphosphate; EST, expressed sequence tag; GREM, genomic repeat expression monitor; mRNA, messenger RNA; MOS, mirror orientation selection; NGSCC, nonmethylated genomic sites coincidence cloning; PEER, primer extension enrichment reaction; PCR, polymerase chain reaction; PERT, phenol emulsion reassociation technique; RACE, rapid amplification of cDNA ends; RDA, representative differences analysis; RFLP, restriction fragment length polymorphism; RNAi, interfering RNA; RT, reverse transcription; SAGE, serial analysis of gene expression; SH, subtractive hybridization; SNP, single nucleotide polymorphism; SSH, suppression subtractive hybridization; YAC, yeast artificial chromosome.

TABLE OF CONTENTS

1. Introduction 2
2. Cloning the Differences: Subtractive Hybridization 4
   2.1 Birth of a Method 4
   2.2 PCR-assisted Subtractive Hybridization 8
   2.3 First Worldwide Success: Representational Differences Analysis 10
   2.4 Further Improvements: Suppression Subtractive Hybridization, Polymerase Chain Reaction Suppression Effect, and Normalization of cDNA Libraries 13
   2.5 Covalently Hybridized Subtraction, Primer Extension Enrichment Reaction, and other Promising Approaches in Subtractive Hybridization 18
3. Finding Common DNA: Coincidence Cloning 20
4. Hybridization in Solution for the Recovery of Genomic Polymorphisms 22
5. Conclusions 24
References 25

1. INTRODUCTION

Watson–Crick hybridization of complementary sequences in nucleic acids is one of the most important fundamental processes necessary for molecular recognition *in vivo* (Watson and Crick 1953), as well as nucleic acid identification and isolation *in vitro* (Southern 1992). The use of experimental techniques based on DNA hybridization in solution is advantageous for many applications, starting from representative complementary DNA (cDNA) library construction for expressed sequence tag (EST) sequencing to the identification of evolutionary conserved sequences, differentially expressed genes, or genomic deletions. Unique characteristics of many such techniques make them powerful competitors for well-known approaches that are appreciated worldwide like microarray hybridization and competitive genomic hybridization. Examples are subtractive hybridization (SH), which allows one to efficiently find differences in genomic DNAs or in cDNA samples; coincidence cloning (CC), which, on the contrary, makes it possible to identify sequences which are common for all samples under comparison; and cDNA normalization, which is used for the smoothing of rare and frequent transcript content in cDNA libraries, thus being extremely useful for representative EST library construction. Moreover, several techniques deal with the large-scale DNA polymorphism recovery, including identification of single nucleotide polymorphisms (SNPs).

Nucleic acid hybridization in solution has few general advantages over hybridization with solid carrier-immobilized nucleic acids: faster hybridization kinetics, better discrimination of proper hybrids, and their availability for further polymerase chain reaction (PCR) amplification and cloning. Among such in-solution hybridization methods, SH is undoubtedly the most popular technique. Analysis of differential gene expression requires application of global approaches that represent a leading tool in postgenomic studies and include transcriptome and proteome analysis as well as methods allowing population-wide sequence and functional polymorphism analysis. Central to these new technologies are DNA chips designed for quantitative and qualitative uses (Brown and Botstein 1999) (see Chapter 11). Although they are now very useful and widely distributed, many popular DNA microarray techniques share a number of shortcomings:

1. The analysis is limited by a number of cDNAs or synthetic oligonucleotides applied on the chip. This number is usually significantly lower than the total gene quantity of the organisms under study. It creates, therefore, the problem that many genes escape such an analysis.
2. General transcriptome-wide chip techniques in their actual state hardly distinguish between different gene splice forms.
3. The expression of genes transcribed at low levels cannot be detected by using standard microarray approaches.
4. cDNA-based chips do not differentiate between many gene family members and/or between many transcripts containing repetitive DNA.
5. Microarray chips lack many natural RNAi cDNAs or synthetic oligonucleotides and therefore cannot be used for comprehensive studies of gene expression regulation at the level of RNA interference by small interfering RNAs.

However, most of these concerns can be effectively addressed by using specific variants of microchip technology, thus making microarrays a truly universal technique (reviewed in Chapter 11). Probably, the most important disadvantage of closed systems such as microarrays is that they require preliminary genomic sequence information in order to identify differentially expressed transcripts.

Open systems have the flexibility of identifying uncatalogued sequences: alternatively, differences in gene expression between two samples can be compared directly by methods such as differential display (Liang and Pardee 1992), differential cloning techniques (Sagerstrom et al. 1997), and combinations of these (Pardinas et al. 1998; Yang et al. 1999). These approaches have been successfully used to identify genes differentially expressed in two populations that exhibit large changes in expression levels, or genes that are expressed at high concentrations in terms of number of copies per cell.

However, these techniques have a low efficiency of identifying rare genes that are differentially expressed (Martin and Pardee 2000). This problem is exacerbated when the change in expression level of rare transcripts is small. Since genes expressed at low levels also play a role in establishing differentiated phenotypes, their identification is essential for a complete mechanistic understanding of cellular changes. The major advantage of SH lies in the ability to identify differentially expressed genes, irrespective of the level of expression, in the absence of sequence information. In addition to preparation of differential cDNA libraries, SH is also extremely useful for identification of genomic DNA differences (Diatchenko et al. 1996, 1999; Ermolaeva et al. 1996; Akopyants et al. 1998; Bogush et al. 1999).

2. CLONING THE DIFFERENCES: SUBTRACTIVE HYBRIDIZATION

2.1 Birth of a Method

SH was first used as early as 1966 by Bautz and Reilly (1966) to purify phage T4 mRNA. Recently, a number of groups have employed variations of the technique, both to clone cDNAs derived from mRNAs that undergo up- or down-regulation (cDNA subtraction), and to identify genomic deletions (genomic subtraction). This approach became well known since 1984 when Palmer and Lamar applied SH for the construction of mouse recombinant DNA libraries, enriched in Y chromosome sequences (Lamar and Palmer 1984). Since that date SH has significantly evolved, thus becoming one of the most important and effective tools in molecular biology. This truly universal approach is being used for a diverse set of applications like cloning and characterization of new genes, recovery of tissue-specific, malignancy-specific, or organism-specific transcripts, identification of genes differentially expressed at different stages of embryo development, cancer progression, regeneration, for the recovery of genes up- or down-regulated in response to external or internal stimuli, etc. (Cekan 2004; Ying 2004). The method is useful for the genome-wide comparison of bacterial DNAs (Bogush et al. 1999), for isolating species-specific loci (Buzdin et al. 2002; Buzdin et al. 2003), and polymorphic markers in both eukaryotic and prokaryotic genomes (Bogush et al. 1999; Nadezhdin et al. 2001). In addition, SH was shown effective for DNA subcloning from yeast artificial chromosomes (YACs) into smaller vectors (Zeschnigk et al. 1999), for mapping of genomic rearrangements associated with cancer or chromosome abnormalities, even for filling in extended gaps in large-scale sequencing projects (Frohme et al. 2001).

However, being such a powerful instrument for molecular biology and biomedicine, SH usefulness is still underestimated. In this section, I have tried to elucidate all major techniques dealing with the subtraction of nucleic acids. As stated above, SH became sound in 1984 when Palmer and Lamar proposed a simple idea of a separation of hybrid molecules: double-stranded homohybrids of the "tracer" or "tester" DNA (a sample containing differential sequences to be identified), from heterohybrids tracer–driver and homohybrids driver–driver ("driver" is a sample containing reference nucleic acid sequences). SH is aimed at the isolation of a fraction of tracer-specific sequences absent from driver. The idea was that tracer and driver DNA would have different sequences on their termini (Lamar and Palmer 1984). The authors wanted to create a mouse recombinant DNA library enriched in Y chromosome sequences (Figure 1). Female mouse DNA (driver) was fragmented by sonication, whereas male DNA (tracer) was digested with *MboI* restriction endonuclease. Tracer DNA was mixed with 100-fold weight excess of a driver, denatured, and allowed to reanneal. Only reassociated tracer–tracer homoduplexes had sticky ends at both termini and could be ligated into the plasmid vector pBR322 digested by *BamHI* restriction endonuclease.

Soon afterwards, this principle was successfully employed for cloning of human DNA fragments absent in patients with Duchenne muscular dystrophy (Kunkel et al. 1985). Simultaneously, other research teams started using similar approaches for the recovery of messenger RNAs distinguishing analyzing samples (Chien et al. 1984; Kavathas et al. 1984), specific for certain cell type, tissue, or organism (Figure 2). Using a poly(A) + fraction of tracer RNA, cDNA first strands were synthesized, initial RNA was then degraded by the addition of NaOH, so that only cDNA first strands complementary to tracer mRNA remained in solution. The tracer was then mixed with taken in 100-fold or more weight excess of driver, which was a poly(A) + RNA fraction from another sample. In the resulting mixture, tracer fragments were either hybridized with the excess of driver complementary strands, or remained in a single-stranded form. The latter single-stranded fraction, which was enriched in tracer-specific sequences, was purified from driver and tracer–driver hybrids on a hydroxyapatite column (which provides column binding by double-stranded nucleic acids). Using purified single-stranded tracer, cDNA second strands are synthesized and the resulting double-stranded cDNA is ligated in either expression (if an additional round of subtraction is needed) or cloning vector (Chien et al. 1984; Kavathas et al. 1984). To increase hybridization rate, chemical accelerators such as phenol could be added to the hybridizing mixture (Travis and Sutcliffe 1988).

However, this approach did not become popular, probably due to three serious limitations: (1) great amounts of mRNA are needed; (2) the technique is extremely laborious; and (3) RNA degradation may cause severe problems at many stages. Recently, the first problem was solved by cloning tracer and driver

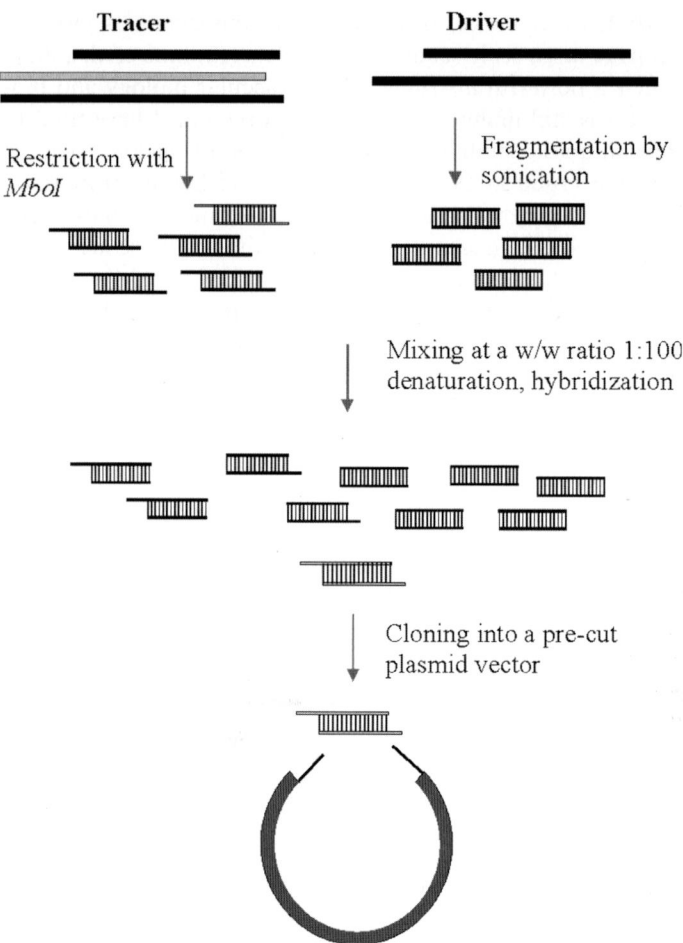

Figure 1. Genomic DNA subtraction scheme, proposed by Palmer and Lamar, based on tracer–tracer hybrids cloning using unique restriction sites.

cDNAs into special single- or double-stranded expression vectors. RNA was produced in *Escherichia coli*, thus making it possible to obtain large amounts for the hybridization (Palazzolo and Meyerowitz 1987; Kuze et al. 1989; Rubenstein et al. 1990). The third barrier was waived in part when driver mRNA was replaced by double-stranded cDNA (the use of single-stranded first strand cDNA was less cost-efficient). However, the latter improvement created a new problem: how to separate tracer–tracer duplexes from tracer–driver and driver–driver after hybridization?

In 1986, Welcher et al. (1986) created the first biotin–streptavidin subtraction system: biotinylated primers were used for driver cDNA synthesis, and the

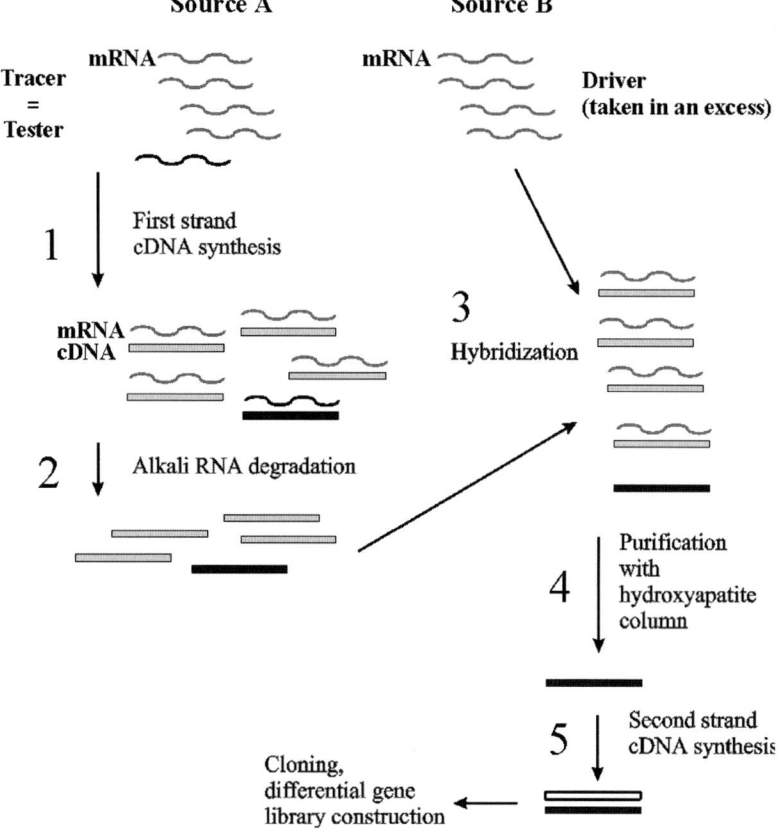

Figure 2. cDNA/mRNA subtraction scheme, utilizing alkali RNA degradation at stage 2.

subtraction products were incubated with streptavidin-coated copper granules. As a result, tracer–tracer duplexes remained in solution, whereas the granules bound all other hybridization products, except single-stranded tracer. This method of subtracted product separation became quite popular (some details being modified). In 1993, magnetic beads replaced copper granules as a solid-phase carrier for conjugated streptavidin in tracer–tracer hybrid purification (Lopez-Fernandez and del Mazo 1993; Sharma et al. 1993).

Overall, these early techniques for cDNA subtraction generally involved one or two rounds of hybridization and used (−) mRNA to drive hybridization to (+) cDNA tracer. However, the preparation of (−) mRNA in large amounts is not always a practical proposition; consequently, for less-abundant sequences, the concentration of driver is likely to be too low to drive hybridization to completion. The degree of enrichment is limited by the driver to tracer ratio, and a single round of hybridization will only enrich adequately those upregulated messages that are rare in the (–) population, but highly abundant in the (+) population. Sequences that are only moderately abundant even after upregulation, or which are upregulated only to a limited extent, will still be obscured by a background of common sequences. Furthermore, the amount of cDNA remaining after hybridization can be tiny and the problem of cloning successfully such minute quantities of cDNA is not trivial.

2.2 PCR-assisted Subtractive Hybridization

Regardless of the method improvements mentioned above, SH is too laborious a process: in a period from 1984 to 1989, the use of SH was described in only 29 research papers (Figure 3). It was obviously PCR that made SH an easy, inexpensive, and widely used technique. Indeed, the researcher could now obtain substantial amounts of tracer and driver DNA in an easy inexpensive way with minute amounts of starting material. Moreover, the use of the PCR solved the problem of cloning of the subtraction products (Hla and Maciag 1990; Timblin

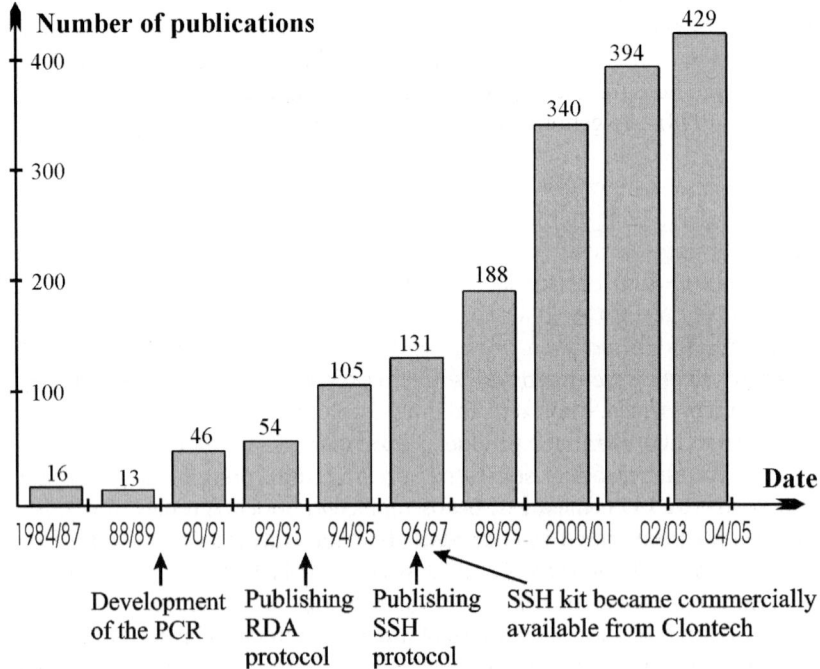

Figure 3. Citation dinamics of the subtractive hybridization approach in the literature. Publications in MedLine – indexed peer-reviewed journals were explored using PubMed server at National Center for Biotechnology Information (NCBI, http://www.ncbi.nlm.nih.gov/). The integral figure of 16 publications is shown here for the period 1984–1987.

Nucleic acids hybridization: potentials and limitations 9

et al. 1990; Hara et al. 1991). Consequently, since 1990 PCR is incorporated in essentially all subtraction protocols and the interest of the scientific community to SH has increased: the number of published SH applications per year is increasing by three- to fourfold (Figure 3).

PCR has also provided a solution to another important problem: before PCR one could not perform an effective subtraction for rare transcripts as their concentrations were small and the reassociation rate during SH was negligible; such transcripts, therefore, were escaping analysis. PCR made it possible to obtain unlimited amounts of DNA for hybridization, thus enabling detection of such infrequent RNAs (Hla and Maciag 1990; Timblin et al. 1990; Hara et al. 1991; Herfort and Garber 1991; Wang and Brown 1991). In addition to gene expression assays, since 1991 SH is being used for the recovery of differential sequences in bacterial genomes (Cook and Sequeira 1991), which is actually one of the most important applications of SH: identification of differences in genomic DNAs of virulent versus nonvirulent strains, between disease-causing bacterial species and their harmless relatives, is extremely important for both creating new diagnostic markers and targeting bacterial genes for new drugs development (Cook and Sequeira 1991; Cruz-Reyes and Ackers 1992).

However, genomic DNA subtraction efficiency depends greatly on the complexities of the DNAs under comparison, as learned from both experimental studies (Wieland et al. 1990; Clapp et al. 1993) and mathematical models simulating SH (Sverdlov and Ermolaeva 1994; Milner et al. 1995; Ermolaeva et al. 1996; Cho and Park 1998). As genome size increases beyond 5×10^8 bp (complexity comparable with that of arabidopsis or drosophila genomes), the kinetics of hybridization start to become an increasingly important factor limiting enrichment of the target (Milner et al. 1995). Mammalian genomes are too complex to reach sufficiently high reassociation rate values, and only major differences (like presence or absence of Y chromosome or extended deletions) can be isolated in such a way. To enhance the kinetics of hybridization, increased hybridization times, higher driver concentrations, greater driver to tracer ratios, longer DNA fragments, and the use of techniques that enhance the rate of reassociation, e.g. phenol emulsion reassociation technique (PERT) (Kohne et al. 1977; Laman et al. 2001) or solvent exclusion (Barr and Emanuel 1990), may be effective. Major considerations on the kinetical requirements for the effective subtraction of complex genomic mixtures and related formulas are given in more detail in Chapter 10.

In addition to poor reassociation rates, one more problem appears when complex eukaryotic DNAs are being compared: repetitive sequences (which form, for example, >40% of mammalian DNA: Lander et al. 2001; Venter et al. 2001) reassociate significantly faster than unique genomic sequences (Milner et al. 1995), and the resulting differential libraries are greatly enriched in repeats (Rubin et al. 1993). This creates a serious obstacle to attempts of direct higher eukaryotic genomic DNA comparison by means of SH. However, as reviewed by Sverdlov (1993) and Sverdlov and Ermolaeva (1994), an adequate reassociation rate theoretically could be obtained for complex genomic mixtures as well, if single-stranded tracer and driver are used, as learned from the authors' mathematical model and preliminary experimental results (Sverdlov and Ermolaeva 1993; Ermolaeva and Sverdlov 1996). Unfortunately, to my knowledge, since 1996 when this idea was published, it was never used in practice for genome-wide comparisons of higher eukaryotic DNAs. So, until now the problem of poor SH applicability to complex nucleic acid mixtures remains unsolved. It should be mentioned here that all above considerations regarding SH reassociation kinetics are in fact more general, being true and actual for all the family of methods based on nucleic acids hybridization in solution as well.

2.3 First Worldwide Success: Representational Differences Analysis

Lisitsyn and Wigler 1993) reported a new SH-based approach termed "representative differences analysis" (RDA), which made the idea of SH quite popular (and which is actively in use until now; see Figure 3). RDA was applied first to the comparison of two mammalian genomes and cloning of the differential sequences. Kinetical limitations are waived here because of the random simplifications of the comparing genomic mixtures due to either the use of nonfrequent-cutter restriction enzymes or PCR selection effect.

In Figure 4, tracer and driver genomic DNAs are digested with the restriction endonuclease, and different oligonucleotide adapters are ligated to the fragmented DNAs. The resulting ligation mixtures are further subjects to 50–100 cycles of PCR amplification with adapter-specific primers. Due to well-known "PCR selection" effect, different fragments in complex genomic mixtures are PCR-amplified with very different efficiencies, primarily depending on the fragment size. Therefore, the major parts of initial DNA fragments are greatly underrepresented or lost in the final amplicons, which contain only 2–10% of the initial fragment diversity. The amplicons are further treated with mung bean nuclease to degrade 3′-terminal adapter sequences (which could cause a problem at the next step due to adapter–adapter cross-hybridization), tracer is then mixed with the excess of a driver, denatured, and allowed to hybridize. The hybridization mixture is then incubated with DNA polymerase to fill in the 3′-termini, followed by a PCR with the primer specific to the tracers' adapter. Only the tracer–tracer duplexes are amplified exponentially, whereas other hybridization products are amplified only linearly or not amplified at all. The exponentially amplified double-stranded duplexes can be easily cloned into a plasmid vector and sequenced. Alternatively, additional rounds of subtraction may be performed to increase library enrichment in differential sequences, presented solely in tracer. This technique is inexpensive, fast, and relatively easy; it permits working with small amounts of the starting material (genomic DNA or cDNA). It is not surprising, therefore, that RDA became quite popular for both transcription analyses and genomic marker recovery (Lisitsyn et al. 1994a, b, 1995; Ayyanathan et al. 1995; Drew and Brindley 1995; Lisitsyn and Wigler 1995; Schutte et al. 1995) (Figure 3).

Nucleic acids hybridization: potentials and limitations 11

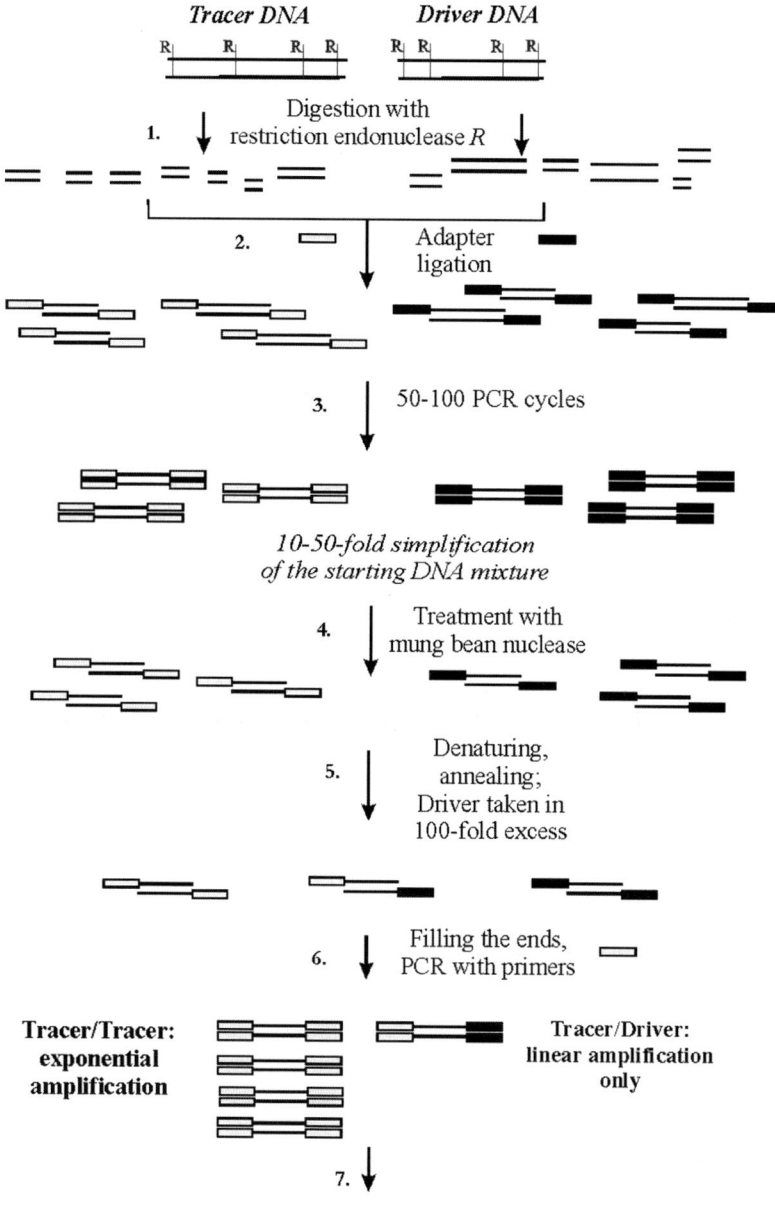

Figure 4. Representative differences analysis (RDA) application for identifying differential genomic DNA sequences. Initial genomic DNA mixture simplification is provided here by 50–100 cycles of PCR amplification at stage 3.

However, RDA has an obvious shortcoming: only small part of a transcriptome or genome is analyzed, whereas the majority (90–98%) escapes analysis. RDA still cannot provide genome-wide comparisons of the DNAs as its results are very fragmentary. Another limitation of the RDA use is the high background of the false-positive signals when cDNA libraries are compared: when the differences in transcription spectra are small, only a small number of tracer–tracer duplexes will be formed, and linear amplification of tracer–driver hybrids will create serious problems in the enriched subtracted library construction (Ayyanathan et al. 1995). This figure is even more pessimistic when such rare differential genes are poorly transcribed. Nevertheless, RDA is widely in use until now; some studies employing RDA are even entitled "genome/transcriptome-wide", which in fact is not the case, as we know now. To my opinion, RDA is advantageous for the recovery of differential marker sequences in DNAs under comparison.

With the increasing interest in SH and its applications, few reliable mathematical models of SH appeared and even two computational programs for the SH simulation *in silico* were published in 1995 (Ermolaeva and Wagner 1995; Milner et al. 1995). Below I briefly describe few successful and original experimental SH applications. The authors of the paper (Sallie 1995) proposed a new modification of the SH comprising the hybridization of subtracted products with filter-immobilized cosmid libraries of the candidate genomic loci. Such a use of SH, to the opinion of the authors, enabled unambiguous mapping of the genomic loci containing differentially transcribed genes. Authors of the next paper (Chen et al. 1995) applied RDA to the more efficient subcloning of the YAC fragments into smaller vectors, suitable for the insert sequencing. Genomes of YAC-containing yeasts (tracer) and those lacking YAC (driver) were fragmented and subtracted; after few cycles of subtraction, the products were cloned and sequenced. An overwhelming majority of the inserts contained sequences from the YAC. This application of the SH, although obviously interesting, did not become popular, probably due to the removal of YACs from large-scale sequencing strategies in favor of more stable bacterial artificial chromosomes (BACs).

A modification of the RDA was used to map cancer-specific deletions (Zeschnigk et al. 1999). Products of the subtraction of cancerous and normal genomes were hybridized with the ordered YAC library, thus making it possible to directly map differential sequences (no complete human genome sequence was available in databases at that time) and, therefore, to identify cancer-specific deletions. In more recent publication (Frohme et al. 2001), RDA was employed to fill the gaps during sequencing of the genome of *Xilella fastidiosa*. Fragments with already defined primary structure were subtracted from *Xilella* genome, differential sequences were hybridized with the complete *X. fastidiosa* genomic clone library and the positive clones (those containing differential sequences which were not sequenced before) were sequenced.

Another interesting approach was the SH-based technique for the restriction fragment length polymorphisms recovery, termed "RFLP subtraction" (Rosenberg et al. 1994). Genomic DNAs under comparison were digested by restriction endonucleases and loaded separately on agarose gels. Following electrophoresis separation, specific zones (e.g. containing fragments from 100 to 500 base pairs (bp) in length) were excised for both tracer and driver, the DNAs were eluted from the agarose, and then used for the SH, resulting in amplicons enriched in differential fragments presenting in that zone in one sample DNA but absent from another one. The technique was proven to be very effective for the recovery of new RFLPs, genetic markers of a universal usefulness. This approach was significantly improved when the authors (Sasaki et al. 1994) performed subtraction directly in the gel. Both digested samples under comparison, one (driver) in a 100-fold weight excess over another one (tracer), were loaded on *the same* track of the gel and separated by electrophoresis. The gel was treated with NaOH to denature DNA, and then neutralized to hybridize denatured DNAs directly in the gel. Finally, the hybridized DNA was eluted from the agarose and was PCR-amplified to select for tracer–tracer duplexes. The authors obtained very high enrichment values, close to the theoretical maximum possible enrichment. This might be explained by very high local concentrations of the fragments of each type in gel which were significantly greater than those in solution. Unfortunately, this approach, which might be a perfect alternative to most popular SH techniques such as RDA and suppression subtractive hybridization (SSH), is very laborious.

2.4 Further Improvements: Suppression Subtractive Hybridization, Polymerase Chain Reaction Suppression Effect, and Normalization of cDNA Libraries

It may be seen from Figure 3 that the development of a new method called "suppression subtractive hybridization" (SSH) (Diatchenko et al. 1996; Jin et al. 1997) and further release of the corresponding kit from Clontech in 1996 caused a revolution in the use of the SH. The technique became more robust, more reproducible, and easier to perform. This clearly enhanced interest was accompanied by some "qualitative" changes in publications citing the SH: in 1984–1991 every third paper described any SH modification or improvement (at least from the point of view of the authors), in 1992–1995 every fifth, in 1998–1999 every tenth, and in 2000–2005 (when SSH was published and the kit became commercially available and appreciated) as rare as every 62nd paper. This might suggest that the popularization of SSH and RDA "killed" the creativity of the authors who were finally satisfied by the existing SH techniques. Probably, this means that these methods cannot be further improved, or that there is no need to improve them for most applications: SH became a routine reproducible procedure.

The principal advantage of the SSH over other SH techniques is the greatly reduced background of false-positive clones. In other methods, this background is caused by a linear amplification of the tracer–driver hybrids which is reduced to a minimum in SSH (described in detail in Chapter 3). Another advantage of the SSH is the custom normalization of cDNA libraries which can be performed in order to equalize the concentrations of different transcripts located in the libraries. Such equalization is needed to avoid discrimination of rare transcripts during the subtraction. Normalization of cDNA libraries can be used for purposes other than SH as well, e.g. for the construction of representative EST libraries. This independent group of methods is described in Chapter 5.

Both advantages of the SSH are based on the "PCR suppression" effect (for detailed description, see Chapter 2). Approximately 40 nt long GC-rich linkers (termed "suppression adapters") are ligated to a fragmented double-stranded DNA (dsDNA). Further treatment with DNA polymerase builds the second strand of the adapters so that the initial DNA fragments are flanked by inverted GC-rich 40-nucleotide sequences. The methods' rationale is that the primers that are complementary to suppression adapters cannot efficiently anneal and initiate the PCR alone (Figure 5) due to significantly stronger intramolecular base pairing of the inverted repeats, which eliminates available

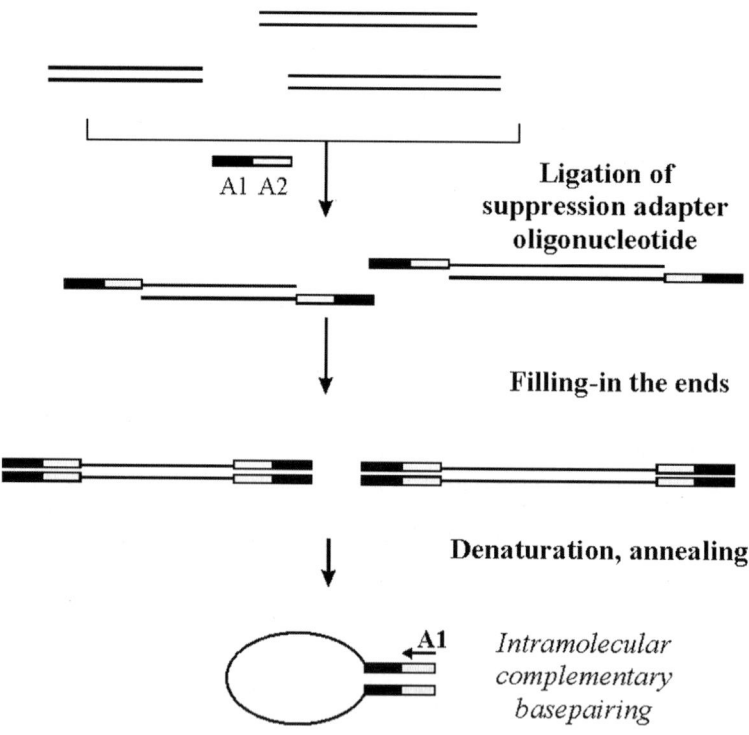

Figure 5. The principle of PCR suppression effect. GC-rich inverted repeats (suppression adapters) base pair intramoleculary, thus preventing annealing of shorter primer oligonucleotides, designed to the adapter sequence. The strength of such an approach is that it greatly reduces background amplification when adapter-specific PCR primers are used.

Nucleic acids hybridization: potentials and limitations

primer binding sites for the PCR (Chapter 2). The use of the PCR suppression effect prevents the background PCR amplification with adapter-specific primers.

Figure 6 depicts the schematic representation of the SSH and its application to identifying differences between two bacterial genomes. The DNAs (both tracer and driver) were digested with restriction endonucleases, and the tracer

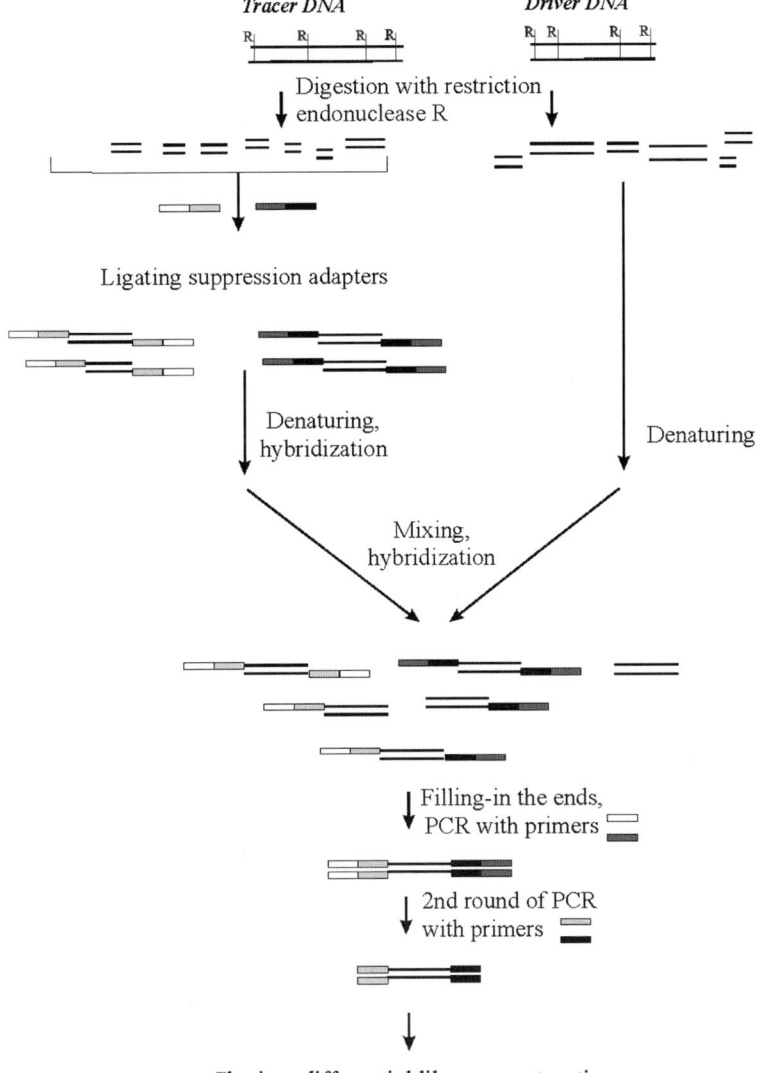

Figure 6. Suppression subtractive hybridization (SSH) scheme, adopted for the comparison of two bacterial genomes. When complex genomes like mammalian DNAs are compared, other approaches must be used.

DNA was then subdivided into two portions (tracer A and tracer B), and two different suppression adapters were ligated separately to these different portions. Nothing was ligated to driver DNA. Both tracer fractions were mixed with driver, denatured, and allowed to hybridize. In the subsequent PCR amplification with primers complementary to both suppression adapters used, only the tracer A or tracer B duplexes, enriched in tracer-specific sequences, could be exponentially amplified further (Akopyants et al. 1998).

Importantly, SSH is also used for comparing transcriptomes (Diatchenko et al. 1996), resulting in high-quality differential cDNA libraries (Chapter 3). In this application, PCR suppression effect made it possible to solve in part one of the most important general problems of cDNA analysis – loss of the rare transcripts from the resulting libraries. The method developed by Sergey Lukyanov's team for normalization of the cDNA libraries in solution (Figure 7; see also Chapter 5) permits an easy and effective smoothing of the concentrations of rare and abundant cDNAs. Two different suppression adapters are ligated to two portions of the fragmented double-stranded cDNA, separately denatured, and then allowed to reanneal for a short time. At this stage, mostly highly abundant sequences hybridize with each other. These fractions are then mixed and allowed to hybridize again without melting. Those cDNAs which did not form duplexes during the first hybridization may hybridize now to form double-stranded molecules. The hybridized cDNA ends are built-in and the mixture is further PCR-amplified with primers complementary to suppression adapters used (Figure 7). This results in exponential amplification of only those duplexes which were formed during the second, but not the first, hybridization. The final amplicon is, therefore, enriched in rare transcript replicas. Such a strategy is a good alternative to a more sophisticated laborious approach (Bonaldo et al. 1996), which utilizes a special subtraction of highly abundant transcripts from the total cDNA pools to enrich the libraries in rare cDNAs.

Although the proportion of background false-positive clones after the use of SSH (multiple rounds of subtraction may be used) is usually low, further improvement was reported recently (Rebrikov et al. 2000). In a technique termed "mirror orientation selection" (MOS) (Figure 8), the number of nondifferential clones is reduced based on the observation that the background, which is caused by reassociation of nontracer-specific molecules, appears just by chance, and each type of such background duplexes is presented by a small number of molecules, compared to the "proper", tracer-specific sequences. As the SSH products (Figure 8) harbor adapter sequences, flanking the tracer DNA in *both* tracer orientations (products A and A′ in the figure) rather than the removal of one adapter, denaturation and subsequent reannealing, followed by filling the ends, may result in tracer fragments flanked by the second adapter sequence from both sides. Using single-primer PCR, such tracer fragments can be exponentially amplified (Figure 8). However, this is not the case for background fragments, as they appear in the resulting SSH amplicons by chance, and each type is presented by very low concentrations (whereas the number of such

Nucleic acids hybridization: potentials and limitations

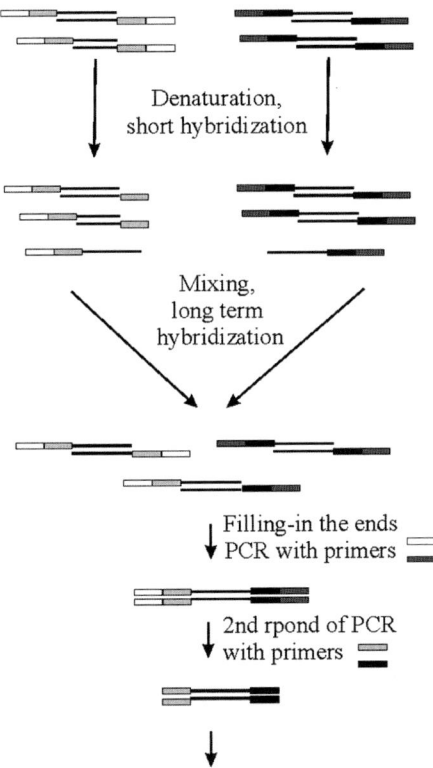

Figure 7. cDNA library normalization scheme using PCR suppression effect. Normalization results in smoothing concentrations of abundant and rare transcripts, which is important for higher transcript repertoire representation in the resulting differential cDNA libraries and for many other applications like EST sequencing.

types may be enormous), and the probability that they will form hybrids carrying adapter sequences at both termini is negligible in most cases (Figure 8, right panel).

Also, SSH may be used in combination with differential display (Pardinas et al. 1998) and microchip hybridization (Yang et al. 1999). The latter seems to be a very promising approach, as it gives an integral picture of a spaciotemporal differential gene expression. Importantly, the direct use of microarray hybridization usually discriminates rare transcripts, whereas preliminary SSH, especially with a stage of cDNA normalization, may greatly enhance both the sensitivity and reproducibility of the results.

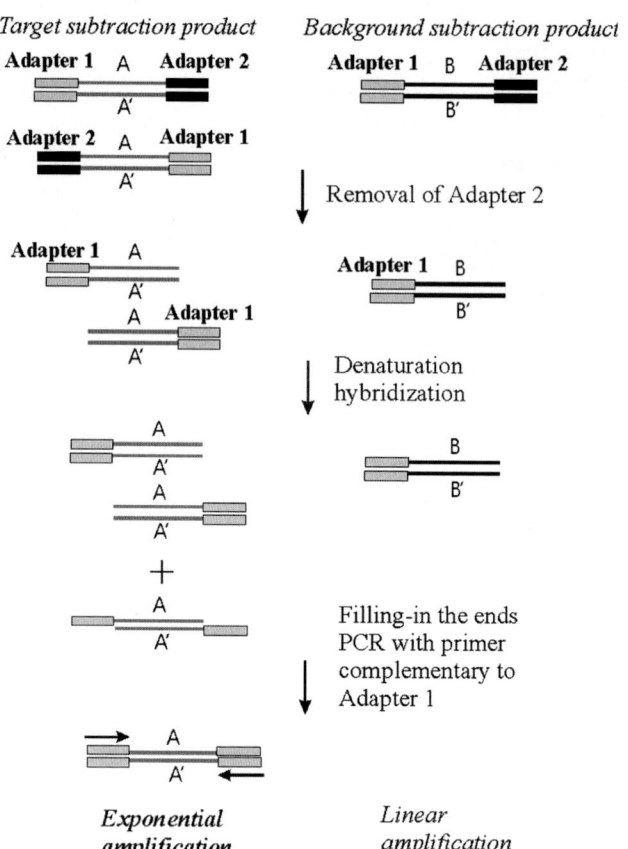

Figure 8. Mirror-oriented selection (MOS) scheme. MOS is aimed at reducing background produced by pseudoselective amplification of occasional unrelated sequences. MOS was found useful for both cDNA and low to medium complexity genomic DNA comparisons.

2.5 Covalently Hybridized Subtraction, Primer Extension Enrichment Reaction, and Other Promising Approaches in Subtractive Hybridization

In Sections 2.2–2.4, I described two most popular techniques based on SH which stand SH among the most effective, timely, and universal approaches in molecular biology like differential display, large-scale shotgun sequencing, EST sequencing, and serial analysis of gene expression (SAGE) (Carulli et al. 1998). As the number of published studies utilizing SH increases, regardless of the lack of serious popularized advances in its methodology during the last 5 years, the potential of SH remains high. In this section, some recent original and promising SH-based techniques will be reviewed. In Shao-Yao Ying's laboratory, a novel approach termed "covalently hybridized subtraction" (CHS) was proposed (Ying and Lin 1999) which utilizes covalent binding of tracer and driver after hybridization. To this end, driver DNA (termed "subtracter" by the authors) was chemically modified so that DNA strands in all posthybridizational duplexes with driver DNA were covalently bound and, therefore, could not be further PCR-amplified, in contrast to tracer–tracer hybrids. This important modification of the SH is described in Chapter 7. Another interesting approach created to improve RDA technique is based on the protection of 3'-termini of tracer DNA by alpha-thiodeoxyribonucleotides (Kuvbachieva and Goffinet 2002). After hybridization with a nonmodified driver, the resulting mixture was treated with a mixture of *Exo*III nuclease that degrades unprotected 3'-ends and mung bean nuclease that digests single-stranded DNA (ssDNA). Only tracer–tracer hybrids remained in solution, which were further PCR-amplified, cloned, and sequenced.

In the new technique termed "primer extension enrichment reaction" (PEER), Ganova-Raeva et al. (2006) utilized a new original rationale of using short fragments of tracer for a primer extension reaction on the template of driver DNA. To this end, double-stranded tracer DNA was converted into small fragments by extensive endonuclease cleavage and then tagged by ligation to a specially designed adapter. The 3'-end of the adapter incorporates a recognition site for a class IIS restriction endonuclease. The fragments are cleaved with the IIS *Mme* I enzyme to create oligonucleotides with unique sequence at the 3'-end derived from the tracer and a 5'-end derived from the adapter. These adapter-tagged oligonucleotides are annealed to the driver DNA template and extended in the presence of biotinylated ddNTPs. This event blocks any further extension and allows the removal of the biotinylated molecules from the reaction by use of streptavidin-coated magnetic beads. Primers that share driver sequences are blocked and removed leaving only primers with unique sequences that can only be found in the tracer. In the presence of initial full-length tracer DNA, these oligonucleotides can prime an extension reaction from the fragments unique to the tracer (target capture). This step converts the tagged primers into DNA templates suitable for PCR amplification by oligonucleotides containing only the adapter sequences. The final step is expected to generate collection of fragments of different sizes that may be cloned and sequenced. The authors demonstrated that at least for some applications PEER was significantly more sensitive and effective than other variations of SH, including SSH. This very promising technique and its comparison with other methods of differential gene expression screening is described in more detail in Chapter 6.

In concluding this section I should mention a few shortcomings shared among the majority of the SH-based techniques. First, PCR amplification can bias representations of the amplified DNAs due to the *PCR selection* effect, when too many cycles are used (see Section 2.3). Second, SH sometimes does not differentiate between members of evolutionary young gene families sharing high nucleotide sequence identity. The same problem appears at both cDNA and genomic DNA levels when genomic repeats are studied. The last but not the least limitation is that the "small" differences in gene expression, which are less than one order of magnitude, are usually hardly detectable in SH-based techniques. However, a number of SH modifications have been proposed to partly or completely address these shortcomings (see Chapters 2, 3, 5–7, and 10).

3. FINDING COMMON DNA: COINCIDENCE CLONING

In contrast to SH, which is aimed at the recovery of differential sequences residing in a tracer sample but absent from a driver sample, the approach termed "coincidence cloning" (CC; Chapter 8) was developed to find DNA fragments which are common to the samples under study. The approach is based on cloning identical nucleotide sequences belonging to different fragmented genomic DNA or cDNA pools, while discarding sequences that are not common to both (Devon and Brookes 1996). By comparing genomic DNA fragments with fragments of any similarly fragmented locus (cloned in the form of BACs, cosmids, etc.), one can select and identify the genomic fragments belonging to this locus.

To this end, both fragmented DNAs under comparison are specifically tagged (e.g. by ligating different terminal adapter oligonucleotides), mixed, denatured, and hybridized, followed by the isolation of duplexes having both specific tags (i.e. "heterohybrid" products derived from both samples, which are common to both tagged DNA mixtures). The former step is the key stage of the whole procedure, as an efficient isolation of proper hybrids provides construction of CC libraries, truly enriched in common sequences. Early versions of the CC technique were not very efficient and, therefore, have not been widely used. Their most serious disadvantage was rather low selectivity, so that the resulting libraries of the fragments contained large amounts of sequences unique to one of the two sets of DNA fragments under comparison. To avoid this, Azhikina and colleagues were the first to exploit the technique of selective PCR suppression (PCR suppression effect was mentioned in Section 2.4 and will be described in detail in Chapter 2), which strongly increased the efficiency of CC (Azhikina et al. 2004, 2006; Azhikina and Sverdlov 2005).

Figure 9 represents a simple model of the use of CC for isolation of evolutionary conserved sequences shared by comparing genomes, reported by Chalaya et al. (2004). Genomic DNAs of human and of New World monkey marmoset *Callithrix pigmaea* were digested with frequent-cutter restriction endonuclease, and two different sets of suppression adapters were ligated to them. Samples were then mixed, denatured, and allowed to reanneal, followed by filling of the ends with DNA polymerase (Figure 9) and treatment with mismatch-specific nucleases (Chalaya et al. 2004). These enzymes recognize improperly matched dsDNAs and cut such "wrong" hybrids, thus clearly enhancing hybridization specificity (see Chapter 10). At the next stage, hybridization products are subjected to PCR with primers specific to the suppression adapters used, so that only human–*C. pigmaea* hybrid molecules are amplified. As a

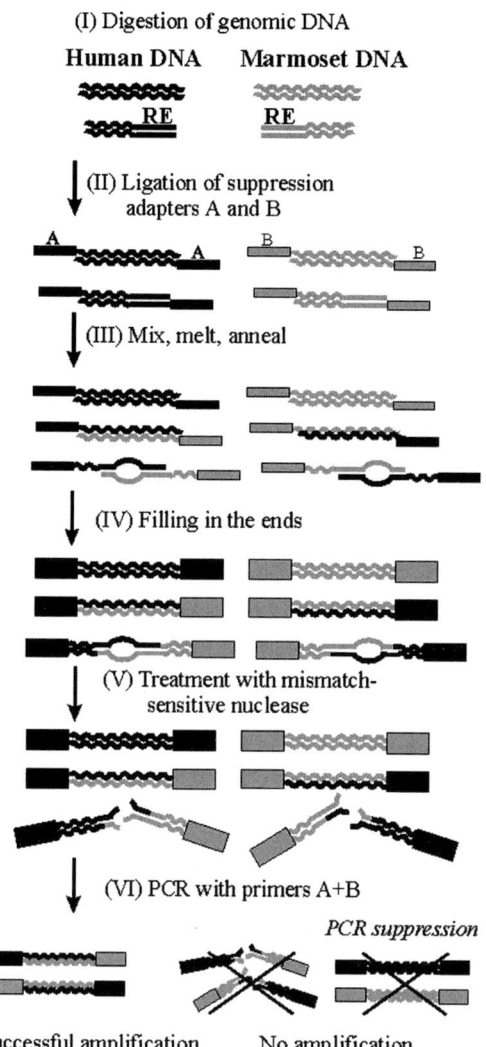

Figure 9. Schematic representation of mispaired DNA rejection (MDR) technique, whose rationale is the specific enzymatic degradation of mismatched hybrids (stage V), which results in significantly lower background mutual hybridization of the two comparing DNA samples. This approach has been demonstrated to be efficient for complex genomic mixtures like primate DNA.

result, the authors managed to create a genomic library highly enriched in evolutionary conserved sequences shared by human and *C. pigmaea* genomes.

Another successful application of the CC is the new technique called "non-methylated genomic sites coincidence cloning" (NGSCC), which results in a set of sequences that are derived from the genomic locus of interest and contain an unmethylated CpG site. The technique is based on the initial fragmentation with a methyl-sensitive restriction enzyme. To simplify the DNA sets to be compared, they can be additionally digested with a frequent cutter that is not sensitive to methylation of its target site, e.g. *Alu*I. As a result, the lengths of the fragments can be restricted to a size that is optimal for subsequent PCR amplifications, usually up to 1.5 kb. Different suppression adapters are then ligated to sticky ends produced by methyl-sensitive restriction enzymes and to blunt ends created by *Alu*I. Further PCR amplification with primers specific to both adapters used results in the amplicon of genomic fragments having unmethylated CpG site at one terminus and *Alu*I restriction site in another.

This amplicon is further hybridized to a new-suppression-adapter-ligated fragmented DNA from the genomic locus of interest (the authors analyzed methylation profiles of an ~1 Mb-long human genomic locus *D19S208-COX7A1* from chromosome 19). In the following nested PCR, only those unmethylated CpG-containing fragments that match to D19S208–COX7A1 genomic locus were amplified. Sequencing of the resulting libraries derived from initial genomic DNAs from healthy and cancerous tissues enabled authors to create the first large-scale comprehensive tissue- and cancer-specific methylation map for that locus (Azhikina and Sverdlov 2005). Recently, the same group of authors combined NGSCC with SAGE, thus creating a new technique termed "RIDGES", which is significantly more informative than NGSCC, as its outcome is 10- to 20-fold more information about methylation sites per one sequenced clone (Azhikina et al. 2006).

However, the use of CC is not restricted to genomic DNA analysis. In particular, a recently published technique termed "genomic repeat expression monitor" (GREM) utilizes CC of preamplified 3′-terminal genomic flanking regions of the repetitive elements with the set of cDNA 5′-terminal parts, which results in the construction of a hybrid genomic DNA or cDNA library, enriched in promoter-active repeats, thus making it possible to create a comprehensive genome-wide map of such repetitive elements (Buzdin et al. 2006a, b). These and other applications of the CC are described in detail in Chapter 8.

4. HYBRIDIZATION IN SOLUTION FOR THE RECOVERY OF GENOMIC POLYMORPHISMS

Unlike SH, which generally deals with finding relatively long differential DNA fragments, this group of methods is aimed at the identification of very small, single nucleotide-scale differences between the comparing DNA samples. The study of such mutations reveals the normal functions of genes, proteins, noncoding RNAs, the causes of many malignancies, and the variability of responses among individuals. A plethora of SNPs are not deleterious by themselves, but are linked to phenotypes associated with diseases and drug responses, thus providing a great opportunity for their use in large-scale association and population studies. Moreover, SNPs are increasingly recognized as important diagnostic markers for the detection of drug-resistant strains of hazardous microorganisms.

An impressive number of research groups working in this field managed to identify 10 million SNPs in recent years. However, this figure seems negligible compared to the real number of SNPs and other mutations present in the genomes. The ideal method for mutation and SNP recovery would detect mutations in large fragments of DNA and position them to single base-pair accuracy, and would be sensitive, precise, and robust. Currently, the need in mutation detection is reflected by the plethora of chemical, enzymatic, bioinformatical, and physically based techniques. Many mutation discovery methods quickly and effectively indicate the presence of a mutation in a sample region, but fail to resolve its characterization and localization; another family of methods permits precise mutation mapping, but in a more laborious and expensive way. The group of novel approaches for mutation detection based on DNA hybridization in solution, which combines high performance, cost-efficiency, reliability, and detailed mutation characterization, will be reviewed in Chapter 9.

At present, mutation discovery through Sanger sequencing is often thought of as the "gold standard" for mutation detection. This perception is distorted due to the fact that this is the *only* method of mutation identification, but this does not mean it is the best for mutation detection. The fact that many scanning methods detect 5–10% of mutant molecules in a wild-type environment immediately indicates that these methods are advantageous over sequencing, at least for some purposes. Using bioinformatical approaches, a large number of mutations (mostly SNPs) were recently discovered.

However, these methodologies require prior knowledge of target sequences, normally obtained through DNA sequencing, and mutation recovery in such case is usually performed by multiple sequence alignment of publicly available sequence data. Recent studies indicate that only a small percentage of mutations can be discovered using this approach and, in particular, that SNPs with low frequency are often missed. It is clear now that high-throughput methods for detecting these variations are needed for in-population screening for complex genetic diseases in which extended genomic loci, large genes, and/or several genes may be affected. To meet the need of these studies, several groups of approaches have been developed. All of them are based on the rationale that mutation-containing DNA molecule will form mismatches at the mutation site when hybridized to the reference wild-type DNA (Figure 10). Thus, when mutant and wild-type DNA are hybridized together, two complementary mismatches are formed. Therefore, the detection and correct position of such mismatches is the key for mutation recovery.

Such mispaired nucleotides can be identified directly or indirectly using very different chemical, enzymatic, or physical approaches, which offer excellent detection efficiencies coupled with high throughput and low unit cost. It should be noted that definition of the mutational change obviously requires a sequencing step, at least to confirm the results. But in this case sequencing is targeted, not a "fishing expedition" in which the region where mutation occurred is unknown and plenty of sequencing work is absolutely required. As a result, these methods are able to cut the costs of detecting a mutation by one order of magnitude or more. Briefly, chemical approaches utilize chemical cleavage or modification of the mispaired nucleotides, enzymatic techniques employ enzymatic recognition of the

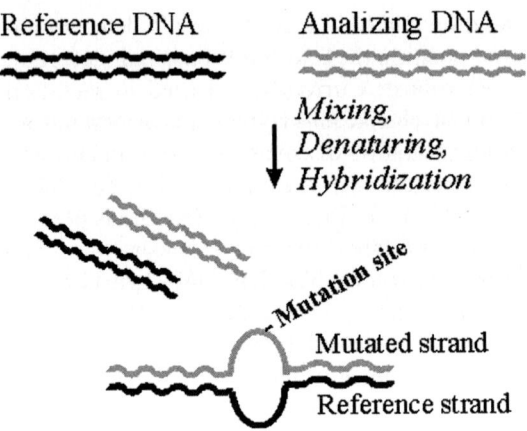

Figure 10. When hybridized, mutated and reference wild-type DNAs form heteroduplexes having mispaired regions, corresponding to the mutation sites.

mismatch (with further binding, cleavage, modification, or ligation of the DNA at the mispaired nucleotides), whereas physical methods look for a physical difference between the mutant strand and wild-type strands of DNA, being based either on physical isolation of imperfectly matched DNA hybrids (like electrophoretic separation) or on finding differences in mismatched versus perfect DNA hybrid physical peculiarities. All these approaches utilize nucleic acids hybridization in solution, and are described in more detail in Chapter 9, in comparison with each other and with direct sequencing-based approaches.

5. CONCLUSIONS

In this chapter I have briefly illustrated how the methods based on nucleic acids hybridization in solution could be helpful for a number of applications, and have provided a short overview underlining the methods' principles for each major technique, except for microarray hybridization, which is thoroughly reviewed in Chapter 11. A more detailed description will be provided in the following chapters of this book: Chapter 2 – PCR suppression effect; Chapter 4 – use of stem-loop oligonucleotides; Chapters 3, 6, and 7 – important modern variations of SH; Chapter 5 – normalization of the cDNA libraries; Chapter 8 – CC; Chapter 9 – hybridization-based mutation detection; and Chapter 10 – current attempts to improve the hybridization specificity.

A wide spectrum of experimental tasks covered by these approaches includes finding differential sequences in both genomic DNAs and cDNAs using microarrays, SH or ordered differential display, genome walking, multiplex PCR, cDNA library construction starting from small amount of total RNA, rapid amplification of cDNA ends (RACE), effective smoothing of the concentrations of rare and abundant transcripts in cDNA libraries, recovery of promoter-active repeats and differentially methylated genomic DNA, identification of common DNA in genomic or cDNA sources, new gene mapping, finding evolutionary conserved sequences and both single-nucleotide and extended mutation discovery or large-scale monitoring. Of course, there is no panacea, an ideal method that would solve all technical problems that researchers face, but a combination of the above approaches will be likely fruitful for conducting successful research. The international team of the authors of this book has tried both to elucidate the current state of the art in hybridization techniques and to help the readers in choosing an appropriate method for performing an experiment in the most efficient way. We hope the book will be useful to all those interested in the modern life science methodologies.

REFERENCES

Akopyants NS, Fradkov A, Diatchenko L, Hill JE, Siebert PD, Lukyanov SA, Sverdlov ED, Berg DE (1998) PCR-based subtractive hybridization and differences in gene content among strains of *Helicobacter pylori*. Proc Natl Acad Sci USA 95:13108–13113

Ayyanathan K, Francis VS, Datta S, Padmanaban G (1995) Development of specific DNA probes and their usage in the detection of Plasmodium vivax infection in blood. Mol Cell Probes 9:239–246

Azhikina T, Gainetdinov I, Skvortsova Y, Batrak A, Dmitrieva N, Sverdlov E (2004) Non-methylated Genomic Sites Coincidence Cloning (NGSCC): an approach to large scale analysis of hypomethylated CpG patterns at predetermined genomic loci. Mol Genet Genomics 271:22–32

Azhikina T, Gainetdinov I, Skvortsova Y, Sverdlov E (2006) Methylation-free site patterns along a 1-Mb locus on Chr19 in cancerous and normal cells are similar. A new fast approach for analyzing unmethylated CCGG sites distribution. Mol Genet Genomics 275:615–622

Azhikina TL, Sverdlov ED (2005) Study of tissue-specific CpG methylation of DNA in extended genomic loci. Biochemistry (Mosc) 70:596–603

Barr FG, Emanuel BS (1990) Application of a subtraction hybridization technique involving photoactivatable biotin and organic extraction to solution hybridization analysis of genomic DNA. Anal Biochem 186:369–373

Bautz EK, Reilly E (1966) Gene-specific messenger RNA: isolation by the deletion method. Science 151:328–330

Bogush ML, Velikodvorskaya TV, Lebedev YB, Nikolaev LG, Lukyanov SA, Fradkov AF, Pliyev BK, Boichenko MN, Usatova GN, Vorobiev AA, Andersen GL, Sverdlov ED (1999) Identification and localization of differences between *Escherichia coli* and *Salmonella typhimurium* genomes by suppressive subtractive hybridization. Mol Gen Genet 262:721–729

Bonaldo MF, Lennon G, Soares MB (1996) Normalization and subtraction: two approaches to facilitate gene discovery. Genome Res 6:791–806

Brown PO, Botstein D (1999) Exploring the new world of the genome with DNA microarrays. Nat Genet 21:33–37

Buzdin A, Khodosevich K, Mamedov I, Vinogradova T, Lebedev Y, Hunsmann G, Sverdlov E (2002) A technique for genome-wide identification of differences in the interspersed repeats integrations between closely related genomes and its application to detection of human-specific integrations of HERV-K LTRs. Genomics 79:413–422

Buzdin A, Ustyugova S, Gogvadze E, Lebedev Y, Hunsmann G, Sverdlov E (2003) Genome-wide targeted search for human specific and polymorphic L1 integrations. Hum Genet 112:527–533

Buzdin A, Kovalskaya-Alexandrova E, Gogvadze E, Sverdlov E (2006a) At least 50% of human-specific HERV-K (HML-2) long terminal repeats serve in vivo as active promoters for host non-repetitive DNA transcription. J Virol 80:10752–10762

Buzdin A, Kovalskaya-Alexandrova E, Gogvadze E, Sverdlov E (2006b) GREM, a technique for genome-wide isolation and quantitative analysis of promoter active repeats. Nucleic Acids Res 34:e67

Carulli JP, Artinger M, Swain PM, Root CD, Chee L, Tulig C, Guerin J, Osborne M, Stein G, Lian J, Lomedico PT (1998) High throughput analysis of differential gene expression. J Cell Biochem Suppl 30–31:286–296

Cekan SZ (2004) Methods to find out the expression of activated genes. Reprod Biol Endocrinol 2:68

Chalaya T, Gogvadze E, Buzdin A, Kovalskaya E, Sverdlov ED (2004) Improving specificity of DNA hybridization-based methods. Nucleic Acids Res 32:e130

Chen H, Pulido JC, Duyk GM (1995) MATS: a rapid and efficient method for the development of microsatellite markers from YACs. Genomics 25:1–8

Chien Y, Becker DM, Lindsten T, Okamura M, Cohen DI, Davis MM (1984) A third type of murine T-cell receptor gene. Nature 312:31–35

Cho TJ, Park SS (1998) A simulation of subtractive hybridization. Nucleic Acids Res 26:1440–1448

Clapp JP, McKee RA, Allen-Williams L, Hopley JG, Slater RJ (1993) Genomic subtractive hybridization to isolate species-specific DNA sequences in insects. Insect Mol Biol 1:133–138

Cook D, Sequeira L (1991) The use of subtractive hybridization to obtain a DNA probe specific for Pseudomonas solanacearum race 3. Mol Gen Genet 227:401–410

Cruz-Reyes JA, Ackers JP (1992) A DNA probe specific to pathogenic *Entamoeba histolytica*. Arch Med Res 23:271–275

Devon RS, Brookes AJ (1996) Coincidence cloning. Taking the coincidences out of genome analysis. Mol Biotechnol 5:243–252

Diatchenko L, Lau YF, Campbell AP, Chenchik A, Moqadam F, Huang B, Lukyanov S, Lukyanov K, Gurskaya N, Sverdlov ED, Siebert PD (1996) Suppression subtractive hybridization: a method for generating differentially regulated or tissue-specific cDNA probes and libraries. Proc Natl Acad Sci USA 93:6025–6030

Diatchenko L, Lukyanov S, Lau YF, Siebert PD (1999) Suppression subtractive hybridization: a versatile method for identifying differentially expressed genes. Methods Enzymol 303:349–380

Drew AC, Brindley PJ (1995) Female-specific sequences isolated from Schistosoma mansoni by representational difference analysis. Mol Biochem Parasitol 71:173–181

Ermolaeva OD, Sverdlov ED (1996) Subtractive hybridization, a technique for extraction of DNA sequences distinguishing two closely related genomes: critical analysis. Genet Anal 13:49–58

Ermolaeva OD, Wagner MC (1995) SUBTRACT: a computer program for modeling the process of subtractive hybridization. Comput Appl Biosci 11:457–462

Ermolaeva OD, Lukyanov SA, Sverdlov ED (1996) The mathematical model of subtractive hybridization and its practical application. Proc Int Conf Intell Syst Mol Biol 4:52–58

Frohme M, Camargo AA, Czink C, Matsukuma AY, Simpson AJ, Hoheisel JD, Verjovski-Almeida S (2001) Directed gap closure in large-scale sequencing projects. Genome Res 11:901–903

Ganova-Raeva L, Zhang X, Cao F, Fields H, Khudyakov Y (2006) Primer Extension Enrichment Reaction (PEER): a new subtraction method for identification of genetic differences between biological specimens. Nucleic Acids Res 34:e76

Hara E, Kato T, Nakada S, Sekiya S, Oda K (1991) Subtractive cDNA cloning using oligo(dT)30-latex and PCR: isolation of cDNA clones specific to undifferentiated human embryonal carcinoma cells. Nucleic Acids Res 19:7097–7104

Herfort MR, Garber AT (1991) Simple and efficient subtractive hybridization screening. Biotechniques 11:598, 600, 602–604

Hla T, Maciag T (1990) Isolation of immediate-early differentiation mRNAs by enzymatic amplification of subtracted cDNA from human endothelial cells. Biochem Biophys Res Commun 167:637–643

Jin H, Cheng X, Diatchenko L, Siebert PD, Huang CC (1997) Differential screening of a subtracted cDNA library: a method to search for genes preferentially expressed in multiple tissues. Biotechniques 23:1084–1086

Kavathas P, Sukhatme VP, Herzenberg LA, Parnes JR (1984) Isolation of the gene encoding the human T-lymphocyte differentiation antigen Leu-2 (T8) by gene transfer and cDNA subtraction. Proc Natl Acad Sci USA 81:7688–7692

Kohne DE, Levison SA, Byers MJ (1977) Room temperature method for increasing the rate of DNA reassociation by many thousandfold: the phenol emulsion reassociation technique. Biochemistry 16:5329–5341

Kunkel LM, Monaco AP, Middlesworth W, Ochs HD, Latt SA (1985) Specific cloning of DNA fragments absent from the DNA of a male patient with an X chromosome deletion. Proc Natl Acad Sci USA 82:4778–4782

Kuvbachieva AA, Goffinet AM (2002) A modification of representational difference analysis, with application to the cloning of a candidate in the Reelin signalling pathway. BMC Mol Biol 3:6

Kuze K, Shimizu A, Honjo T (1989) A new vector and RNase H method for the subtractive hybridization. Nucleic Acids Res 17:807

Laman AG, Kurjukov SG, Bulgakova EV, Anikeeva NN, Brovko FA (2001) Subtractive hybridization of biotinylated DNA in phenol emulsion. J Biochem Biophys Methods 50:43–52

Lamar EE, Palmer E (1984) Y-encoded, species-specific DNA in mice: evidence that the Y chromosome exists in two polymorphic forms in inbred strains. Cell 37:171–177

Lander ES, Linton LM, Birren B, Nusbaum C, Zody MC, Baldwin J, Devon K, et al. (2001) Initial sequencing and analysis of the human genome. Nature 409:860–921

Liang P, Pardee AB (1992) Differential display of eukaryotic messenger RNA by means of the polymerase chain reaction. Science 257:967–971

Lisitsyn N, Wigler M (1993) Cloning the differences between two complex genomes. Science 259:946–951

Lisitsyn N, Wigler M (1995) Representational difference analysis in detection of genetic lesions in cancer. Methods Enzymol 254:291–304

Lisitsyn NA, Leach FS, Vogelstein B, Wigler MH (1994a) Detection of genetic loss in tumors by representational difference analysis. Cold Spring Harb Symp Quant Biol 59:585–587

Lisitsyn NA, Segre JA, Kusumi K, Lisitsyn NM, Nadeau JH, Frankel WN, Wigler MH, Lander ES (1994b) Direct isolation of polymorphic markers linked to a trait by genetically directed representational difference analysis. Nat Genet 6:57–63

Lisitsyn NA, Lisitsina NM, Dalbagni G, Barker P, Sanchez CA, Gnarra J, Linehan WM, Reid BJ, Wigler MH (1995) Comparative genomic analysis of tumors: detection of DNA losses and amplification. Proc Natl Acad Sci USA 92:151–155

Lopez-Fernandez LA, del Mazo J (1993) Construction of subtractive cDNA libraries from limited amounts of mRNA and multiple cycles of subtraction. Biotechniques 15:654–6, 658–659

Martin KJ, Pardee AB (2000) Identifying expressed genes. Proc Natl Acad Sci USA 97:3789–3791

Milner JJ, Cecchini E, Dominy PJ (1995) A kinetic model for subtractive hybridization. Nucleic Acids Res 23:176–187

Nadezhdin EV, Vinogradova TV, Sverdlov ED (2001) Interspecies subtractive hybridization of cDNA from human and chimpanzee brains. Dokl Biochem Biophys 381:415–418

Palazzolo MJ, Meyerowitz EM (1987) A family of lambda phage cDNA cloning vectors, lambda SWAJ, allowing the amplification of RNA sequences. Gene 52:197–206

Pardinas JR, Combates NJ, Prouty SM, Stenn KS, Parimoo S (1998) Differential subtraction display: a unified approach for isolation of cDNAs from differentially expressed genes. Anal Biochem 257:161–168

Rebrikov DV, Britanova OV, Gurskaya NG, Lukyanov KA, Tarabykin VS, Lukyanov SA (2000) Mirror orientation selection (MOS): a method for eliminating false positive clones from libraries generated by suppression subtractive hybridization. Nucleic Acids Res 28:e90

Rosenberg M, Przybylska M, Straus D (1994) "RFLP subtraction": a method for making libraries of polymorphic markers. Proc Natl Acad Sci USA 91:6113–6117

Rubenstein JL, Brice AE, Ciaranello RD, Denney D, Porteus MH, Usdin TB (1990) Subtractive hybridization system using single-stranded phagemids with directional inserts. Nucleic Acids Res 18:4833–4842

Rubin CM, Leeflang EP, Rinehart FP, Schmid CW (1993) Paucity of novel short interspersed repetitive element (SINE) families in human DNA and isolation of a novel MER repeat. Genomics 18:322–328

Sagerstrom CG, Sun BI, Sive HL (1997) Subtractive cloning: past, present, and future. Annu Rev Biochem 66:751–783

Sallie R (1995) Isolation of candidate genes by subtractive and sequential (Boolean) hybridization: an hypothesis. Med Hypotheses 45:142–146

Sasaki H, Nomura S, Akiyama N, Takahashi A, Sugimura T, Oishi M, Terada M (1994) Highly efficient method for obtaining a subtracted genomic DNA library by the modified in-gel competitive reassociation method. Cancer Res 54:5821–5823

Schutte M, da Costa LT, Hahn SA, Moskaluk C, Hoque AT, Rozenblum E, Weinstein CL, Bittner M, Meltzer PS, Trent JM, et al. (1995) Identification by representational difference analysis of a homozygous deletion in pancreatic carcinoma that lies within the BRCA2 region. Proc Natl Acad Sci USA 92:5950–5954

Sharma P, Lonneborg A, Stougaard P (1993) PCR-based construction of subtractive cDNA library using magnetic beads. Biotechniques 15:610, 612

Southern EM (1992) Detection of specific sequences among DNA fragments separated by gel electrophoresis. 1975. Biotechnology 24:122–139

Sverdlov ED (1993) Subtractive hybridization-a technique for extracting DNA sequences, discriminating between two closely-related genomes. Mol Gen Mikrobiol Virusol 6:3–12

Sverdlov ED, Ermolaeva OD (1993) Subtractive hybridization. Theoretical analysis, and a principle of the trap. Bioorg Khim 19:1081–1088

Sverdlov ED, Ermolaeva OD (1994) Kinetic analysis for subtractive hybridization of transcripts. Bioorg Khim 20:506–514

Timblin C, Battey J, Kuehl WM (1990) Application for PCR technology to subtractive cDNA cloning: identification of genes expressed specifically in murine plasmacytoma cells. Nucleic Acids Res 18:1587–1593

Travis GH, Sutcliffe JG (1988) Phenol emulsion-enhanced DNA-driven subtractive cDNA cloning: isolation of low-abundance monkey cortex-specific mRNAs. Proc Natl Acad Sci USA 85:1696–1700

Venter JC, Adams MD, Myers EW, Li PW, Mural RJ, Sutton GG, Smith HO, et al. (2001) The sequence of the human genome. Science 291:1304–1351

Wang Z, Brown DD (1991) A gene expression screen. Proc Natl Acad Sci USA 88:11505–11509

Watson JD, Crick FH (1953) Molecular structure of nucleic acids; a structure for deoxyribose nucleic acid. Nature 171:737–738

Welcher AA, Torres AR, Ward DC (1986) Selective enrichment of specific DNA, cDNA and RNA sequences using biotinylated probes, avidin and copper-chelate agarose. Nucleic Acids Res 14:10027–10044

Wieland I, Bolger G, Asouline G, Wigler M (1990) A method for difference cloning: gene amplification following subtractive hybridization. Proc Natl Acad Sci USA 87:2720–2724

Yang GP, Ross DT, Kuang WW, Brown PO, Weigel RJ (1999) Combining SSH and cDNA microarrays for rapid identification of differentially expressed genes. Nucleic Acids Res 27:1517–1523

Ying SY (2004) Complementary DNA libraries: an overview. Mol Biotechnol 27:245–252

Ying SY, Lin S (1999) High-performance subtractive hybridization of cDNAs by covalent bonding between specific complementary nucleotides. Biotechniques 26:966–968, 970–972, 979 passim

Zeschnigk M, Horsthemke B, Lohmann D (1999) Detection of homozygous deletions in tumors by hybridization of representational difference analysis (RDA) products to chromosome-specific YAC clone arrays. Nucleic Acids Res 27:e30

CHAPTER 2

SELECTIVE SUPPRESSION OF POLYMERASE CHAIN REACTION AND ITS MOST POPULAR APPLICATIONS

SERGEY A. LUKYANOV*, KONSTANTIN A. LUKYANOV, NADEZHDA G. GURSKAYA, EKATERINA A. BOGDANOVA, ANTON A. BUZDIN

Shemyakin-Ovchinnikov Institute of Bioorganic Chemistry, Russian Academy of Sciences, Moscow 117997, Miklukho-Maklaya 16/10
*Corresponding author
Phone: +(7495) 3307029; Fax: +(7495) 3307056; E-mail: luk@ibch.ru*

Abstract:	This chapter is devoted to methods based on suppression polymerase chain reaction (PCR) effect (cDNA library construction starting from a small amount of total RNA; suppression subtractive hybridization, SSH; ordered differential display, ODD; Marathon cDNA RACE; genome walking; variants of coincidence cloning, CC; normalization of cDNA libraries; multiplex PCR, mPCR; to *in vitro* cloning). Taken together, these approaches allow one to analyze complex DNA samples, from searching sequences of interest to determining complete structures of the respective genes.
Keywords:	Selective PCR suppression, PCR suppression, suppression of the PCR, PS, pan handle, suppression adapter, inverted repeats, annealing temperature, primer concentration, suppression sequence, preparation of full-size cDNA, small amount of biological material, selective amplification, differential display, ordered differential display, targeted differential display, shortcoming, disadvantage, normalized cDNA libraries, evolutionary conserved sequences, search for promoter sites, chromosome walking, *in vitro* cloning, multiplex PCR.
Abbreviations:	cDNA, complementary DNA; dNTP, deoxyribonucleotidetriphosphate; ITR, inverted terminal repeat; mRNA, messenger RNA; ODD, ordered differential display; PAGE, polyacrylamide gel electrophoresis; PCR, polymerase chain reaction; PS, PCR suppression effect; RACE, rapid amplification of cDNA ends; RT, reverse transcription; SSP, selective suppression of PCR; SSH, suppression subtractive hybridization; TGDD, targeted genomic differential display.

TABLE OF CONTENTS

1. Introduction 30
2. Selective PCR Suppression 30
3. Preparation of ITR-Containing DNA Samples 32
4. Strategy of Employment of SSP 33
5. SSP-based Procedures 33
   5.1 Construction of cDNA Libraries from a Small Amount
       of Biological Material 35
   5.2 Detection of Differentially Expressed Genes 36
   5.3 Search for 5'- and 3'-Terminal Fragments of cDNA 40
   5.4 Search for Promoter Sites (Chromosome Walking) 40
   5.5 *In vitro* Cloning 41
   5.6 Multiplex PCR 42
6. Detailed Protocols 43
   6.1 Materials 43
   6.2 Methods 45
References 48

1. INTRODUCTION

The most important processes in various biological systems (cell differentiation and morphogenesis during embryonic development and regeneration, apoptosis or malignization of cells, etc.) are under the control of specific regulatory genes. To understand the underlying molecular mechanisms, the genes involved in their regulation should be revealed and studied. At present, the majority of methods of molecular biology involved in unraveling such problems are based on polymerase chain reaction (PCR), which has made it possible to work with small amounts of biological material. However, the use of PCR requires information on the sequence of the DNA under study. When the sequence is partially or completely unknown, the PCR often encounters difficulties.

The selective suppression of PCR (SSP) phenomenon discovered in this laboratory (Launer et al. 1994) generated a number of highly effective, mutually complementary methods of finding and analyzing new functionally important DNA and RNA sequences when information, complete or partial, on their primary structure is absent. The use of SSP enables one to skip labor-intensive and not very effective methods of physical fractionation of DNA and makes the methods for search and analysis of genetic sequences easier, more rapid, and more reproducible.

2. SELECTIVE PCR SUPPRESSION

SSP consists in inhibiting the amplification of DNA molecules flanked by inverted terminal repeats (ITRs) in PCR with a primer corresponding to the external part of the ITR, provided that the primer is considerably shorter than the ITR (Figure 1).

Selective suppression of polymerase chain reaction

Restriction endonuclease-treated DNA or cDNA

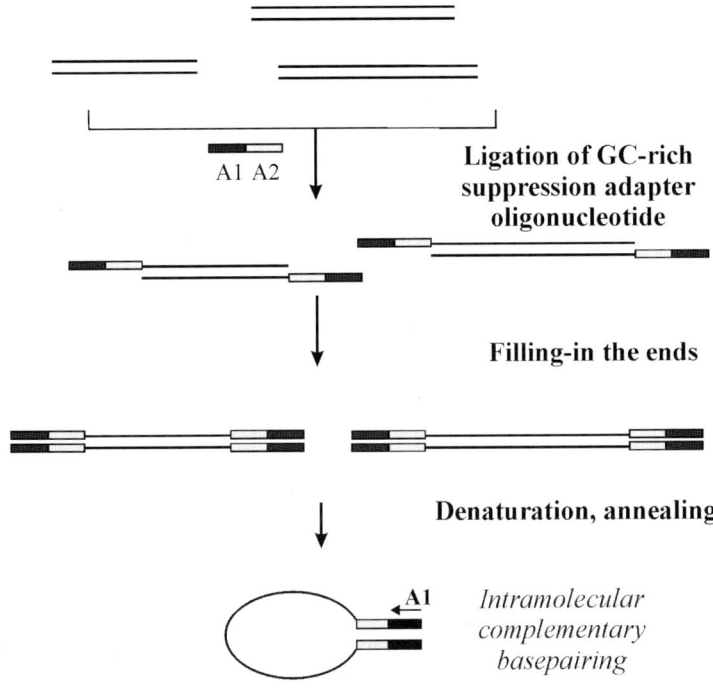

Figure 1. Schematic representation of the PCR suppression effect. DNA molecules flanked by inverted terminal repeats form intramolecular terminal duplexes, thus preventing terminal adapter-specific primer annealing and, consequently, inhibiting the PCR.

The principle of PCR suppression is that the target DNA is designed to adopt a hairpin form. In general, in all its applications, PS allows the amplification of wanted sequences and simultaneously suppresses the amplification of unwanted ones. Pan handle-like stem-loop DNA constructs for PS are created by ligation of long GC-rich adapters to DNA or cDNA restriction fragments (Figure 1) (Launer et al. 1994). As a result, each single-stranded DNA (ssDNA) fragment is flanked by terminal inverted repeats (i.e. by self-complementary ends). During PCR, on denaturing and annealing, the self-complementary ends of each single strand form duplex stems, converting each fragment into a large pan handle-like stem-loop structure. The formation of stable duplex structures at the fragment ends makes the PCR with the adapter-primer (A-primer) alone relatively inefficient, because the intramolecular annealing of the complementary termini is kinetically favored and more stable than the intermolecular annealing of shorter A-primers. This effect is therefore called PCR suppression (Launer et al. 1994).

In the presence of both A-primers and target-primers (T-primers, targeted at the gene-specific sequences in the single-stranded loops), however, PCR is efficient. The T-primer anneals to its target and is used by DNA polymerase to initiate DNA synthesis. The newly synthesized product has two termini, which are not complementary and, thus, cannot fold into a stem-loop structure. This fragment is, therefore, not subject to the PS effect, and is efficiently amplified. Consequently, only the fragments containing the target are exponentially amplified by PCR, while the background fragments without the target remain inert.

The inhibition of amplification depends on many parameters, of which the following are the major ones:

1. The difference in the annealing temperatures of ITR and the amplification primer. The ratio of the length and GC content of the whole ITR and the sequence corresponding to the primer affects the SSP effectiveness considerably. The use of ITR 40–50 bp long with an increased GC content in its internal (suppression) part and of an amplification primer corresponding to the external 20–25 bp segment of the ITR appears to be optimum.
2. The length of the ITR-containing DNA molecule. The longer the molecule, the less probable is the encounter and intramolecular hybridization of its termini. Thus, SSP effect does not inhibit, or inhibits sparingly, amplification of very long DNA (the threshold value depends on the conditions and is usually 6–8 kbp).
3. Primer concentration in PCR. SSP depends on competition between intramolecular hybridization and primer annealing. Therefore, SSP is more effective at low-primer concentrations.

SSP can be used to suppress amplification of an unwanted DNA fraction, which requires suitable long ITR, i.e. suppression sequences, to be introduced into the DNA.

3. PREPARATION OF ITR-CONTAINING DNA SAMPLES

Two main methods for the attachment of suppression sequences to DNA molecules were developed:
1. Ligation of double-stranded DNA (dsDNA) fragments with a pseudodouble-stranded, or so-called suppression, adapter (Siebert et al. 1995)
2. PCR with a long (suppression) primer whose 3′-terminal part is also present within the DNA fragments (Launer et al. 1994)

The first method is highly versatile, since it does not require any special sequences to be introduced into DNA. For example, a DNA sample for ligation can be obtained by the synthesis of double-stranded cDNA or by the treatment of cDNA or genomic DNA with restriction endonucleases. Apparently, blunt-end (rather than staggered-end) DNA fragments are optimum because, in this case, blunt-end suppression adapters can be universally used in ligation. The second technique is effective when an amplified DNA sample is being dealt with so that it harbors known sequences. This can be exemplified by cDNA samples obtained through addition of a homopolymer sequence to the first cDNA strand followed by PCR with a T-primer or C- and T-primers.

4. STRATEGY OF EMPLOYMENT OF SSP

SSP can be used in the analysis of complex mixtures of DNA fragments (cDNA or fragmented genomic DNA) to suppress an undesirable DNA fraction while retaining exponential amplification of target sequences. By and large, SSP allows the selection of asymmetrically flanked DNA molecules from mixtures with symmetrically flanked ones. Before the discovery of SSP, such a problem could not be solved using PCR. Three major schemes of SSP employment can be outlined.

The first scheme (Figure 2A) is based on the addition of one suppression sequence to all DNA molecules, followed by amplification with two primers of which one corresponds to the external part of the suppression sequence while the other is complementary to the target DNA (gene-specific or oligo(dT)-containing primer). In the course of PCR, DNA molecules bearing the sequence of the second primer are selected. This scheme underlies methods such as the search for the genomic DNA or cDNA sequences belonging to a known fragment (Siebert et al. 1995; Chenchik et al. 1996), the construction of cDNA libraries on the basis of low amounts of total RNA (Lukyanov et al. 1997), and mRNA ordered differential display (ODD) (Matz et al. 1997).

The second scheme (Figure 2B) is based on an addition of two different suppression sequences to DNA followed by amplification with primers corresponding to their external parts. In the course of PCR, asymmetrically flanked DNA molecules – bearing different suppression sequences at their ends – are selected. This scheme underlies the *in vitro* cloning method (Lukyanov et al. 1996).

The third scheme (Figure 2C) is a sophisticated variant of the previous one. Two DNA samples are supplied with different 5'-terminal (but not 3'-terminal) suppression sequences. The samples are then denatured and hybridized together. After filling in the 3'-termini, PCR is carried out with primers corresponding to the external parts of the suppression sequences. As in the second scheme, asymmetrically flanked DNA molecules are selected, but in this case each of these DNA must anneal with the complementary strand from the alternative sample. Such selection of heteroduplexes enables subtractive hybridization (SH) of cDNA (Launer et al. 1994), cDNA normalization (Gurskaya et al. 1996), modified coincidence cloning (Azhikina et al. 2004; Chalaya et al. 2004; Buzdin et al. 2006), and a search for evolutionarily conserved cDNA.

5. SSP-BASED PROCEDURES

Currently, studies on the effects of gene expression on various biological processes often employ the following strategy: (1) construction of cDNA libraries from biological samples under study; (2) screening of these libraries for

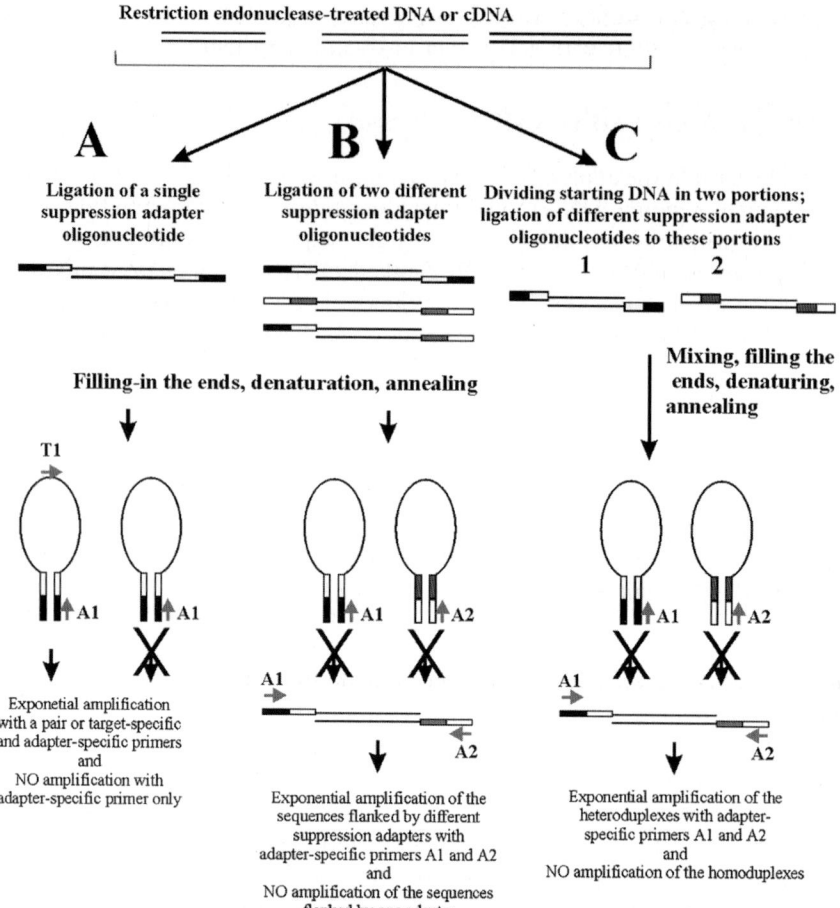

Figure 2. Outline of the major strategies using PCR suppression effect. (A) Single suppression adapter is ligated to the sample; (B) two different suppression adapters are ligated; (C) the sample is subdivided in two portions, which are further ligated to two different suppression adapters.

genes (more exactly, their fragments) differentially expressed or interesting in other aspects; and (3) preparation of full-size cDNA and genomic copies of the genes of interest. All these stages can be carried out using SSP-based methods that are considered in detail in section 5.1.

PS effect has already found applications in a variety of methods (reviewed in Luk'ianov et al. (1999)). In this book, suppression SH, multiplex PCR (mPCR), and targeted genomic differential display (TGDD) will be discussed:.

5.1 Construction of cDNA Libraries from a Small Amount of Biological Material

cDNA libraries are needed for a wide range of problems related to the functional and structural aspects of gene expression. The methods for their construction from relatively large amounts of biological material were developed long ago and are now routine genetic engineering procedures. However, when biological material is scarce and therefore poly(A) + RNA is insufficiently available (this is the standard situation nowadays), the conventional methods are of no avail.

The advent of PCR gave impetus to new methods of constructing cDNA libraries from small amounts of total RNA. The use of PCR, however, requires information on at least a partial sequence of the DNA to be amplified. Since mRNAs of the overwhelming majority of genes contain a poly(A) sequence, PCR can be carried out with an oligo(dT)-containing primer (T-primer). Exponential amplification needs an artificial sequence to be attached to all cDNA molecules for annealing of a second primer.

Several methods for the introduction of such a sequence have been described: (1) addition of a homopolymer sequence to the 3'-terminus of the first strand of cDNA using terminal deoxynucleotidyl transferase (tailing) (Frohman et al. 1988); (2) ligation of a synthetic single-stranded oligonucleotide to the first strand of cDNA (Edwards et al. 1991) or to mRNA (Liu and Gorovsky 1993) using T4 RNA ligase; and (3) ligation of a double-stranded oligonucleotide adapter to double-stranded cDNA using T4 DNA ligase (Akowitz and Manuelidis 1989).

Ligation of double-stranded molecules using T4 DNA ligase is more effective and reproducible than tailing or ligation of single-stranded substrates using T4 RNA ligase (Chenchik et al. 1996). However, the method (Akowitz and Manuelidis 1989) using T4 DNA ligase has the substantial drawback that it can be applied only to poly(A) + RNA. If cDNA is synthesized on the basis of total RNA, even the use of a T-primer does not prevent the synthesis of a large excess of background cDNA on ribosomal RNA as a template. On the basis of SSP, a method was developed that involves T4 DNA ligase-mediated ligation of a suppression adapter to double-stranded cDNA, followed by PCR with a T-primer and a primer corresponding to the external part of the adapter (Figure 2A) (Lukyanov et al. 1997).

Such PCR is accompanied by selective amplification of the cDNA fraction comprising the T-primer structure. At the same time, SSP inhibits amplification of the rest of cDNA, which is adapter-flanked from both sides. Thus, a cDNA library can be constructed based on a small amount of total RNA without isolating poly(A) + RNA fraction. We have shown the potential of this procedure on a model system (Lukyanov et al. 1997). The libraries so constructed were subsequently used to reveal and analyze functionally important expressing sequences in various biological systems (Bogdanova et al. 1997, 1998; Kazanskaya et al. 1997). This method enables representative, essentially full-size, cDNA libraries to be constructed from small (10–100 ng) amounts of total RNA that are of as high a quality as libraries produced by conventional methods from 5 to 10 mg of poly(A) + RNA.

5.2 Detection of Differentially Expressed Genes

Understanding the molecular mechanisms of biological processes requires the search for, and study of, genes expressed differentially in these processes. The identification methods for such genes are based on the detection of mRNA molecules that are diversely represented in various tissues and at various stages of a biological process. Changes in composition and content of cellular mRNA can be analyzed in several modes. Apart from microarray hybridizations, two strategies are most widely used: differential display and SH.

5.2.1 *Differential display*

The differential display technique was proposed in 1992 (Liang and Pardee 1992) and is currently in great demand in searches for differentially expressed genes. This technique is based on the employment of a short oligonucleotide primer that has a low annealing temperature and can direct PCR amplification of a limited pool of cDNA fragments. Comparative polyacrylamide gel electrophoresis (PAGE) analysis of such cDNA samples allows the identification of differentially represented cDNA fragments. Use of an arbitrary primer, however, precludes systematic comparison of the samples by all mRNA types: characteristic sets of DNA fragments are random, and the overwhelming majority of these fragments correspond to the types of transcripts most abundant in the original samples.

In addition, use of oligonucleotide primers with a low annealing temperature is accompanied by nonspecific amplification and leads to numerous artifacts. A totally different approach, involving a systematic comparison of samples by all mRNA types and based on the separation of 3′-terminal restriction fragments of cDNA, was proposed by Ivanova and Belyavsky (1995). Its drawbacks are its limited sensitivity and labor-intensive character. We have developed a method of mRNA ODD that is also based on an analysis of 3′-terminal restriction fragments of cDNA (Matz et al. 1997). However, to select these fragments, our approach uses PCR rather than physical separation (Figure 3). This results in selective amplification of 3′-terminal fragments of cDNA (from T-primer to the nearest restriction site), which substantially simplifies the method and enhances its sensitivity. The effectiveness of the ODD method was exemplified by revealing sequences differentially expressed along the anterior–exterior axis of planarian (Matz et al. 1997).

In TGDD, the PS effect is used to amplify a set of genomic or cDNA fragments with a single primer targeting repetitive sequences. Because of PS, the vast amount of other genomic fragments that do not have the targeted repeat remains unamplified. The PCR product consists of many fragments, each containing part of the

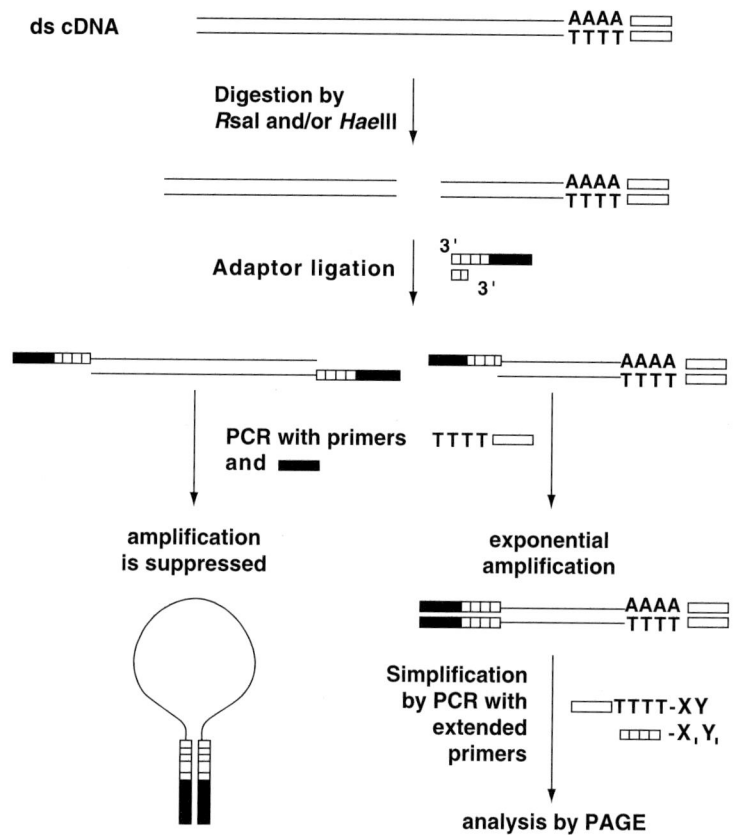

Figure 3. Schematic view of the ordered differential display (ODD) approach, which takes advantage of the PCR suppression effect for the specific amplification of cDNA 3′ regions.

repeat and one of the flanking regions. After resolution of the PCR products on a sequencing gel, a stable pattern of fragments specific for each genome or cDNA is obtained, with any differences revealed by their different mobility (Figure 4) (Broude et al. 1997, 1999; Lavrentieva et al. 1999; Vinogradova et al. 2002). Single-nucleotide polymorphisms, insertions, and deletions were detected when differential fragments were isolated and subjected to sequence analysis (Broude et al. 1999). In TGDD, the complexity of the displayed fragments is controlled by the 3′-anchoring of the primers, essentially in the same way as in the amplified fragment length polymorphism approach (Vos et al. 1995) and in the ordered cDNA differential display (Matz et al. 1997). This allows nonoverlapping sets of genomic fragments to be surveyed with different 3′-anchors in different reactions.

In contrast to other PCR-based display methods in which repeated sequences have been targeted (Weising et al. 1995; Donohue et al. 1997), TGDD shows high

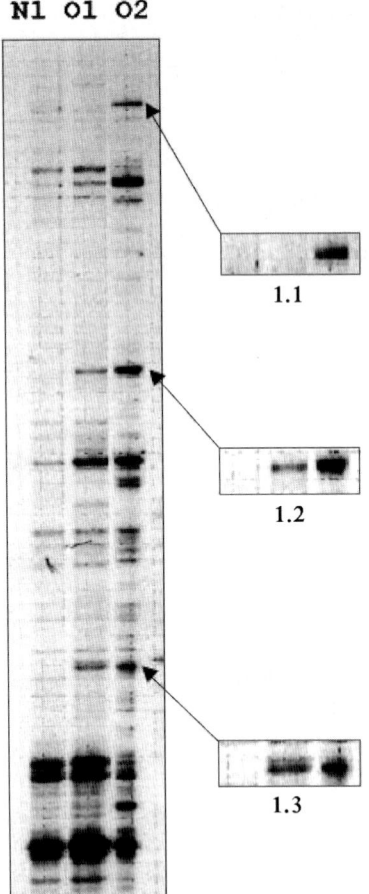

Figure 4. Representative electrophoregram showing results of the PS-assisted differential display of the endogenous retrovirus – driven transcripts in a normal (N1) and two cancerous tissues (O1 and O2). The fragments 1.1, 1.2, and 1.3 are abundant in O1 and O2 samples, but are hardly detectable in N1. The differential fragments are then removed from gels and sequenced, with their differential status to be confirmed by any independent approach like RT-PCR or Northern blot.

specificity: ~90% of the clones obtained the targets (Broude et al. 1999; Lavrentieva et al. 1999). TGDD was applied for the genome-wide recovery of polymorphic human genomic loci containing endogenous retroviral inserts (Lavrentieva et al. 1999) and demonstrated excellent performance, with high specificity, robustness, and high discrimination power. It can be adapted for the display of any type of repeats and its application is especially promising in cases where repeat instability is anticipated. Indeed, TGDD modifications were recently successfully employed for the identification of differentially expressed human endogenous retroviruses (Vinogradova et al. 2002) and for the recovery of polymorphic insertions of other mammalian repetitive elements – human L1 retrotransposons (Badge et al. 2003).

Shortcomings of the method should also be mentioned. These are common to all PCR-based methods with simultaneous amplification of a multitude of genomic fragments: (1) there is biased amplification of certain fragments, which leads to overrepresentation of some of the fragments and the loss of others during amplification; and (2) the complete digestion of all the DNA samples is paramount for reliable DNA comparisons.

5.2.2 Subtractive hybridization of cDNA

SH of cDNA is a process of exhaustive hybridization of two cDNA samples named *driver* and *tracer* that is meant to reveal sequences (targets) that are present in tracer (also frequently called tester) but are absent from, or at a lower level in, driver. SH involves hybridization of tracer with excess driver, followed by the separation of hybrid molecules from target molecules.

The use of SH of cDNA resulted in the detection of a considerable number of functionally important genes involved in embryonic development, cell differentiation, tumor transformation, and metastasizing. However, the low representativeness of the resulting enriched libraries, moderate enrichment degrees, and labor-intensity of procedures for the purification of the enriched fraction considerably hinders the use of this approach when the amount of biological material is limited and in the identification of genes whose transcripts are sparingly presented in the cell (1–10 copies per cell).

Our cDNA SH procedure (Diatchenko et al. 1996, 1999) allows one to overcome the problem of searches for rare transcripts by introducing a step of normalization of concentrations of various transcripts in the cDNA sample under study. This is achieved by using SSP (Figure 2C), which led to this method being named suppression subtractive hybridization (SSH). The effectiveness of this procedure was confirmed in model experiments with the use, as a target, of an exogenous viral DNA added to tracer in predetermined concentrations (Launer et al. 1994). This method, which will be described in detail in Chapter 3, was employed in revealing sequences differentially expressed in various biological processes, such as activation of immune response in a culture of immunocompetent cells (Gurskaya et al. 1996), alteration of the metastatic potential of tumor cells, or regeneration of planarian (Bogdanova et al. 1998), and for constructing human tissue-specific cDNA libraries (Diatchenko et al. 1996). The important advantages of SSH are: (1) it incorporates two hybridization steps, leading to efficient normalization of cDNA concentrations; (2) it requires only one subtraction round; and (3) it does not require physical separation of single-stranded and double-stranded fractions. The frequency of false positives is low and ~90% of clones are different (Diatchenko et al. 1996). As a result, SSH has become one of the most popular and efficient methods for subtractive expression studies (von Stein et al. 1997; Zhang and DuBois 2001).

Currently, the SSH technique is widely used (e.g. Chu et al. 1997; Hudson et al. 1997; Mueller et al. 1997; Yokomizo et al. 1997). Elements of the proposed technique of SH were successfully employed in the solution of some other problems, for example, constructing normalized cDNA libraries (Lukyanov et al. 1996; Luk'ianov et al. 1999) and revealing evolutionarily conserved sequences (Chalaya et al. 2004).

5.3 Search for 5'- and 3'-Terminal Fragments of cDNA

One of the most important and technically difficult objectives associated with the characterization of genes is the preparation of full-size cDNA. Conventional methods for detection of genetic sequences (screening of cDNA libraries, cloning of conserved genes using PCR with degenerated oligonucleotide primers, identification of differentially expressed genes using mRNA differential display, or cDNA SH) usually allow identification of only a fragment of cDNA. To clone full-size cDNA more rapidly and effectively, a number of methods jointly referred to as rapid amplification of cDNA ends (RACE) were recently proposed for *in vitro* amplification of the ends of cDNA (for review, see Schaefer 1995).

The majority of RACE methods now known are based on the introduction into the 3'-end of the first cDNA strand of an additional nucleotide sequence, which subsequently serves as the annealing site for a PCR primer. However, the problem of amplification suppression of nonspecific sequences upon use of the RACE strategy for rare genes is even more complicated than in constructing cDNA libraries. SSP makes it possible to overcome this difficulty. The cDNA sample is ligated with a suppression adapter and then amplified with a primer corresponding to the external part of the suppression adapter and a gene-specific primer (Figure 5; see also a more generalized scheme in Figure 2A). Only molecules containing the annealing site for the specific primer, i.e. 5'- or 3'-terminal sequences (depending on the direction of the specific primer), are thereby amplified. Amplification of the remainder of the molecules is prevented by SSP. The effectiveness of this method was tested in model experiments on known genetic sequences (Chenchik et al. 1996; Ackerman et al. 1997; Fleury et al. 1997; Loftus et al. 1997; Meyerson et al. 1997; Yang et al. 1997) and is currently widely used in many laboratories.

5.4 Search for Promoter Sites (Chromosome Walking)

Analysis of the genomic organization of the sequences isolated and cloning of regulatory regions of genes is important in the structure-functional analysis of genes. However, existing PCR-based techniques of chromosome walking are rather labor-intensive and ineffective. We proposed a novel technique of PCR walking on genomic DNA based on SSP (Siebert et al. 1995). To produce an ITR-containing DNA sample, genomic DNA is treated with a restriction endonuclease and ligated with a suppression adapter. Isolation of regulatory regions of the gene under study is carried out in subsequent PCR on the resulting DNA sample with gene-specific and adapter primers according to the scheme for isolation of full-size cDNA. This technique has also been widely used (e.g. Johansson and Karlsson 1996; Rosenberg and Dyer 1996; Takenoshita et al. 1996; Chong et al. 1997; Morii et al. 1997; Wade et al. 1997).

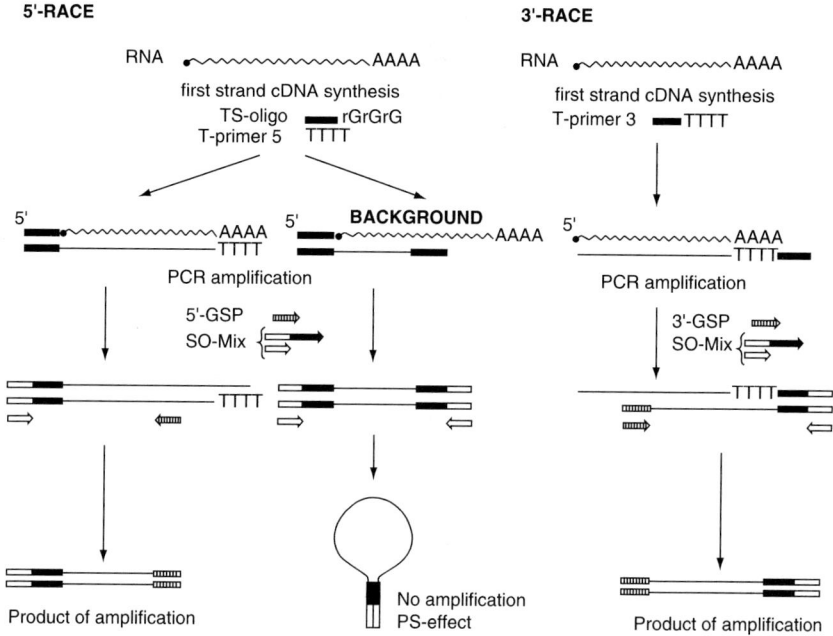

Figure 5. Rapid amplification of the cDNA ends (RACE) technique based on the PS effect. This fast and cost-effective procedure permits reliable cloning and identification of both 3′ and 5′ cDNA terminal regions.

5.5 *In Vitro* Cloning

Although the majority of classical genetic engineering methods were recently improved or displaced by more effective PCR-based techniques, traditional methods retained their leading role in the molecular cloning of DNA (cloning in bacterial, phage, and other *in vivo* systems). Based on SSP, a new method, named *in vitro* cloning, was proposed (Lukyanov et al. 1996), which allows PCR amplification and then sequencing of individual DNA molecules of unknown sequence without *in vivo* cloning.

The method of *in vitro* cloning includes the following stages (Figure 2B):
1. Simultaneous ligation of dsDNA fragments with two suppression adapters
2. Multiple dilution of the resulting sample to adjust the DNA content in the volume to be used in amplification to single molecules
3. PCR amplification of single DNA molecules using primers complementary to the external parts of the adapters.

The resulting PCR products, named *in vitro* clones, correspond to single DNA molecules. Owing to SSP, these clones are necessarily flanked with sequences of different adapters, which allows one to sequence the cloned DNA fragment by any method suitable for PCR products. *In vitro* cloning can be used in solving a wide range of problems of molecular biology in the place of conventional *in vivo* cloning. This method is especially convenient when no more than several dozen clones are needed. We used this approach in the differential screening of cDNA libraries constructed by SH (Luk'ianov and Luk'ianov 1997). In addition, we developed a protocol for rapid preparation of a panel of overlapping subclones to sequence long-stretched (5 kb) DNA fragments (Fradkov et al. 1998).

5.6 Multiplex PCR

PCR suppression was also used to develop a new strategy for mPCR amplification. In mPCR, multiple DNA targets are amplified simultaneously in one tube (Edwards and Gibbs 1994). Amplification of each target in conventional PCR requires two gene-specific primers. At a high level of multiplexing, it is often difficult to avoid primer interactions and achieve efficient and uniform target synthesis. In spite of numerous studies aimed at developing an effective strategy for mPCR (Shuber et al. 1995; Henegariu et al. 1997) and minimizing primer–primer interactions (Brownie et al. 1997), mPCR still presents a challenge (Broude 2002).

Although high multiplexing levels were achieved in some studies (e.g. Fan et al. 2000), they remain exceptions. Typically, a routine mPCR does not exceed 5- to 10-plex. Suppression PCR requires one gene-specific primer per amplicon and one primer, which is common for all targets. Therefore, an n-plex PCR would require only n primers instead of *2n* in conventional PCR (Figure 6). As expected, PS-based mPCR allowed efficient amplification of several targeted sequences (Broude et al. 2001; Broude 2002), and only simple adjustment of conditions was necessary to amplify simultaneously 30 DNA targets of different length from different human chromosomes (Broude 2002).

Additionally, PS-based mPCR exhibited excellent specificity and provided allele specificity in a multiplex format (Broude et al. 2001; Broude 2002). Although the PS approach includes two additional steps compared with conventional PCR (digestion of genomic DNA with a restriction enzyme and ligation with the adapters), this does not create many problems. It was demonstrated that it is possible to use one restriction enzyme for multiple target amplifications, so that a single DNA sample can be used in many experiments. Thus, application of the suppression approach for mPCR seems to offer several advantages over traditional methods, such as higher levels of multiplexing, higher specificity, simpler primer design, and primer cost savings (Broude 2002).

Thus, the discovery of SSP has resulted in a number of mutually complementary procedures of analysis of complex DNA samples: these include construction of cDNA libraries from small amounts of biological material, SH and differential display of mRNA for revealing differentially expressed sequences, fast cloning of full-size cDNA, chromosome walking for the cloning of promoter and other genomic regions, mPCR, *in vitro* cloning, and other applications.

Selective suppression of polymerase chain reaction

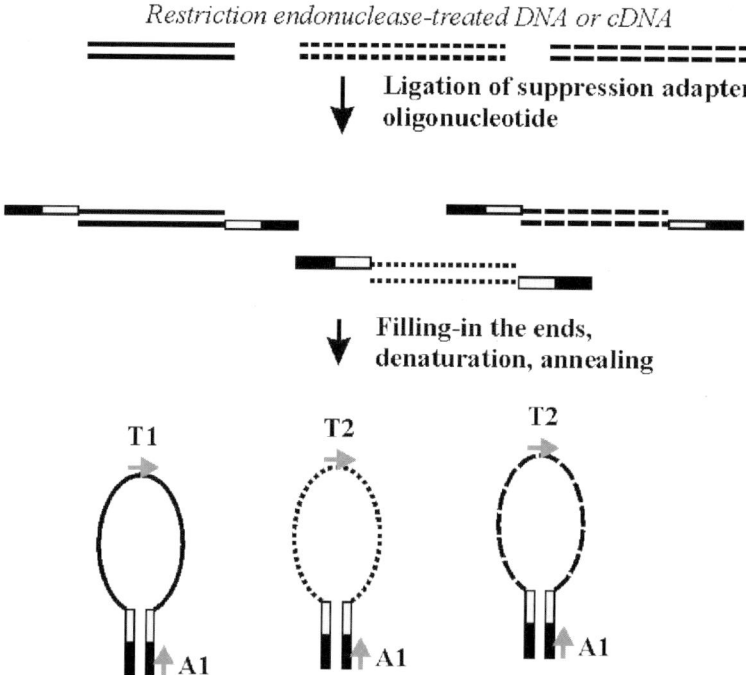

Figure 6. Scheme for the multiplex PCR amplification with only one specific primer used per one locus amplified (instead of two primers in conventional PCR).

6. DETAILED PROTOCOLS

6.1 Materials

6.1.1 Oligonucleotides

Many adapter oligonucleotides were shown to be efficient for the selective PCR suppression. We will provide here the protocols for direct suppression adapter ligation, as the direct adapter ligation is more uniform and widely used technique than the introduction of adapter sequences by PCR. The full-size adapter is usually a 40–45 nt long sequence with an inner (3'-terminal) part highly enriched in GC content. We recommend the following adapter oligonucleotides to be used for any of the above PS applications (whenever possible, oligonucleotides should be HPLC- or gel-purified):
Adapters and related oligonucleotides:
Set 1
A1A2 (adapter oligonucleotide, 44 nt long)

5'-*CTAATACGACTCACTATAGGGCTCGAGCGGCCGCCCGGGCAGGT*-3'
A3 (oligonucleotide complementary to the adapter 3'-terminal part, 10 nt long)
3'-*GGCCCGTCCA*-5'
A1 ("outer" PCR primer, 22 nt long)
5'-*CTAATACGACTCACTATAGGGC*-3'
A2 ("inner" primer for nested PCR, 22 nt long)
5'-*TCGAGCGGCCGCCCGGGCAGGT*-3'

Set 2
B1B2 (adapter oligonucleotide, 43 nt long)
5'-*TGTAGCGTGAAGACGACAGAAAGGGCGTGGTGCGGAGGGCGGT*-3'
B3 (oligonucleotide complementary to the adapter 3'-terminal part, 11 nt long)
3'-*GCCTCCCGCCA*-5'
B1 ("outer" PCR primer, 21 nt long)
5'-*TGTAGCGTGAAGACGACAGAA*-3'
B2 ("inner" primer for nested PCR, 22 nt long)
5'-*AGGGCGTGGTGCGGAGGGCGGT*-3'

Set 3
C1C2 (adapter oligonucleotide, 44 nt long)
5'-*AGCAGCGAACTCAGTACAACAAGTCGACGCGTGCCCGGGCTGGT*-3'
C3 (oligonucleotide complementary to the adapter 3'-terminal part, 11 nt long)
3'-*GGGCCCGACCA*-5'
C1 ("outer" PCR primer, 21 nt long)
5'-*AGCAGCGAACTCAGTACAACA*-3'
C2 ("inner" primer for nested PCR, 23 nt long)
5'-*AGTCGACGCGTGCCCGGGCTGGT*-3'

Note: In principle, all these adapter oligonucleotide sets can be used for any of the selective PCR suppression-based techniques; however, in some particular cases adapter and related nested PCR primer sequences may interfere with genomic or cDNA-specific oligonucleotides designed by the user. To avoid this problem, one should make a preliminary *in silico* analysis of oligonucleotide compatibility in terms of formation of extended cross-hybridization patterns and primer dimers. Such an analysis will make it possible to choose the best appropriate adapter oligonucleotide set among those presented here or among any other suppression adapters. For some specific applications, custom adapter design is needed, e.g. introduction of restriction endonuclease recognition sites. Another important issue is that one should pay attention to new restriction sites that may appear upon adapter ligation, if any further manipulations with restriction enzymes are needed.

The recommended annealing temperature for any outer or inner nested PCR primer varies between 60°C and 67°C. We advise 65°C as the default T_m, or as the starting T_m if any further fine tuning of the PCR conditions is required. Please note that the cycling parameters in this protocol have been optimized using the MJ Research PTC-200 DNA thermal cycler. For a different type of thermal cycler, the cycling parameters must be optimized for that machine.

Selective suppression of polymerase chain reaction 45

6.1.2 Buffers and enzymes

We exemplify here the PS adapter ligation to genomic DNA or cDNA sample. In reality, it is up to the user to choose the restriction endonuclease (*Rsa* I in this example) and the ligation enzyme (here T4 DNA ligase). We recommend blunt-end adapter ligation as a more universal approach. If sticky-end making restriction endonuclease is used to process genomic DNA or cDNA, Klenow fragment or other DNA polymerase enzyme may be employed to produce blunt-ends.

1. Rsa I restriction endonuclease (10 U/ml, New England Biolabs).
2. 10X Rsa I restriction buffer: 100 mM Bis-Tris propane/HCl, pH 7.0, 100 mM $MgCl_2$, and 1 mM dithiothreitol (DTT).
3. T4 DNA ligase (3 U/µl, New England Biolabs).
4. T4 10X DNA ligation buffer.
5. T4 DNA polymerase (3 U/µl, New England Biolabs).
6. 10X buffer for T4 DNA polymerase.
7. 10 mM each dNTP (Amersham Pharmacia Biotech, Piscataway, NJ).
8. TN buffer: 10 mM Tris-HCl, 10 mM NaCl.
9. 10X PCR buffer.

6.1.3 Starting material

Genomic DNA or double-stranded cDNA can be used for the ligation of suppression adapters. For example, in the case of suppression SH, 2 µg of genomic DNA or RNA is required per experiment. Most commonly used methods for isolation of RNA and genomic DNA are appropriate. Nevertheless, the quality of DNA or RNA is very important for successful experiment. Whenever possible, samples being compared should be purified side by side utilizing the same reagents and protocol. If genomic DNA is used as a starting material, the next step is restriction endonuclease digestion. If RNA is used as a starting material, the next step will be the cDNA synthesis, including two major steps: first strand cDNA synthesis and second strand cDNA synthesis, using commercially available kits.

6.2 Methods

6.2.1 Rsa I digestion

This step generates shorter, blunt-ended genomic DNA or double-stranded cDNA fragments optimal for several PS applications.
1. Add the following reagents into the sterile 1.5 ml tube:
   – 1.5 µg genomic DNA or double-stranded cDNA
   – 5.0 µl 10X *Rsa* I restriction buffer
   – 1.5 µl *Rsa* I (10 U/µl)
   – to 50 µl deionized sterile water
2. Mix and incubate at 37°C for 2–4 h

3. Analyze 5 μl of the digest mixture on a 1.5% agarose gel along with undigested DNA to analyze the efficiency of *Rsa* I digestion
Note: continue the digestion during electrophoresis and terminate the reaction only after you are satisfied with the results of the analysis.
4. Add 2.5 μl of 0.2 M EDTA to terminate the reaction
5. Perform phenol:chloroform extraction and ethanol precipitation
6. Dissolve each pellet in 5–10 μl of water and store at –20°C.

6.2.2 Adapter annealing

This step is needed to properly anneal 40–45 nt long adapter on a 8–12 nt long oligonucleotide complementary to the adapter 3′-terminal part (e.g. adapter A1A2 on oligonucleotide A3). This will create a "pseudo double-stranded" adapter which may be ligated by DNA ligase to digested genomic DNA or cDNA.

1. In PCR tube, prepare an adapter annealing mix of the following components:
    – 22.5 μl of 100 μM adapter oligonucleotide (e.g. A1A2, B1B2, C1C2)
    – 22.5 μl of 100 μM short complementary oligonucleotide (e.g. A3, B3, C3)
    – 5 μl 10X PCR buffer
2. Incubate the annealing mixture in a thermal cycler at 65°C for 10 min. Do not remove the samples from the thermal cycler.
3. Immediately commence the following program:
    60°C – 5 min
    55°C – 5 min
    50°C – 5 min
    45°C – 5 min
    40°C – 5 min
    35°C – 5 min
    30°C – 5 min
    25°C – 5 min
4. Store annealed adapter solution at –20°C.

6.2.3 Adapter ligation

1. Dilute 1 μl of each *Rsa* I-digested tracer cDNA from the above section with 5 μl of TN buffer.
2. Prepare a ligation mix of the following components for each reaction:
    – 3 μl sterile water
    – 2 μl 5X ligation buffer
    – 1 μl T4 DNA ligase (3 U/μl)
    – 4 μl annealed adapter solution

Please note that ATP required for ligation is in the T4 DNA ligation buffer (300 μM final).

3. Centrifuge tubes briefly, and incubate at 16°C overnight.
4. Stop ligation reaction by adding 1 μl of 0.2 M EDTA.
5. Heat samples at 72°C for 5 min to inactivate the ligase.

Selective suppression of polymerase chain reaction

6. Briefly centrifuge the tubes. Purify ligated DNA from the excess of adapter oligonucleotides using commercially available PCR product purification kit (e.g. manufactured by Promega or QiaGen). Dissolve the DNA in a final volume of 50 µl. Preparation of your experimental adapter-ligated DNA is now complete.

6.2.4 Ligation efficiency test

The following PCR experiment is recommended to verify that at least 25% of the DNA fragments obtained have adapters on both ends. However, this control experiment is not needed if there are no doubts in annealed adapter ligation efficiency. This experiment is designed to amplify fragments that span the adapter–DNA junctions by adapter-specific A1 primer and a unique gene-specific primer (if cDNA was ligated) or genomic DNA-specific primer (if genomic DNA was used). The PCR products generated using one gene-specific primer and adapter-specific primer should be about the same intensity as the PCR products amplified using two gene-specific primers. When designing primers, pay attention to the location of restriction sites for the enzyme you use (*Rsa* I in our example) in analyzing cDNA or genomic DNA; it is important that the amplified gene-specific fragment has no *Rsa* I restriction site. Here, we provide an example of using primers specific for the gene *GAPDH* (GAPDH 3′ primer 5′-*TCCACCACCCTGTTGCTGTA*-3′, GAPDH 5′ primer 5′-*ACCACAGTCCATGCCATCAC*-3′), which work well for human, mouse, and rat cDNA samples.

1. Dilute 1 µl of ligated cDNA into 200 µl of water.
2. Combine the reagents in two separate tubes as follows:

Component (µl)	Tube 1	2
cDNA (ligated to adapter A1A2)	1	1
GAPDH 3′ primer (10 µM)	1	1
GAPDH 5′ primer (10 µM)	0	1
PCR primer A1 (10 µM)	1	0
Total volume	3	3

3. Prepare a master mix for both reaction tubes. For each reaction, combine the reagents in the following order:

Reagent	Amount per reaction tube (µl) 2	Amount for reactions (µl)
Sterile H$_2$O	18.5	37.0
10X PCR reaction buffer	2.5	5.0
dNTP mix (10 mM)	0.5	1.0
50X Taq polymerase mix	0.5	1.0
Total volume	22.0	44.0

4. Mix thoroughly and briefly centrifuge the tubes.
5. Aliquot 22 µl of master mix into each reaction tube from step 2.

6. Overlay with 50 µl of mineral oil. Skip this step if an oil-free thermal cycler is used.
7. Incubate the reaction mixture in a thermal cycler at 75°C for 5 min to extend the adapters. (Do not remove the samples from the thermal cycler.)
8. Immediately commence 20 cycles of:
   94°C – 30 s
   65°C – 30 s
   68°C (or 72°C, depending on the polymerase used) – 2.5 min
9. Examine the products by electrophoresis on a 2% agarose/EtBr gel.

If no products are visible after 20 cycles, perform 5 more cycles of amplification, and again analyze the product by gel electrophoresis. The number of cycles will depend on the abundance of the specific gene. The efficiency of ligation is estimated to be the ratios of the intensities of the bands corresponding to the PCR products of tube 2 to 1. Low ligation efficiency of 25% or less may substantially reduce the efficiency of the subsequent PS-based procedure. In this case, the ligation reaction should be repeated with fresh samples before proceeding to the next step.

For mouse or rat cDNAs, the PCR products amplified with the *GAPDH* 3' primer and PCR primer A1 will be ~1.2 kb instead of the 0.75 kb band observed for human cDNA (because rat and mouse *GAPDH* cDNAs lack the *Rsa* I restriction site in 340 nt position). However, for the human cDNA (which contains the *Rsa* I site), the presence of a 1.2 kb band suggests that the cDNAs are not completely digested by *Rsa* I. If a significant amount of this longer PCR product persists, the procedure should be repeated from the step of *Rsa* I digestion.

REFERENCES

Ackerman SL, Kozak LP, Przyborski SA, Rund LA, Boyer BB, Knowles BB (1997) The mouse rostral cerebellar malformation gene encodes an UNC-5-like protein. Nature 386:838–842

Akowitz A, Manuelidis L (1989) A novel cDNA/PCR strategy for efficient cloning of small amounts of undefined RNA. Gene 81:295–306

Azhikina T, Gainetdinov I, Skvortsova Y, Batrak A, Dmitrieva N, Sverdlov E (2004) Non-methylated Genomic Sites Coincidence Cloning (NGSCC): an approach to large scale analysis of hypomethylated CpG patterns at predetermined genomic loci. Mol Genet Genomics 271:22–32

Badge RM, Alisch RS, Moran JV (2003) ATLAS: a system to selectively identify human-specific L1 insertions. Am J Hum Genet 72:823–838

Bogdanova E, Matz M, Tarabykin V, Usman N, Shagin D, Zaraisky A, Lukyanov S (1998) Inductive interactions regulating body patterning in planarian, revealed by analysis of expression of novel gene scarf. Dev Biol 194:172–181

Bogdanova EA, Matts MV, Tarabykin VS, Usman N, Luk'ianov SA (1997) Differential gene expression during the reparative regeneration of differing polarities in planarians. Ontogenez 28:132–137

Broude NE (2002) Stem-loop oligonucleotides: a robust tool for molecular biology and biotechnology. Trends Biotechnol 20:249–256

Broude NE, Chandra A, Smith CL (1997) Differential display of genome subsets containing specific interspersed repeats. Proc Natl Acad Sci USA 94:4548–4553

Broude NE, Storm N, Malpel S, Graber JH, Lukyanov S, Sverdlov E, Smith CL (1999) PCR based targeted genomic and cDNA differential display. Genet Anal 15:51–63

Broude NE, Zhang L, Woodward K, Englert D, Cantor CR (2001) Multiplex allele-specific target amplification based on PCR suppression. Proc Natl Acad Sci USA 98:206–211

Brownie J, Shawcross S, Theaker J, Whitcombe D, Ferrie R, Newton C, Little S (1997) The elimination of primer-dimer accumulation in PCR. Nucleic Acids Res 25:3235–3241

Buzdin A, Kovalskaya-Alexandrova E, Gogvadze E, Sverdlov E (2006) GREM, a technique for genome-wide isolation and quantitative analysis of promoter active repeats. Nucleic Acids Res 34:e67

Chalaya T, Gogvadze E, Buzdin A, Kovalskaya E, Sverdlov ED (2004) Improving specificity of DNA hybridization-based methods. Nucleic Acids Res 32:e130

Chenchik A, Diachenko L, Moqadam F, Tarabykin V, Lukyanov S, Siebert PD (1996) Full-length cDNA cloning and determination of mRNA 5′ and 3′ ends by amplification of adaptor-ligated cDNA. Biotechniques 21:526–534

Chong SS, Pack SD, Roschke AV, Tanigami A, Carrozzo R, Smith AC, Dobyns WB, Ledbetter DH (1997) A revision of the lissencephaly and Miller-Dieker syndrome critical regions in chromosome 17p13.3. Hum Mol Genet 6:147–155

Chu ZL, McKinsey TA, Liu L, Gentry JJ, Malim MH, Ballard DW (1997) Suppression of tumor necrosis factor-induced cell death by inhibitor of apoptosis c-IAP2 is under NF-kappaB control. Proc Natl Acad Sci USA 94:10057–10062

Diatchenko L, Lau YF, Campbell AP, Chenchik A, Moqadam F, Huang B, Lukyanov S, Lukyanov K, Gurskaya N, Sverdlov ED, Siebert PD (1996) Suppression subtractive hybridization: a method for generating differentially regulated or tissue-specific cDNA probes and libraries. Proc Natl Acad Sci USA 93:6025–6030

Diatchenko L, Lukyanov S, Lau YF, Siebert PD (1999) Suppression subtractive hybridization: a versatile method for identifying differentially expressed genes. Methods Enzymol 303:349–380

Donohue PJ, Hsu DK, Winkles JA (1997) Differential display using random hexamer-primed cDNA, motif primers, and agarose gel electrophoresis. Methods Mol Biol 85:25–35

Edwards JB, Delort J, Mallet J (1991) Oligodeoxyribonucleotide ligation to single-stranded cDNAs: a new tool for cloning 5′ ends of mRNAs and for constructing cDNA libraries by in vitro amplification. Nucleic Acids Res 19:5227–5232

Edwards MC, Gibbs RA (1994) Multiplex PCR: advantages, development, and applications. PCR Methods Appl 3:S65–75

Fan JB, Chen X, Halushka MK, Berno A, Huang X, Ryder T, Lipshutz RJ, Lockhart DJ, Chakravarti A (2000) Parallel genotyping of human SNPs using generic high-density oligonucleotide tag arrays. Genome Res 10:853–860

Fleury C, Neverova M, Collins S, Raimbault S, Champigny O, Levi-Meyrueis C, Bouillaud F, Seldin MF, Surwit RS, Ricquier D, Warden CH (1997) Uncoupling protein-2: a novel gene linked to obesity and hyperinsulinemia. Nat Genet 15:269–272

Fradkov AF, Lukyanov KA, Matz MV, Diatchenko LB, Siebert PD, Lukyanov SA (1998) Sequence-independent method for in vitro generation of nested deletions for sequencing large DNA fragments. Anal Biochem 258:138–141

Frohman MA, Dush MK, Martin GR (1988) Rapid production of full-length cDNAs from rare transcripts: amplification using a single gene-specific oligonucleotide primer. Proc Natl Acad Sci USA 85:8998–9002

Gurskaya NG, Diatchenko L, Chenchik A, Siebert PD, Khaspekov GL, Lukyanov KA, Vagner LL, Ermolaeva OD, Lukyanov SA, Sverdlov ED (1996) Equalizing cDNA subtraction based on selective suppression of polymerase chain reaction: cloning of Jurkat cell transcripts induced by phytohemaglutinin and phorbol 12-myristate 13-acetate. Anal Biochem 240:90–97

Henegariu O, Heerema NA, Dlouhy SR, Vance GH, Vogt PH (1997) Multiplex PCR: critical parameters and step-by-step protocol. Biotechniques 23:504–511

Hudson C, Clements D, Friday RV, Stott D, Woodland HR (1997) Xsox17alpha and -beta mediate endoderm formation in Xenopus. Cell 91:397–405

Ivanova NB, Belyavsky AV (1995) Identification of differentially expressed genes by restriction endonuclease-based gene expression fingerprinting. Nucleic Acids Res 23:2954–2958

Johansson M, Karlsson A (1996) Cloning and expression of human deoxyguanosine kinase cDNA. Proc Natl Acad Sci USA 93:7258–7262

Kazanskaya OV, Severtzova EA, Barth KA, Ermakova GV, Lukyanov SA, Benyumov AO, Pannese M, Boncinelli E, Wilson SW, Zaraisky AG (1997) Anf: a novel class of vertebrate homeobox genes expressed at the anterior end of the main embryonic axis. Gene 200:25–34

Launer GA, Lukyanov KA, Tarabykin VS, Lukyanov SA (1994) Simple method for cDNA amplification starting from small amount of total RNA. Mol Gen Mikrobiol Virusol 6:38–41

Lavrentieva I, Broude NE, Lebedev Y, Gottesman, II, Lukyanov SA, Smith CL, Sverdlov ED (1999) High polymorphism level of genomic sequences flanking insertion sites of human endogenous retroviral long terminal repeats. FEBS Lett 443:341–347

Liang P, Pardee AB (1992) Differential display of eukaryotic messenger RNA by means of the polymerase chain reaction. Science 257:967–971

Liu X, Gorovsky MA (1993) Mapping the 5′ and 3′ ends of Tetrahymena thermophila mRNAs using RNA ligase mediated amplification of cDNA ends (RLM-RACE). Nucleic Acids Res 21:4954–4960

Loftus SK, Morris JA, Carstea ED, Gu JZ, Cummings C, Brown A, Ellison J, Ohno K, Rosenfeld MA, Tagle DA, Pentchev PG, Pavan WJ (1997) Murine model of Niemann-Pick C disease: mutation in a cholesterol homeostasis gene. Science 277:232–235

Luk'ianov KA, Gurskaia NG, Bogdanova EA, Luk'ianov SA (1999) Selective suppression of polymerase chain reaction. Bioorg Khim 25:163–170

Luk'ianov KA, Luk'ianov SA (1997) In vitro cloning of DNA fragments by one polymerase chain reaction. Bioorg Khim 23:882–887

Lukyanov K, Diatchenko L, Chenchik A, Nanisetti A, Siebert P, Usman N, Matz M, Lukyanov S (1997) Construction of cDNA libraries from small amounts of total RNA using the suppression PCR effect. Biochem Biophys Res Commun 230:285–288

Lukyanov KA, Matz MV, Bogdanova EA, Gurskaya NG, Lukyanov SA (1996) Molecule by molecule PCR amplification of complex DNA mixtures for direct sequencing: an approach to in vitro cloning. Nucleic Acids Res 24:2194–2195

Matz M, Usman N, Shagin D, Bogdanova E, Lukyanov S (1997) Ordered differential display: a simple method for systematic comparison of gene expression profiles. Nucleic Acids Res 25:2541–2542

Meyerson M, Counter CM, Eaton EN, Ellisen LW, Steiner P, Caddle SD, Ziaugra L, Beijersbergen RL, Davidoff MJ, Liu Q, Bacchetti S, Haber DA, Weinberg RA (1997) hEST2, the putative human telomerase catalytic subunit gene, is up-regulated in tumor cells and during immortalization. Cell 90:785–795

Morii E, Jippo T, Tsujimura T, Hashimoto K, Kim DK, Lee YM, Ogihara H, Tsujino K, Kim HM, Kitamura Y (1997) Abnormal expression of mouse mast cell protease 5 gene in cultured mast cells derived from mutant mi/mi mice. Blood 90:3057–3066

Mueller CG, Rissoan MC, Salinas B, Ait-Yahia S, Ravel O, Bridon JM, Briere F, Lebecque S, Liu YJ (1997) Polymerase chain reaction selects a novel disintegrin proteinase from CD40-activated germinal center dendritic cells. J Exp Med 186:655–663

Rosenberg HF, Dyer KD (1996) Molecular cloning and characterization of a novel human ribonuclease (RNase k6): increasing diversity in the enlarging ribonuclease gene family. Nucleic Acids Res 24:3507–3513

Schaefer BC (1995) Revolutions in rapid amplification of cDNA ends: new strategies for polymerase chain reaction cloning of full-length cDNA ends. Anal Biochem 227:255–273

Shuber AP, Grondin VJ, Klinger KW (1995) A simplified procedure for developing multiplex PCRs. Genome Res 5:488–493

Siebert PD, Chenchik A, Kellogg DE, Lukyanov KA, Lukyanov SA (1995) An improved PCR method for walking in uncloned genomic DNA. Nucleic Acids Res 23:1087–1088

Takenoshita S, Hagiwara K, Nagashima M, Gemma A, Bennett WP, Harris CC (1996) The genomic structure of the gene encoding the human transforming growth factor beta type II receptor (TGF-beta RII). Genomics 36:341–344

Vinogradova T, Leppik L, Kalinina E, Zhulidov P, Grzeschik KH, Sverdlov E (2002) Selective differential display of RNAs containing interspersed repeats: analysis of changes in the transcription of HERV-K LTRs in germ cell tumors. Mol Genet Genomics 266:796–805

von Stein OD, Thies WG, Hofmann M (1997) A high throughput screening for rarely transcribed differentially expressed genes. Nucleic Acids Res 25:2598–2602

Vos P, Hogers R, Bleeker M, Reijans M, van de Lee T, Hornes M, Frijters A, Pot J, Peleman J, Kuiper M, et al. (1995) AFLP: a new technique for DNA fingerprinting. Nucleic Acids Res 23:4407–4414

Wade DP, Puckey LH, Knight BL, Acquati F, Mihalich A, Taramelli R (1997) Characterization of multiple enhancer regions upstream of the apolipoprotein(a) gene. J Biol Chem 272:30387–30399

Weising K, Atkinson RG, Gardner RC (1995) Genomic fingerprinting by microsatellite-primed PCR: a critical evaluation. PCR Methods Appl 4:249–255

Yang WP, Levesque PC, Little WA, Conder ML, Shalaby FY, Blanar MA (1997) KvLQT1, a voltage-gated potassium channel responsible for human cardiac arrhythmias. Proc Natl Acad Sci USA 94:4017–4021

Yokomizo T, Izumi T, Chang K, Takuwa Y, Shimizu T (1997) A G-protein-coupled receptor for leukotriene B4 that mediates chemotaxis. Nature 387:620–624

Zhang Z, DuBois RN (2001) Detection of differentially expressed genes in human colon carcinoma cells treated with a selective COX-2 inhibitor. Oncogene 20:4450–4456

CHAPTER 3

SUPPRESSION SUBTRACTIVE HYBRIDIZATION

SERGEY A. LUKYANOV*, DENIS REBRIKOV, ANTON A. BUZDIN

Shemyakin-Ovchinnikov Institute of Bioorganic Chemistry, Russian Academy of Sciences, Moscow 117997, Miklukho-Maklaya 16/10
*Corresponding author
Phone: +(7495) 3307029; Fax: +(7495) 3307056; E-mail: luk@ibch.ru*

Abstract: Suppression subtractive hybridization (SSH) is a widely used method for separating DNA molecules that distinguish two closely related DNA samples. Two of the main SSH applications are cDNA subtraction and genomic DNA subtraction. To our knowledge, SSH is one of the most powerful and popular methods for generating subtracted cDNA or genomic DNA libraries. It is based primarily on a suppression polymerase chain reaction (PCR) technique (described narrowly in Chapter 3) and combines normalization and subtraction in a single procedure. The normalization step equalizes the abundance of DNA fragments within the target population, and the subtraction step excludes sequences that are common to the populations being compared. This dramatically increases the probability of obtaining low-abundance differentially expressed cDNAs or genomic DNA fragments and simplifies analysis of the subtracted library. SSH technique is applicable to many comparative and functional genetic studies for the identification of disease, developmental, tissue-specific, or other differentially expressed genes, as well as for the recovery of genomic DNA fragments distinguishing the samples under comparison. This chapter provides an insight into SSH practical use and contains detailed protocol for generation of subtracted cDNAs (which is the most frequent SSH application) and differential screening of the resulting subtracted cDNA library. As shown in many examples, the SSH technique may result in over 1000-fold enrichment for rare sequences in a single round of subtractive hybridization. Finally, we discuss the characteristics of cDNA-subtracted libraries, the nature and level of background nondifferentially expressed clones in the libraries, as well as procedure for rapid identification of truly differentially expressed cDNA clones.

Keywords: Differentially regulated genes, suppression polymerase chain reaction (PCR) effect, enrichment, hybridization time, high complexity, false positive, mirror orientation selection, mirror-oriented selection (MOS), protocol, cap switch, tagging RNA 5′-ends, highly abundant cDNA, primer annealing site, adapter sequence, physical separation, mathematical model, background, restriction endonuclease, heat denaturation, reannealing, target sequence, technical comments, differential screening, efficiency of SSH, Northern blot, level of enrichment, random clones, removal of the adapter sequences, size of cDNA fragments, disadvantage, drawback, equalization.

Abbreviations: cDNA, complementary DNA; dNTP, deoxyribonucleotidetriphosphate; mRNA, messenger RNA; MOS, mirror orientation selection; PCR, polymerase chain reaction; RACE, rapid amplification of cDNA ends; RT, reverse transcription; SSH, suppression subtractive hybridization.

TABLE OF CONTENTS

1. Introduction 54
2. The Principle of Suppression Subtractive Hybridization 55
3. The Principle of Mirror-Oriented Selection 57
4. Technical Comments and Considerations 59
   4.1 Normalization Step and the Efficiency of SSH 59
   4.2 Differential Screening of the Subtracted Libraries 61
   4.3 Starting RNA Materials 61
   4.4 Size of cDNA Fragments 62
5. Detailed Protocols: Materials 63
   5.1 Oligonucleotides 63
   5.2 Buffers and Enzymes 63
6. Detailed Protocols: Methods 65
   6.1 Preparation of Subtracted cDNA or Genomic DNA Library 65
   6.2 Mirror Orientation Selection 70
   6.3 Cloning of Subtracted cDNAs 74
   6.4 Differential Screening of the Subtracted cDNA Library 74
7. Notes 77
References 83

1. INTRODUCTION

Subtractive hybridization methods are valuable tools for identifying differentially regulated genes important for cellular growth and differentiation. Over the last decade, numerous subtractive hybridization techniques have been developed (many of which are described in this book, see Chapters 1–3, 6, and 7) and used to isolate significant genes in many systems (Sargent and Dawid 1983; Hedrick et al. 1984; Hara et al. 1991; Wang and Brown 1991; Hubank and Schatz 1994). However, while having some advantages many of them require either tedious, complicated procedures or large amounts of starting material, thereby reducing their overall utility. Hence, these can be greatly improved if the procedures can be streamlined and/or used with minute amounts of starting material. Suppression subtractive hybridization (SSH) is a widely used method for separating DNA molecules that distinguish two closely related DNA samples of either cDNA or genomic DNA nature (Luk'ianov et al. 1994; Diatchenko et al.

1996; Gurskaya et al. 1996; Akopyants et al. 1998). In particular, the SSH protocol, which we describe here, normalizes (equalizes) sequence abundance among the target cDNA population, eliminates any intermediate step(s) for physical separation of single-stranded cDNAs and double-stranded cDNAs, requires only one round of subtractive hybridization, and can achieve greater than 1000-fold enrichment for differentially expressed cDNAs (Rebrikov et al. 2004). The SSH method is based on a suppression polymerase chain reaction (PCR) effect ((Lukyanov et al. 1995; Siebert et al. 1995), see Chapter 2) and combines normalization and subtraction in a single procedure (Gurskaya et al. 1996). The normalization step equalizes the abundance of DNA fragments within one round of subtraction (Diatchenko et al. 1996; Gurskaya et al. 1996; Jin et al. 1997). Nevertheless, in practice, not all differentially expressed genes are equally enriched by SSH. The level of enrichment of a particular cDNA depends on its original abundance, the ratio of its concentration in the samples being subtracted, and the number of other differentially expressed genes. Other factors, such as the complexity of a starting material, hybridization time, and ratio of two samples being subtracted, play a very important role in SSH's success in a given application. For instance, the high complexity of mammalian genomic DNA makes SSH application very difficult. Likewise, some cDNA subtractions are also challenging because of the nature of the starting samples. Subtracted libraries generated using complex samples may contain very high background. An especially challenging problem is the inclusion of "false positive" clones that generate a differential signal in a primary screening procedure, but are not confirmed by subsequent detailed analysis. To overcome this problem, a simple procedure called mirror orientation selection (MOS) can be used to substantially decrease the number of background clones (Rebrikov et al. 2000, 2004).

In this chapter, we describe the SSH technique for generating subtracted cDNA or genomic DNA libraries. Detailed protocols for cDNA synthesis, subtractive hybridization, PCR amplification, library generation, and differential screening analysis are provided. We also describe the MOS procedure that substantially decreases the number of background clones in SSH-generated libraries. Finally, we show an example of SSH- and MOS-subtracted library.

2. THE PRINCIPLE OF SUPPRESSION SUBTRACTIVE HYBRIDIZATION

Figure 1 presents a brief overview of the SSH procedure (application to subtraction of cDNA samples). SSH includes several steps. First, cDNA is synthesized from the two types of tissues or cell populations being compared. The cDNA population in which specific transcripts are to be found is called tracer cDNA (or tester cDNA), and the reference cDNA population is called driver cDNA. For cDNA synthesis, the conventional method described by Gubler and Hoffman including poly(A) + RNA isolation (Gubler and

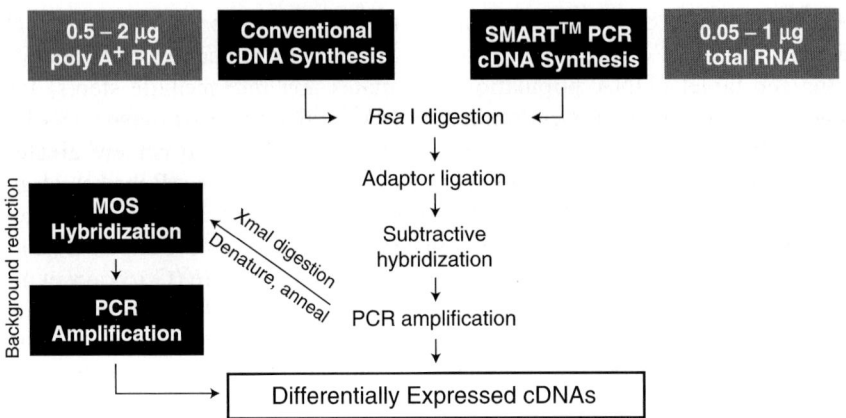

Figure 1. Brief overview of the suppression subtractive hybridization (SSH). For better results, mirror-oriented selection (MOS) procedure can be recruited as the option.

Hoffman 1983) can be used. If enough poly(A)+ RNA is not available, the "Cap Switch" amplification technology (SMART, BD Biosciences Clontech) or the FirstChoice RLM-RACE Kit (Ambion), exactly tagging RNA 5′-ends with synthetic oligonucleotides, can be used to preamplify high-quality cDNA from total RNA (Maruyama and Sugano 1994). At the second step, double-stranded cDNAs are synthesized independently from the tracer and driver, and are further digested with a four-base-cutting restriction enzyme that yields blunt ends, such as *Rsa* I or *Alu* I (Rebrikov et al. 2004). The tracer cDNA is then subdivided into two portions (1 and 2) and each is ligated to a different double-stranded adapter (adapters 1 [Ad1] and 2R [Ad2R]). The ends of the synthetic oligonucleotide-derived adapters are not phosphorylated, so only one strand of each adapter becomes covalently attached to the 5′-ends of the cDNAs. The molecular events that occur during subtractive hybridization and selective amplification of differentially expressed genes are illustrated in Figure 2. In the first hybridization, an excess of driver cDNA is added to each sample of tracer cDNA.

The samples are then heat-denatured and allowed to anneal. Figure 2 shows the type A, B, C, and D molecules generated in each sample. During this first hybridization step the subset of single-stranded tracer molecules (fraction A) is normalized, which means that concentrations of high and low abundance cDNAs become roughly equal. Normalization occurs because the annealing process generating homohybrid (B) and heterohybrid (C) cDNAs is faster for more abundant molecules, due to the second order of hybridization kinetics, than annealing of the less abundant cDNAs that remain single-stranded (A). By controlling the extent of the hybridization, the single-stranded forms of highly abundant cDNAs can then be reduced to the same levels as those of less abundant ones, thereby normalizing the representation of tracer cDNA population (Hames and Higgins 1985). At the same time, the population of type A molecules is significantly enriched for differentially expressed sequences, because common for tracer and driver samples nontarget cDNAs form type C molecules with the driver. During the second hybridization, the two samples from the first hybridization are combined and annealed further with additional freshly denatured driver. Under these conditions, only single-stranded type A tracer cDNAs are able to reassociate and form (B), (C), and new (E) hybrids (Figure 2). Type E hybrids are double-stranded tracer molecules with different single-stranded ends, one of which corresponds to Ad1 and another to Ad2R.

Freshly denatured driver is added to further enrich fraction E in differentially expressed sequences. The entire population of molecules is then subjected to two rounds of PCR to selectively amplify the differentially expressed sequences (Rebrikov et al. 2004). Prior to the first PCR, adapter ends are filled in, thus creating the complementary primer binding sites needed for amplification (Figure 2). Type A and D molecules lack primer annealing sites and cannot be amplified. Type B molecules form stem-loop pan handle-like structures that suppress amplification (Lukyanov et al. 1995; Siebert et al. 1995). Type C molecules have only one primer annealing site and can be amplified only at a linear rate. Only type E molecules, which have different adapter sequences at their ends and, thus, two different primer annealing sites, can be amplified exponentially.

Differentially expressed sequences are greatly enriched in type E fraction, and therefore in the subtracted cDNA pool. This method does not involve any physical separation of single-stranded molecules from double-stranded hybrids. The comprehensive mathematical model for the formation of fraction E molecules, as well as the rate of enrichment, has been described previously (Gurskaya et al. 1996). Although SSH technique in its standard form was shown to be very effective for many objects, in some cases, e.g. when the samples under comparison had relatively small number of differential sequences, there was a high background in the SSH-generated subtracted libraries, and another approach termed MOS could be helpful to significantly reduce the background.

3. THE PRINCIPLE OF MIRROR-ORIENTED SELECTION

The MOS technique is based on the rationale that, after PCR amplification during SSH, each kind of background molecule has only one orientation relative to the adapter sequences (Rebrikov et al. 2004). This directionality corresponds to the orientation of the progenitor molecule. On the contrary, the target DNA fragments are involved in PCR amplification owing to efficient enrichment in the SSH procedure. As a result, each specific sequence has many progenitors and is represented by both orientations relatively to adapter sequences (Rebrikov et al. 2000, 2004). The procedure includes removing adapter-derived primer NP1

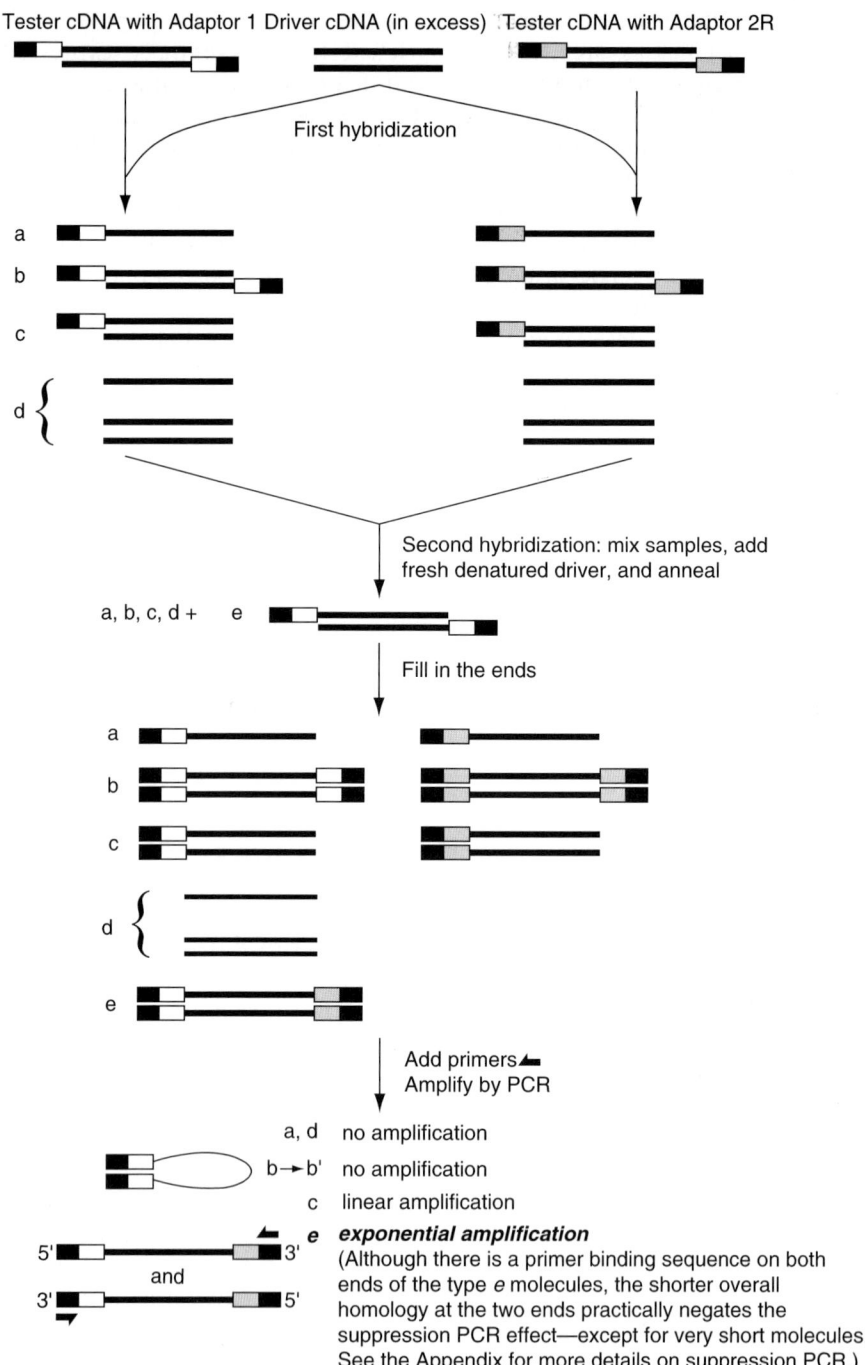

Figure 2.

(Continued)

in secondary PCR of SSH, (Figure 3) by restriction endonuclease (*Xma* I in this example), heat denaturation and reannealing of the SSH sample (Figure 3). Some of the newly formed hybrids from target DNAs bear adapter 2R (represented by adapter 2R-derived primer NP2R in secondary PCR of SSH, Figure 3) at both termini. These molecules are generated as a result of hybridization of molecules with "mirror" orientation of adapters 1 and 2. Thus, they can primarily (if not exclusively) be derived from a fraction of target DNA. Finally, the 3'-ends are filled in and PCR with primers corresponding to NP2R (also called NP2Rs or MOS PCR primer) is performed. In this reaction only molecules bearing NP2R at both termini can be amplified exponentially. Therefore, the final PCR products are enriched in target sequences.

4. TECHNICAL COMMENTS AND CONSIDERATIONS

4.1 Normalization Step and the Efficiency of SSH

The SSH technique has been demonstrated to be efficient for generating cDNAs highly enriched for differentially expressed genes of both high and low abundance. The high level of enrichment of cDNAs for rare transcripts has been achieved by the inclusion of a normalization step in the subtraction procedure, as evidenced by Northern blot analysis of random cDNA clones from many subtracted cDNA libraries (Siebert et al. 1995).

Using the mathematical model of the subtraction procedure (Ermolaeva et al. 1996; Gurskaya et al. 1996) and SUBTRACT program (Ermolaeva and Wagner 1995), it was calculated that the rare specific transcripts can be enriched by greater than 1000 during one round of subtraction. This conclusion was supported in a model experiment with artificial targets added to cDNA libraries (Diatchenko et al. 1996). In practice, the level of enrichment for a particular

---

Figure 2. cont'd. Overall scheme for suppression subtractive hybridization. There are six steps in this procedure. (1) double-stranded cDNAs are synthesized from tracer and driver mRNAs and digested with *Rsa* I to generate optimal fragments for hybridization reactions. (2) adapters Ad1 and Ad2R are ligated to two separate populations of the tracer cDNAs which are then (3) mixed with 30X excess driver cDNAs. The mixtures are processed in the first hybridization to normalize and enrich differentially expressed sequences among ss tracer molecules. (4) The reactions from the first hybridization are mixed and processed for a second hybridization in the presence of additional driver ss cDNAs, resulting in combination of different hybrid types. (5) The ends of the respective hybrids are filed in to generate cDNA fragments from differentially expressed genes that can be preferentially amplified by PCR using appropriate primers. (6) Two rounds of PCR are performed to preferentially amplify differentially expressed genes. Solid lines represent the *Rsa* I-digested tracer or driver cDNA. Solid boxes represent the outer part of the adapter Ad1 and Ad2 longer strands and corresponding PCR primer P1 sequence. Clear boxes represent the inner part of the adapters Ad1 longer strand and corresponding nested PCR primers NP1. Shaded boxes represent the inner part of the adapter Ad2R longer strand and corresponding nested PCR primer NP2R. Note that after filling in the recessed 3'-ends with DNA polymerase, type a, b, and c molecules having adapter 2 are also present, but are not shown.

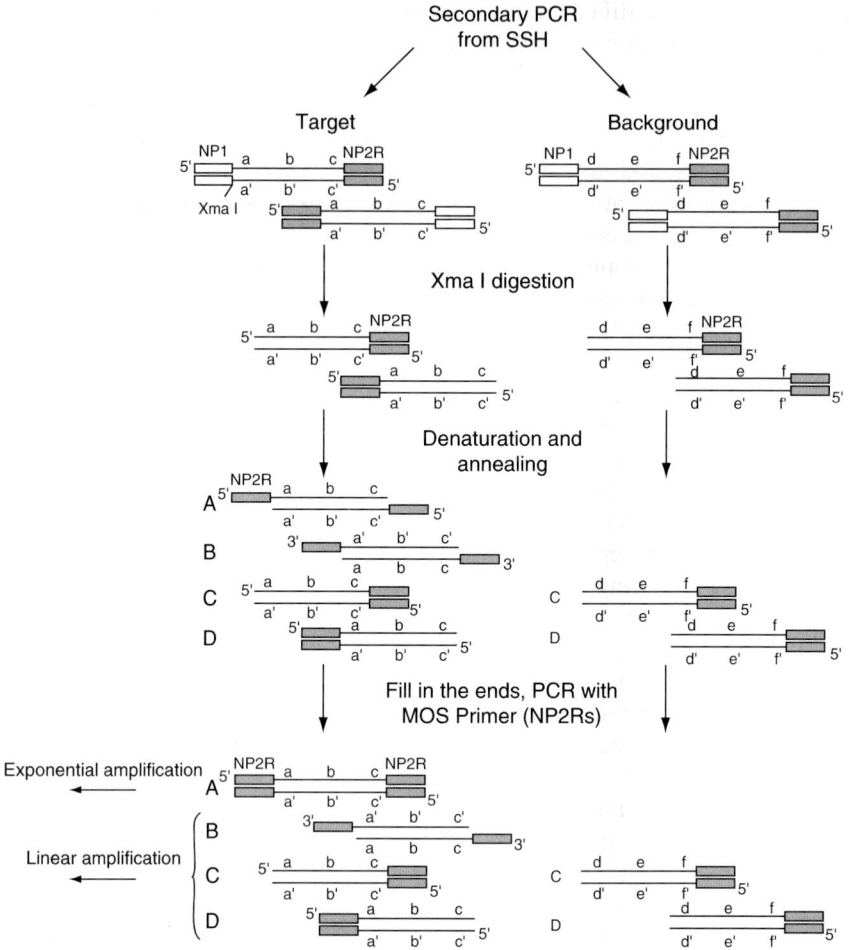

Figure 3. Schematic diagram of the mirror-oriented selection (MOS) procedure. The method is based on the assumption that each background molecule has only one orientation with respect to the Ad1 (represented by NP1) and Ad2R (represented by NP2R) adapters used in suppression subtractive hybridization (SSH), whereas truly differentially expressed target cDNA fragments are represented by both sequence orientations. MOS PCR primer is a shortened variant of the NP2R primer used in SSH.

gene depends greatly on its original abundance, the ratio of its concentrations in driver and tracer, and a number of other differentially expressed genes. With the incorporation of a normalization step in subtraction procedure, the highest enrichment level can be obtained for differentially expressed mRNAs exhibiting low abundance and/or large differences in expression levels in the tracer and driver RNA populations. However, as with other subtractive hybridization techniques, the efficiency of SSH is lower in experiments designed to detect mRNAs that show only moderate (e.g. twofold to fourfold) differences between the tracer and driver populations. Nevertheless, these types of differentially expressed genes can still be identified by the extensive screening of subtracted libraries (Rebrikov et al. 2004; Gurskaya et al. 1996).

4.2 Differential Screening of the Subtracted Libraries

Although SSH method greatly enriches for differentially expressed genes, and MOS approach application significantly reduces SSH background, the subtracted sample will still contain some cDNAs that correspond to mRNAs common to both the tracer and driver samples, depending somewhat on the quality of RNA purification and the performance of the particular subtraction. However, it mainly arises when very few mRNA species are differentially expressed in tracer and driver. In general, fewer differentially expressed mRNAs and less quantitative difference in expression lead to higher background – even if one obtains a good enrichment for differentially expressed cDNAs. When background is expected to be high, identification of differentially expressed cDNAs by picking random clones from the subtracted library for Northern blot analysis can be time consuming. To solve this problem, the incorporation of a differential screening step is an efficient and desirable means to minimize background before embarking on Northern blot analysis (Rebrikov et al. 2004).

There are two approaches for differentially screening the subtracted library. The first is to hybridize the subtracted library with [α^{32}P]dCTP-labeled cDNA probes synthesized directly from the tracer and driver mRNAs (Hedrick et al. 1984). Clones corresponding to differentially expressed mRNAs will hybridize only with the tracer probe, and not with the driver probe. Although this approach is widely used, it has one major disadvantage: only cDNA molecules corresponding to highly abundant mRNAs (i.e. mRNAs which constitute more than ~0.2% of the total cDNA in the probe) can hybridize with the clones (Hames and Higgins 1985; Wang and Brown 1991). Clones corresponding to low-abundance differentially expressed mRNAs will not be detected by this screening procedure.

To avoid missing low-abundance differentially expressed genes, we recommend a second approach, in which the subtracted library is hybridized with forward- and reverse-subtracted cDNA mixtures. Clones representing mRNAs that are truly differentially expressed will hybridize only with the forward probe; clones that hybridize with the reverse probe most probably represent background (Wang and Brown 1991; Lukyanov et al. 1996). It should be noted, that the removal of the adapter sequences from the ends of the cDNA molecules generated by the SSH procedure, is critical for reducing the background caused by hybridization of the adapters in the subtracted cDNA probes to those immobilized on the nylon filters.

4.3 Starting RNA Materials

Normally, 2–4 mg of poly(A)+ RNA for both the tracer and driver are needed for a comprehensive subtraction scheme using both forward and reverse SSH. The resulting PCR products can be used for subtracted cDNA library construction and differential screening experiments. However, in some cases, such amount of poly(A)+ RNAs may be difficult to obtain. To circumvent this problem, an amplification step for both the driver and tracer poly(A)+ RNA can be incorporated to generate sufficient quantities of both cDNA samples (Gurskaya et al. 1996) before initiating the SSH procedure. Alternatively, total RNA can be used successfully as starting material for the preamplification (Lukyanov et al. 1997; Zhumabayeva et al. 2001a, b). However, preamplification of either poly(A)+ or total RNA samples invariably increases the background in the final PCR products and may result in the loss of some sequences. The utilization of preamplifiation should be minimized whenever it is possible.

4.4 Size of cDNA Fragments

For an efficient SSH procedure, the starting tracer and driver cDNAs have to be cleaved into smaller fragments. A four-base cutter, *Rsa* I, is recommended for this purpose since it generates optimal fragments (average ~600 bp) for SSH. Although this step may not be a desirable manipulation for obtaining full-length differentially expressed cDNAs, dividing each cDNA into multiple fragments has two important advantages. First, long DNA fragments may form complex networks which prevent the formation of appropriate hybrids needed to position two independent adapters, Ad1 and Ad2R, at the ends of the target molecules. Second, small cDNA fragments provide better representation of individual genes, since cDNAs derived from related but distinct members of gene families may cross-hybridize with each other, thereby eliminating them from the final subtracted cDNA products (Ko 1990). Dividing the cDNAs into smaller and different portions increases the possibility that a particular differentially expressed member of a gene family will contain a smaller fragment that is sufficiently different from other homologous members and can be enriched in the final subtracted cDNA mixture (Wang and Brown 1991; Hubank and Schatz 1994). Once a small cDNA fragment is cloned and sequenced, numerous approaches, such as 5'- or 3'-rapid amplification of cDNA ends (RACE) (see Chapter 2), can be used to quickly obtain corresponding full-length cDNAs (reviewed in Luk'ianov et al. (1999)).

Another drawback of the SSH technique is the typical small inserts in the final subtracted cDNA libraries (average ~200 bp). This problem is generated by several factors related to the SSH procedure, such as more efficient hybridization of shorter fragments, preferential amplification of these fragments by PCR, and higher efficiency of subcloning of short fragments in plasmid vectors than those of longer ones. To minimize this undesirable selection for shorter fragments, several modifications of the SSH procedure can be incorporated to increase the representation of larger inserts in the final subtracted products, such as size selection of the subtracted cDNA products before subcloning. To minimize this problem in the current of SSH protocol described here, we use adapters with identical sequences for the first 22 nucleotides at their 5'-ends (Ad 1 and Ad2R), instead of being completely different in sequence as in the original description (Ad1 and Ad2) (Siebert et al. 1995). These sequence changes introduce short complimentary inverted terminal repeats on the end of the cDNA molecule which carry different adapters on their ends and allow the primary amplification to be carried out with a single PCR primer. This introduces weak suppression PCR effect during the primary amplification since the length of complementary part is equal of the length of the primer. In this condition the amplification of very short (less than 200 bp) DNA fragments significantly diminished (Jin et al. 1997) and the risk of nonspecific amplification also decreases (Rebrikov et al. 2004; Takarada 1994.

These equalization steps are not intended to eliminate the shorter fragments which may represent truly differentially expressed cDNAs, but instead are designed to balance representation of different fragment sizes, thereby increasing the complexity of the subtracted cDNA library.

5. DETAILED PROTOCOLS: MATERIALS

5.1 Oligonucleotides

The following oligonucleotides are used at a concentration of 10 µM. Whenever possible, oligonucleotides should be HPLC- or gel-purified.
1. cDNA synthesis primers: 5'-*TTTTGTACAAGCT(T)$_{30}$*-3'
2. Adapters:
   (Ad1) – 5'-*CTAATACGACTCACTATAGGGCTCGAGCGGCCGCCCGGG CAGGT*-3'
   3'-*GGCCCGTCCA*-5'
   (Ad2) – 5'-*TGTAGCGTGAAGACGACAGAAAGGGCGTGGTGCGGAGG GCGGT*-3'
   3'-*GCCTCCCGCCA*-5'
   (Ad2R) – 5'-*CTAATACGACTCACTATAGGGCAGCGTGGTCGCGGCCGA GGT*-3'
   3'-*GCCGGCTCCA*-5'
3. PCR primers: (P1) – 5'-*CTAATACGACTCACTATAGGGC*-3'
   (P2) – 5'-*TGTAGCGTGAAGACGACAGAA*-3'
   Nested primer 1 (NP1) – 5'-*TCGAGCGGCCGCCCGGGCAGGT*-3'
   Nested primer 2 (NP2) – 5'-*AGGGCGTGGTGCGGAGGGCGGT*-3'
   Nested primer 2R (NP2R) – 5'-*AGCGTGGTCGCGGCCGAGGT*-3'
4. MOS PCR primer (NP2Rs) – 5'-*GGTCGCGGCCGAGGT*-3'
5. Blocking solution: a mixture of the cDNA synthesis primer, nested primers (NP1 and NP2R), and their respective complementary oligonucleotides (2 mg/ml each).

5.2 Buffers and Enzymes

1. AMV reverse transcriptase (20 U/µl; Life Technologies, Gaithersburg, MD).
2. 5X First strand buffer: 250 mM Tris-HCl, pH 8.5, 40 mM MgCl$_2$, 150 mM KCl, and 5 mM dithiothreitol (DTT).

3. 20X Second strand enzyme cocktail: DNA polymerase I (6 U/µl, New England Biolabs, Beverly, MA).
4. RNase H (0.25 U/µl, Epicentre Technologies, Madison,WI).
5. T4 DNA ligase (3 U/µl, New England Biolabs).
6. 5X Second strand buffer: 500 mM KCl, 50 mM ammonium sulfate, 25 mM $MgCl_2$, 0.75 mM b-NAD, 100 mM Tris-HCl, pH 7.5, 0.25 mg/ml BSA.
7. T4 DNA polymerase (3 U/µl, New England Biolabs).
8. 10X *Rsa* I restriction buffer: 100 mM Bis-Tris propane/HCl, pH 7.0, 100 mM $MgCl_2$, and 1 mM DTT.
9. *Rsa* I (10 U/ml, New England Biolabs).
10. T4 DNA ligase (400 U/µl: contains 3 mM ATP [New England Biolabs]).
11. 5X DNA ligation buffer: 250 mM Tris-HCl, pH 7.8, 50 mM $MgCl_2$, 10 mM DTT, 0.25 mg/ml BSA.
12. 4X Hybridization buffer: 4 M NaCl, 200 mM HEPES, pH 8.3, 4 mM cetyltrimethyl ammonium bromide (CTAB).
13. Dilution buffer: 20 mM HEPES-HCl, pH 8.3, 50 mM NaCl, 0.2 mM EDTA.
14. Advantage cDNA PCR Mix (BD Biosciences Clontech, Palo Alto, CA). This mix contains a mixture of KlenTaq-1 and DeepVent DNA polymerases (New England Biolabs, Beverly, MA) and TaqStart antibody (BD Biosciences Clontech); 10X reaction buffer (40 mM Tricine-KOH (pH 9.2 at 22°C), 3.5 mM $Mg(OAc)_2$, 10 mM KOAc, 75 mg/ml BSA). The TaqStart antibody provides automatic hot start PCR (Kellogg et al. 1994). Alternatively, *Taq* DNA polymerase can be used alone, but five additional thermal cycles will be needed in both the primary and secondary PCR, and the additional cycles may cause higher background. If the Advantage cDNA PCR Mix is not used, manual hot start or hot start with wax beads is strongly recommended to reduce nonspecific DNA synthesis.
15. 10 mM each dNTP (Amersham Pharmacia Biotech, Piscataway, NJ).
16. 20X EDTA/glycogen mix: 0.2 M EDTA; 1 mg/ml glycogen.
17. 4 M NH_4OAc, TN buffer: 10 mM Tris-HCl, 10 mM NaCl.
18. ExpressHyb hybridization solution (BD Biosciences Clontech).
19. Sterile deionized water.

Please note that the cycling parameters in this protocol have been optimized using the MJ Research PTC-200 DNA thermal cycler. For a different type of thermal cycler, the cycling parameters must be optimized for that machine. We recommend performing subtractions in both directions (tracer–driver and in parallel driver–tracer) for each DNA pair being compared. This forward- and reverse-subtracted DNA may be useful for differential screening of the resulting subtracted libraries. We also recommend performing self-subtractions (with both tracer and driver prepared from the same DNA sample) as a control experiment for fast examination of subtraction efficiency (see Note 1). For models such as RNA or DNA injections, or viral infections, it is extremely important to add appropriate DNA into driver sample (see Note 2).

6. DETAILED PROTOCOLS: METHODS

6.1 Preparation of Subtracted cDNA or Genomic DNA Library

6.1.1 RNA and DNA isolation

Two micrograms of genomic DNA or RNA is required per subtraction. Most commonly used methods for isolation of RNA and genomic DNA are appropriate for subtraction experiments (Chomczynski and Sacchi 1987). Nevertheless, the quality of DNA or RNA is very important for successful experiment. Whenever possible, samples being compared should be purified side by side utilizing the same reagents and protocol. Alternatively, commercially available kits from different vendors can be used for RNA and DNA isolation.

If genomic DNA is used as a starting material, the next step is *Rsa* I or *Alu* I digestion (Section 6.1.3). If RNA is used as a starting material, the next step is cDNA synthesis (Section 6.1.2).

Note: For simplicity, the term "cDNA" will be used throughout the protocol, but the protocol is suitable for genomic DNA subtraction with no changes in the amount of any reagents required to perform subtraction.

6.1.2 cDNA synthesis

There are two steps involved in cDNA synthesis: first strand cDNA synthesis and second strand cDNA synthesis. During first strand cDNA synthesis, AMV reverse transcriptase synthesizes cDNA using poly(A)+ RNA as a template. During second strand cDNA synthesis, DNA polymerase I uses first strand cDNA as a template. The following protocol is recommended for generating a subtracted library from 2 µg of poly(A)+ RNA. If enough poly(A)+ RNA is not available, the switch mechanism at the 5′-end of RNA templates (SMART) amplification technology (BD Biosciences Clontech), or FirstChoice RLM-RACE kit (Ambion) can be used to preamplify high-quality cDNA from total RNA (11) (see Note 3).

First strand cDNA synthesis

Perform this procedure individually with each tracer and driver poly(A)+ RNA sample.

1. For each tracer and driver sample, combine the following components in a sterile 0.5-ml microcentrifuge tube (do not use a polystyrene tube):
   – Poly(A) + RNA (2 µg) to 2–4 µl
   – cDNA synthesis primer (10 µM) to 1 µl
   – If needed, add sterile H_2O to a final volume of 5 µl
2. Incubate the tubes at 70°C in a thermal cycler for 2 min
3. Cool at room temperature for 2 min and briefly centrifuge using a microcentrifuge at maximum rotation speed (we recommend 6000–7000 rpm)
4. Add the following to each reaction tube:
   – 2 µl 5X First strand buffer
   – 1 µl dNTP mixture (each 10 mM)
   – 0.5 µl Sterile H_2O

(Optional: to monitor the progress of cDNA synthesis, dilute 0.5 µl of [α³²P] dCTP (10 mCi/ml, 3000 Ci/mM) with 9 µl of H₂O and replace the H₂O above with 1 µl of the diluted label.)
- 0.5 µl 0.1 M DTT
- 1 µl AMV reverse transcriptase (20 U/µl)

5. Gently vortex and briefly centrifuge the tubes
6. Incubate the tubes at 42°C for 1.5 h in an air incubator
7. Place the tubes on ice to terminate first strand cDNA synthesis and immediately proceed to second strand cDNA synthesis

Second strand cDNA synthesis

1. Add the following components (previously cooled on ice) to the first strand synthesis reaction tubes:
   - 48.4 µl Sterile water
   - 16.0 µl 5X Second strand buffer
   - 1.6 µl dNTP mix (10 mM)
   - 4.0 µl 20X Second strand enzyme cocktail
2. Mix the contents and briefly spin the tubes. The final volume should be 80 µl.
3. Incubate the tubes at 16°C (water bath or thermal cycler) for 2 h.
4. Add 2 µl (6 U) of T4 DNA polymerase. Mix contents well.
5. Incubate the tube at 16°C for 30 min in a water bath or a thermal cycler.
6. Add 4 µl of 0.2 M EDTA to terminate second strand synthesis.
7. Perform phenol:chloroform extraction and ethanol precipitation (see Note 4).
8. Dissolve pellet in 50 µl of TN buffer.
9. Transfer 6 µl to a fresh microcentrifuge tube. Store this sample at ?20°C until after *Rsa* I digestion. This sample will be used for agarose gel electrophoresis to estimate yield and size range of the ds cDNA-synthesized products.

6.1.3 *Rsa I digestion*

Perform the following procedure with each experimental double-stranded tracer and driver cDNA. This step generates shorter, blunt-ended double-stranded cDNA fragments optimal for subtractive hybridization.

1. Add the following reagents into the tube from Section 1.2.2, step 8:
   - 43.5 µl double-stranded cDNA
   - 5.0 µl 10X *Rsa* I restriction buffer
   - 1.5 µl *Rsa* I (10 U/µl)
2. Mix and incubate at 37°C for 2–4 h.
3. Use 5 µl of the digest mixture and analyze on a 2% agarose gel along with undigested cDNA (Section 6.1.2, step 9 or Section 6.1.1 for genomic DNA) to analyze the efficiency of *Rsa* I digestion.

Note: Continue the digestion during electrophoresis and terminate the reaction only after you are satisfied with the results of the analysis.

4. Add 2.5 µl of 0.2 M EDTA to terminate the reaction.
5. Perform phenol:chloroform extraction and ethanol precipitation (see Note 4–6).
6. Dissolve each pellet in 6 µl of TN buffer (see Note 7) and store at −20°C.

Driver cDNA preparation is now complete.

Suppression subtractive hybridization 67

6.1.4 Adapter ligation

It is strongly recommended that you perform subtractions in both directions for each tracer or driver cDNA pair. Forward subtraction is designed to enrich differentially expressed transcripts present in tracer but not in driver; reverse subtraction is designed to enrich differentially expressed sequences present in driver but not in tracer. The availability of such forward- and reverse-subtracted cDNAs will be useful for differential screening of the resulting subtracted tracer cDNA library (see Section 6.4).

The tracer cDNAs are ligated separately to Ad1 (tracer 1-1 and 2-1) and Ad2R (tracer 1-2 and 2-2). It is highly recommended that a third ligation of both adapters 1 and 2R to the tracer cDNAs (unsubtracted tracer control 1-c and 2-c) be performed and used as a negative control for subtraction. Please note that the adapters are not ligated to the driver cDNA.

1. Dilute 1 µl of each *Rsa* I-digested tracer cDNA from Section 6.1.3 with 5 µl of TN buffer.
2. Prepare a master ligation mix of the following components for each reaction:
   – 3 µl Sterile water
   – 2 µl 5X Ligation buffer
   – 1 µl T4 DNA ligase (400 U/µl)

Please note that ATP required for ligation is in the T4 DNA ligase (3 mM initial, 300 µM final).

3. For each tracer cDNA mixture, combine the following reagents in a 0.5-ml microcentrifuge tube in the order shown. Pipet the solution up and down to mix thoroughly.

Tube no.	1	2
Component	Tracer 1-1 (µl)	Tracer 1-2 (µl)
Diluted tracer cDNA	2	2
Adapter Ad1 (10 µM)	2	–
Adapter Ad2R (10 µM)	–	2
Master ligation mix	6	6
Final volume	10	10

4. In a fresh microcentrifuge tube, mix 2 µl of tracer 1-1 and 2 µl of tracer 1-2. This is your unsubtracted tracer control 1-c. Do the same for each tracer cDNA sample. After ligation, approximately one third of the cDNA molecules in each unsubtracted tracer control tube will have two different adapters on their ends, suitable for exponential PCR amplification with adapter-derived primers.
5. Centrifuge the tubes briefly and incubate at 16°C overnight.
6. Stop the ligation reaction by adding 1 µl of 0.2 M EDTA.
7. Heat samples at 72°C for 5 min to inactivate the ligase.
8. Briefly centrifuge the tubes. Remove 1 µl from each unsubtracted tracer control (1-c, 2-c, . . .) and dilute into 1 ml of H_2O. These samples will be used for PCR amplification.

Preparation of your experimental adapter-ligated tracer cDNAs 1-1 and 1–2 is now complete.

Perform ligation efficiency test before proceeding to the next section (see Note 8).

6.1.5 Subtractive hybridization

First hybridization

1. For each tracer sample, combine the reagents in the following order:

Component	Hybridization 1.1 (μl)	Hybridization 1.2 (μl)
Rsa I-digested driver cDNA (Section 6.1.3, step 7)	1.5	1.5
Ad1-ligated tracer 1-1 (Section 6.1.4, step 8)	1.5	–
Ad2R-ligated tracer 1-2 (Section 6.1.4, step 5)	–	1.5
4X Hybridization buffer	1.0	1.0
Final volume	4.0	4.0

2. Overlay samples with one drop of mineral oil and centrifuge briefly.
3. Incubate samples in a thermal cycler at 98°C for 1.5 min. Incubate samples at 68°C for 8 h (see Note 9) and then proceed immediately to the second hybridization (see Note 21).

Second hybridization

Repeat the following steps for each experimental driver cDNA.

1. Add the following reagents into a sterile 0.5-μl microcentrifuge tube:
   – 1 μl Driver cDNA (Section 6.1.3, step 6)
   – 1 μl 4X Hybridization buffer
   – 2 μl Sterile water
2. Place 1 μl of this mixture in a 0.5-ml microcentrifuge tube and overlay it with one drop of mineral oil.
3. Incubate in a thermal cycler at 98°C for 1.5 min (see Note 10).
4. Remove the tube of freshly denatured driver from the thermal cycler (see Note 11).
5. To the tube of freshly denatured driver cDNA, add hybridized sample 1.1 and hybridized sample 1.2 (from first hybridization) in that order. This ensures that the two hybridization samples are mixed only in the presence of excess driver cDNA.
6. Incubate the hybridization reaction at 68°C overnight.
7. Add 100 μl of dilution buffer to the tube and mix well by pipetting.
8. Incubate in a thermal cycler at 72°C for 7 min.
9. Store hybridization solution at −20°C (see Note 12).

PCR amplification

Differentially presented DNAs are selectively amplified during the reactions described in this section. Each experiment should have at least four reactions: subtracted tracer cDNAs, unsubtracted tracer control (1-c), reverse-subtracted tracer cDNAs, and unsubtracted driver control for the reverse subtraction (2-c).

PRIMARY PCR

1. Place a 1 µl aliquot of each diluted cDNA sample (i.e. each subtracted sample from Section 6.1.6, step 8, and the corresponding diluted unsubtracted tracer control from Section 6.1.4, step 8) into an appropriately labeled tube (see Note 12).
2. Prepare a master mix for all of the primary PCR tubes plus one additional tube. For each reaction combine the reagents in the order shown:

Reagent	Amount per reaction (µl)
Sterile water	19.5
10X PCR reaction buffer	2.5
dNTP mix (10 mM)	0.5
PCR primer P1 (10 µM)	1.0
50X Advantage cDNA PCR Mix	0.5
Total volume	24.0

3. Place 24 µl aliquot of master mix into each reaction tube prepared in step 1.
4. Overlay with 50 µl of mineral oil. Skip this step if an oil-free thermal cycler is used.
5. Incubate the reaction mixture in a thermal cycler at 75°C for 5 min to extend the adapters (see Note 13). Do not remove the samples from the thermal cycler.
6. Immediately commence 26 cycles of:
   95°C 10 s
   66°C 10 s
   72°C 1.5 min
7. Analyze 4 µl from each tube on a 2% agarose/EtBr gel run in 1X TAE buffer (see Notes 14 and 15).

SECONDARY PCR

1. Dilute 2 µl of each primary PCR mixture in 38 µl of water.
2. Place 1 µl aliquot of each diluted primary PCR product mixture from step 1 into an appropriately labeled tube.
3. Prepare a master mix for the secondary PCR samples plus one additional reaction by combining the reagents in the following order:

Reagent	Amount per reaction (µl)
Sterile water	18.5
10X PCR reaction buffer	2.5
Nested PCR primer NP1 (10 µM)	1.0
Nested PCR primer NP2R (10 µM)	1.0
dNTP mix (10 mM)	0.5
50X Advantage cDNA PCR Mix	0.5
Total volume	24.0

4. Place 24 µl aliquot of master mix into each reaction tube from step 2.
5. Overlay with one drop of mineral oil. Skip this step if an oil-free thermal cycler is used.
6. Immediately commence 10–12 cycles of:
   95°C 10 s
   68°C 10 s
   72°C 1.5 min
7. Analyze 4 µl from each reaction on a 2% agarose/EtBr gel.
8. Store PCR products at −20°C. This PCR product is now enriched for differentially presented DNAs. At this point if you are not going to perform MOS, please go to Section 6.3 (cloning of subtracted library) in this method section.

6.2 Mirror Orientation Selection

The major drawback of SSH is the presence of background clones that represent nondifferentially expressed DNA species in the subtracted libraries. In some difficult cases, the number of background clones may considerably exceed the number of target clones. To overcome this problem, we recommend MOS – a simple procedure that substantially decreases the number of background clones in the libraries generated by SSH (see Note 16).

We recommend the use of MOS in the following cases:
- If the percentage of differentially expressed clones found during differential screening is very low (e.g. 1–5%). The MOS procedure can increase the number of differential clones up to tenfold.
- If most of the differentially expressed clones found are false positive clones (i.e. clones that appear to be differentially expressed in the differential screening procedure, but turn out not to be differentially expressed in the Northern blot or reverse transcription–PCR analysis). The MOS procedure decreases the portion of false positive clones by several folds.
- If the primary PCR in SSH requires more than 30 cycles (but not more than 36 cycles, see Note 15) to generate agarose gel detectable PCR product. If the primary PCR requires more than 30 cycles, the problems described in the previous two items will usually appear. If you want to perform MOS, please follow the following procedure for PCR amplification using the second hybridization solution (Section 6.1.6, step 9).

6.2.1 PCR amplification for MOS

If the complexity of tracer and driver samples is very high or if the difference in gene expression between tracer and driver is very small, one can plan to perform MOS from the beginning of the experiment. In that case, after subtractive hybridization (Section 6.1.5), perform PCR amplification using the following protocol instead of using protocol in Section 6.1.6. If you have already made the

Suppression subtractive hybridization

SSH subtracted library and found high background upon differential screening, you may perform MOS on the SSH-generated library. You can use the hybridization mix generated in Section 6.1.6, step 9) for PCR amplification using the following protocol:

PRIMARY PCR

1. Transfer 10 µl of each diluted second hybridization (from Section 6.1.5) into appropriately labeled tubes (see Note 12)
2. Prepare a master mix for the primary PCR-1. For each reaction, combine the reagents as follows:

Component	Amount per reaction (µl)
Sterile water	92.5
10X PCR buffer	12.5
dNTP mixture (10 mM each)	2.5
PCR primer	15.0
50X polymerase mixture	2.5
Total volume	115

3. Mix well and briefly centrifuge the tube
4. Place 115 µl aliquot of master mix into each reaction tube from step 1
5. Place 125 µl aliquot of final mix into five 0.5 µl PCR tubes (25 µl per tube)
6. Overlay with one drop of mineral oil
7. Incubate the reaction mixture in a thermal cycler at 72°C for 5 min to extend the adapters (see Note 13)
8. Immediately commence thermal cycling (see Note 17 to calculate the number of PCR cycles you need):
   95°C 10 s
   66°C 10 s
   72°C 1.5 min
9. Combine 2 µl of each (of 5) primary PCR-1 product in one tube and add 390 µl of water
10. Place 1 µl aliquot of each diluted primary PCR-1 product mixture from step 9 into an appropriately labeled PCR tube
11. Prepare master mix for primary PCR-2 as follows:

Component	Amount per reaction (µl)
Sterile water	19.5
10X PCR buffer	2.5
dNTP mixture (10 mM each)	0.5
PCR primer	11.0
50X polymerase mixture	0.5
Total volume	24

12. Mix well and briefly centrifuge the tube
13. Place 24 µl aliquot of master mix into each reaction tube from step 10
14. Overlay with one drop of mineral oil
15. Immediately commence thermal cycling:
    – 10 Cycles:
    94°C 30 s
    66°C 30 s
    72°C 1.5 min
16. Analyze 4 µl from each reaction on a 2% agarose/EtBr gel

SECONDARY PCR

1. Dilute 2 µl of each primary PCR-2 product generated in Section 6.2.1, step 16 in 38 µl of water
2. Place 2 µl aliquot of each diluted primary PCR-2 product into an appropriately labeled tube
3. Prepare a master mix for secondary PCR. For each reaction, combine the reagents as follows:

Component	Amount per reaction (µl)
Sterile water	37.0
10X PCR buffer	5.0
dNTP mixture (10 mM each)	1.0
PCR primer NP1	2.0
PCR primer NP2R	2.0
50X polymerase mixture	1.0
Total volume	48.0

4. Mix well and briefly centrifuge the tube
5. Place 48 µl aliquot of master mix into each reaction tube from step 2
6. Overlay with one drop of mineral oil
7. Immediately commence thermal cycling:
    – 10 Cycles:
    95°C 10 s
    68°C 10 s
    72°C 1.5 min
8. Analyze 4 µl from each tube on a 2% agarose/EtBr gel
9. The PCR product of secondary PCR is purified by phenol or chloroform extraction and ethanol precipitation (see Note 4)
10. Dissolve the pellet in 20–40 µl of NT buffer up to concentration 20 ng/µl of DNA
11. Analyze 2 µl of purified PCR product from step 9 on a 2% agarose/EtBr gel
12. Dilute 1 µl of purified PCR product from step 9 in 1.6 ml of water (this will be your undigested control)
13. Store at −20°C

6.2.2 Xma I digestion

1. Add the following reagents into the tube:
   - 12 μl H$_2$O
   - 2 μl 10X *Xma* I restriction buffer
   - 5 μl DNA (Section 6.2.1, secondary PCR, step 10)
   - 1 μl *Xma* I (10 U/μl)
2. Mix and incubate at 37°C for 2 h
3. Add 2 μl of 0.2 M EDTA to terminate the reaction
4. Incubate at 70°C for 5 min to inactivate enzyme
5. Store at −20°C.

6.2.3 MOS hybridization

Combine the following reagents in a fresh 1.5-ml tube:
- 2 μl H$_2$O
- 1 μl *Xma* I-digested DNA
- 1 μl 4X Hybridization buffer

Place 2 μl of this mixture in a 0.5-ml microcentrifuge tube and overlay with one drop of mineral oil.
- Incubate in a thermal cycler at 98°C for 1.5 min
- Incubate in a thermal cycler at 68°C for 3 h
- Add 200 μl of dilution buffer to the tube and mix by pipetting
- Heat in a thermal cycler at 70°C for 7 min.
- Store at −20°C.

6.2.4 MOS PCR amplification

1. Prepare a master mix for all MOS PCR reactions as follows:

Component	Amount per reaction (μl)
Sterile water	19.5
10X PCR buffer	2.5
dNTP mixture (10 mM each)	0.5
MOS PCR primer (NP2Rs)	1.0
50X polymerase mixture	0.5
Total volume	24.0

2. Add 1 μl of each diluted cDNA sample (after hybridization and the corresponding undigested control) to an appropriately labeled tube containing 24 μl of master mix.
3. Overlay with one drop of mineral oil.
4. Incubate the reaction mix in a thermal cycler at 72°C for 2 min to extend the adapters. (Do not remove the samples from the thermal cycler.)
5. Immediately commence thermal cycling:
   - 19 Cycles:
   94°C 30 s

62°C 30 s
72°C 1.5 min
6. Analyze 4 µl from each tube on 2% agarose/EtBr gel.

6.3 Cloning of Subtracted cDNAs

Once a subtracted sample is confirmed to be enriched in cDNAs derived from differentially expressed genes, the PCR products (from Section 6.1.5, secondary PCR or from Section 6.2.4, MOS PCR amplification) can be subcloned using several conventional cloning techniques. The following describes two such methods that are currently used.

6.3.1 T/A cloning

Use 3 µl of the secondary PCR product (Section 6.1.5, step 8) or MOS PCR product (Section 6.2.4, step 6) for cloning with a T/A-based system, such as the Advantage PCR Cloning Kit (Invitrogen), according to the manufacturer's protocol. The library may be transformed into bacteria (electrocompetent cells) by electroporation (1.8 kV) using a pulser (BioRad) and plated onto agar plates containing ampicillin, X-Gal, and IPTG. Recombinant (white colonies) clones are picked and used to inoculate LB medium in 96-well microtiter plates. Bacteria should be allowed to grow at 37°C for 4 h before insert amplification (Section 6.4.1). Typically, 10^4 independent clones from 1 µl of secondary PCR product can be obtained using the above cloning system and electroporation. It is important to optimize the cloning efficiency because a low cloning efficiency will result in a high background.

6.3.2 Site-specific or blunt-end cloning

For site-specific cloning, cleave at the *Eag* I, *Not* I, and *Xma* I (*Sma* I, *Srf* I) sites embedded in the adapter sequences and then ligate the products into an appropriate plasmid vector. Keep in mind that all of these sites might be present in the cDNA fragments. The *Rsa* I site in the adapter sequences can also be used for blunt-ended cloning. Commercially available cloning kits are suitable for these purposes. The number of independent colonies obtained for each library depends on the estimated number of differentially expressed genes, as well as the subtraction and subcloning efficiencies. In general, 500 colonies can be initially arrayed and studied. The complexity of the library can be increased by additional subcloning of secondary PCR products (from Section 6.1.5) or MOS PCR products (from Section 6.2.4).

6.4 Differential Screening of the Subtracted cDNA Library

Two approaches can be utilized for differential screening of the arrayed subtracted cDNA clones; cDNA dot blots and colony dot blots. For colony dot blots, bacterial colonies are spotted on nylon filters, grown on antibiotic

Suppression subtractive hybridization

plates, and processed for colony hybridization. This method is cheaper and more convenient, but it is less sensitive and gives a higher background than PCR-based cDNA dot blots. The cDNA array approach is highly recommended (Section 6.4.2).

6.4.1 Amplification of cDNA inserts by PCR

For high-throughput screening, a 96-well format PCR from one of several thermal cycler manufacturers is recommended. Alternatively, single tubes can be used.
1. Randomly pick 96 white bacterial colonies
2. Grow each colony in 100 μl of LB-amp medium in a 96-well plate at 37°C for at least 2 h (up to overnight) with gentle shaking
3. Prepare a master mix for 100 PCR reactions (see Note 18):

Reagent	Amount per reaction (μl)
10X PCR buffer	2.0
Nested primer NP1*	0.6
Nested primer NP2R*	0.6
dNTP mix (10 mM)	0.4
H_2O	15.0
50X Advantage cDNA PCR Mix	0.4
Total volume	19.0

Alternatively, primers flanking the insertion site of the vector can be used in PCR amplification of the inserts.

4. Place 19 μl aliquot of the master mix into each tube or well of the reaction plate
5. Transfer 1 μl of each bacterial culture (from step 2) to each tube or well containing master mix (see Note 19)
6. Perform PCR in an oil-free thermal cycler with the following conditions:
   – 1 cycle:
   94°C 2 min
   – Then 22 cycles:
   94°C 30 s
   68°C 3 min
7. Analyze 5 μl from each reaction on a 2% agarose/EtBr gel (see Note 20).

6.4.2 Preparation of cDNA dot blots of the PCR products

1. For each PCR reaction, combine 5 μl of the PCR product and 5 μl of 0.6 M NaOH (freshly made or at least freshly diluted from concentrated stock).
2. Transfer 1–2 μl of each mixture to a nylon membrane. This can be accomplished by dipping a 96-well replicator in the corresponding wells of a microtiter dish used in the PCR amplification and spotting it onto a dry nylon

filter. Make at least two identical blots for hybridization with subtracted and reverse-subtracted probes (Section 6.1.4; see Note 22).
3. Neutralize the blots for 2–4 min in 0.5 M Tris-HCl (pH 7.5) and wash in 2X SSC.
4. Immobilize cDNA on the membrane using a UV cross-linking device (such as Stratagene's UV Stratalinker), or bake the blots for 4 h at 68°C.

6.4.3 Differential hybridization with tracer and driver cDNA probes

Label tracer and driver cDNA probes by random-primer labeling using a commercially available kit. The hybridization conditions given here are optimized for BD Biosciences Clontech's ExpressHyb solution; the optimal hybridization conditions for other systems should be determined empirically.

The following four different probes will be used for differential screening hybridization:
1. Tracer-specific subtracted probe (forward-subtracted probe)
2. Driver-specific subtracted probe (reverse-subtracted probe)
3. cDNA probe synthesized directly from tracer mRNA
4. cDNA probe synthesized directly from driver mRNA

See Note 23.
1. Prepare a prehybridization solution for each membrane:
   (a) Combine 50 µl of 20X SSC, 50 µl of sheared salmon sperm DNA (10 mg/ml), and 10 µl of blocking solution (containing 2 mg/ml of unpurified NP1, NP2R, cDNA synthesis primers, and their complementary oligonucleotides).
   (b) Boil the blocking solution for 5 min, then chill on ice.
   (c) Combine the blocking solution with 5 ml of ExpressHyb hybridization solution (BD Biosciences Clontech).
2. Place each membrane in the prehybridization solution prepared in step 1. Prehybridize for 40–60 min with continuous agitation at 72°C.

Note: It is important to add blocking solution in prehybridization solution. Because subtracted probes contain the same adapter sequences as arrayed clones, these probes hybridize to all arrayed clones regardless of the sequences.
3. Prepare hybridization probes:
   (a) Mix 50 µl of 20X SSC, 50 µl of sheared salmon sperm DNA (10 mg/ml) and 10 µl blocking solution, and purified probe (at least 10^7 cpm per 100 ng of subtracted cDNA). Make sure the specific activity of each probe is approximately equal.
   (b) Boil the probe for 5 min, then chill on ice.
   (c) Add the probe to the prehybridization solution.
4. Hybridize overnight with continuous agitation at 72°C.
5. Prepare low-stringency (2X SSC/0.5% SDS) and high-stringency (0.2X SSC/0.5% SDS) washing buffers and warm them up to 68°C.
6. Wash membranes with low-stringency buffer (4X 20 min at 68°C), then wash with high stringency buffer (2X 20 min at 68°C).

Suppression subtractive hybridization

7. Perform autoradiography.
8. If desired, remove probes from the membranes by boiling for 7 min in 0.5% SDS. Blots can typically be reused at least five times.

Note: To minimize hybridization background, store the membranes at –20°C when they are not in use.

7. NOTES

Suppression subtractive hybridization and MOS are available as the custom service from Evrogen (http://www.evrogen.com/).

1. Self-subtraction (with both tracer and driver prepared from one DNA sample) is recommended as the best comprehensive control. In the self-subtracted control, subtracted secondary PCR requires more cycles than unsubtracted secondary PCR. A number of other control experiments may be performed for fast analysis of SSH and MOS experiments (see below).
2. For experimental systems such as transfection, overexpression, mRNA injection, or viral infection using mammalian or viral expression systems, we strongly recommend that you use affecting RNA or DNA sequence for compensation of overexpressed sequence concentration. For example, if you are searching for p53-up-regulated genes in a p53 overexpressed cell line, add *Rsa* I-digested p53 cDNA into *Rsa* I-digested driver sample (about one tenths of driver cDNA concentration) after you prepare adapter-ligated tracer. Adding exogenous DNA/RNA earlier (in RNA sample) or before *Rsa* I digestion may cause disproportion of this material in initial DNAs.
3. We recommend the use of poly(A)+ RNA as starting material. Amplified cDNA should be used as a starting material only when enough RNA is not available. The amplification of two cDNA samples to be subtracted is a crucial procedure and any disproportion during cDNA amplification may cause artifacts in the subtraction results. Some RNA types cannot be amplified because the messages are too long and are not available for subtraction and analysis.
4. Phenol–chloroform extraction and ethanol precipitation:
    (a) Add equal volumes of phenol:chloroform:isoamyl alcohol (25:24:1) and vortex thoroughly.
    (b) Centrifuge the tubes at 14,000 rpm for 10 min.
    (c) Remove the top aqueous layer and transfer to a fresh microcentrifuge tube.
    (d) Add equal volumes of chloroform:isoamyl alcohol (24:1) and vortex thoroughly.
    (e) Centrifuge the tubes at 14,000 rpm for 10 min.
    (f) Remove the top aqueous layer and transfer to a fresh microcentrifuge tube.
    (g) Add 0.5 volume of 4 M NH_4OAc, mix, then add 2.5 volumes of 95% ethanol and vortex thoroughly.
    (h) Centrifuge the tubes at 14,000 rpm for 20 min.
    (i) Remove the supernatant carefully.

(j) Add 200 µl of 80% ethanol.
(k) Centrifuge the tubes at 14,000 rpm for 10 min.
(l) Remove the supernatant carefully.
(m) Air-dry the pellets for 5–10 min.
(n) Dissolve the pellets in appropriate volume of TN buffer.

5. Using glycogen or any type of coprecipitants during DNA precipitation may increase viscosity of DNA solution and prevent DNA hybridization in some cases. We recommend avoiding use of these reagents if possible.
6. We do not recommend using silica matrix-based PCR purification systems at this stage.
7. Water may denature short DNA fragments and may make the adapter ligation difficult. We advise using TN buffer.
8. Ligation efficiency test:
    (a) Place 1 µl aliquot of each undiluted unsubtracted control sample (Section 6.1.4, step 8) into an appropriately labeled 0.5-ml PCR tube.
    (b) Prepare a master mix for all of the reaction tubes. Combine the reagents as follows:

Component	Amount per reaction (µl)
H_2O	19.5
10X PCR buffer	2.5
dNTP mixture (10 mM each)	0.5
PCR primer P1	1.0
50X polymerase mixture	0.5
Total volume	24.0

   (c) Mix and briefly centrifuge the tubes.
   (d) Place 24 µl aliquot of master mix into each of the reaction tubes prepared in step 1.
   (e) Overlay with one drop of mineral oil.
   (f) Incubate the reaction mixture in a thermal cycler at 72°C for 5 min to extend the adapters.
   (g) Immediately commence thermal cycling:
      – 15 Cycles:
        95°C 10 s
        66°C 10 s
        72°C 1.5 min
   (h) Analyze 4 µl from each tube on a 2% agarose/EtBr gel. This PCR product should have a similar pattern to that of *Rsa* I-digested DNA. If PCR products are not visible after 15 cycles, perform three more cycles and again analyze the PCR product. If PCR products are not visible after 21 cycles, the activity of the polymerase mixture needs to be examined. If there is no problem with the polymerase mixture, the ligation reaction should be repeated with fresh samples before proceeding to the next step.

9. Recommended first hybridization times for different DNA samples:

Sample type	First hybridization time (h)
Bacterial genome subtraction	1–3
Eukaryotic genome subtraction	3–5
cDNA subtraction	7–12

10. We recommend that you use two blocks thermal cycler (or two thermal cyclers nearby) for proper and fast operations.
11. We recommend transferring this tube immediately after denaturing (98°C for 1.5 min) into thermal cycler with first hybridization process (68°C) and waiting for 1 min before proceeding to the next step.
12. If hybridization mix was frozen, we recommend the following before proceeding with PCR reactions: mix hybridization samples well by pipetting, heat in a thermal cycler at 72°C for 7 min, then mix again by pipetting and use for PCR.
13. This step "fills in" the missing strand of the adapters and thus creates binding sites for the PCR primers.
14. For some complicated subtractions (with complex tissues or eukaryotic genomes), we recommend performing primary PCR two times one by one. This procedure may significantly reduce background, generated by partial disruption of PCR suppression effect. First, perform primary PCR as described in Section 6.1.5. Then perform another primary PCR as follows:
    (a) Dilute 2 µl of each primary PCR product (from step 7) in 78 µl of water.
    (b) Place 1 µl of each diluted primary PCR product from step 1 into appropriately labeled tube.
    (c) Combine the following reagents to prepare a master mix for each reaction.

Component	Amount per reaction (µl)
10X PCR buffer	2.5
PCR primer P1	1.0
dNTP mix (10 Mm)	0.5
H2O	19.5
50X Advantage cDNA PCR Mix	0.5
Total volume	24.0

(d) Mix well and briefly centrifuge the tube.
(e) Place 24 µl aliquot of master mix into each reaction tube from step 2.
(f) Overlay with one drop of mineral oil.
(g) Immediately commence thermal cycling:
   – 10 Cycles:
   95°C 10 s
   66°C 10 s
   72°C 1.5 min
(h) Analyze 4 µl from each tube on a 2% agarose/EtBr gel, then proceed to secondary PCR (Section 6.1.5).

15. If the SSH primary PCR requires more then 36 cycles, "*in vitro* cloning" will occur. As a result, only false-positive clones may be found during differential screening procedure.
16. To illustrate the utility of combining SSH and MOS for eukaryotic genome comparison, we will describe our efforts to isolate genes that are present in one freshwater planaria warm strain but are absent in another. In this study, we used two closely related strains of freshwater planaria *Girardia tigrina* that reproduce in different ways. Whereas one strain has exclusively asexual reproduction, the other reproduces both sexually and asexually. The genomes of both strains of *G. tigrina* we compared to search for genetic determinants of asexuality. Total DNA from these strains was purified using the procedure described in Section 6.1.1. The SSH and MOS combination was used to isolate genes that are differentially present in each planaria strain. Forward subtraction (AB) was performed using asexual DNA (sample A) as tracer and sexual DNA (sample B) as driver, and the forward-subtracted DNA was enriched for DNA fragments specific to the asexual strain of freshwater planaria. Reverse-subtracted DNA (BA) was enriched in DNA fragments specific to the sexual planaria strain. Self-subtractions were performed for both DNA samples to get a quick idea of subtraction efficiency. Subsequent MOS PCR analysis confirmed that the self-subtractions (as well as undigested controls) require more PCR cycles to generate visible PCR product, indicating that the subtraction was successful (Figure 4).

We anticipated that the differences between tracer and driver DNA would be small, so we proceeded with a differential screening procedure. From each forward- and reverse-subtracted library, 86 randomly selected clones were arrayed (DNA dot blot) onto nylon membranes. DNA dot blots were hybridized to probes prepared from the subtracted and reverse-subtracted libraries. Figure 5 shows typical results of differential screening of a subtracted cDNA libraries obtained using the SSH and MOS combination. Normally, MOS results for DNA or cDNA reveal the following types of clones:

(a) Clones hybzridizing to the one probe only. These correspond to the differentially presented DNA, but must be verified by Southern blot analysis. The signal intensity depends on the copy number in genomic or extrachromosomal DNA.

(b) Clones hybridizing to both subtracted probes with the same efficiency. These clones most probably do not correspond to the differentially presented DNA and thus represent a background.

(c) Clones hybridizing to both subtracted probes with different hybridization efficiencies. In the case of genomic DNA subtraction, these clones may represent genes (DNA fragments) with different number of copies per genome. In the case of cDNA subtraction, these clones do represent differentially expressed clones. In some cDNA subtractions, this difference can be a result of random fluctuation and does not represent differentially

Suppression subtractive hybridization

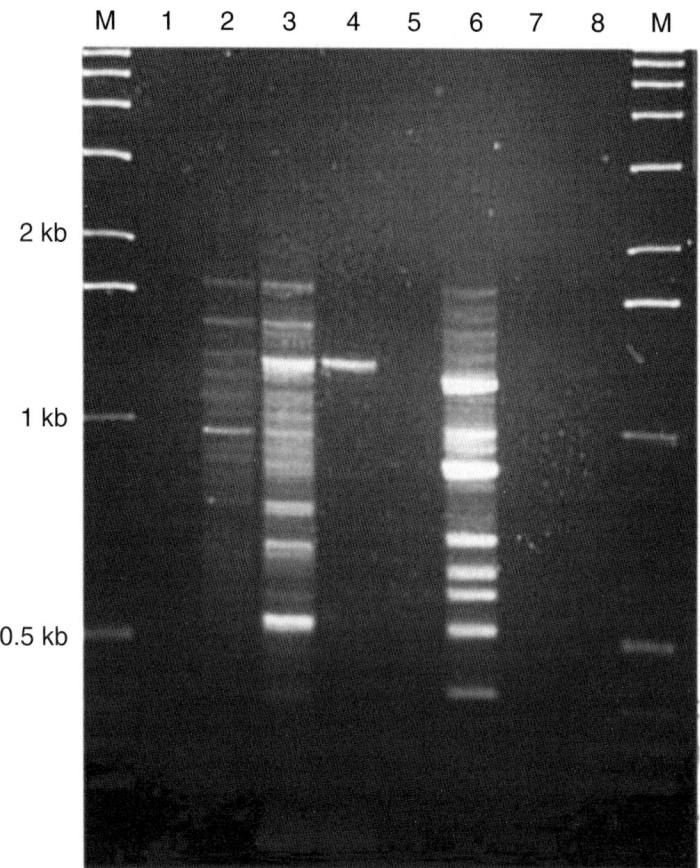

Figure 4. MOS stage significantly improves differential library quality in many cases. Lane 1: MOS PCR product of undigested control of BB self-subtraction. Lane 2: MOS PCR product of BB self-subtraction. Lane 3: MOS PCR product of BA experimental subtraction. Lane 4: MOS PCR product of undigested control of BA experimental subtraction. Lane 5: MOS PCR product of undigested control of AB experimental subtraction. Lane 6: MOS PCR product of AB experimental subtraction. Lane 7: MOS PCR product of AA self-subtraction. Lane 8: MOS PCR product of undigested control of AA self-subtraction.

expressed cDNA. For this reason, it is always recommended to confirm true differential expression of these clones by Northern blot analysis or reverse transcription polymerase chain reaction (RT-PCR).

(d) Clones that do not hybridize noticeably to either hybridization probe. These clones may not contain DNA insertion or may be present at very low concentration in the subtracted probe. (In most cases, they do not represent differentially presented clones.)

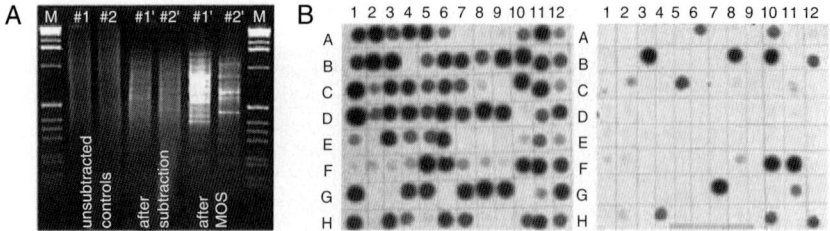

Figure 5. Typical results of suppression subtractive hybridization, MOS, and differential screening of the resulting libraries. (A) MOS reveals real differential bands and eliminates most of false positives. M – 1 kb DNA size marker ladder, #1 and #2 – cDNA samples 1 and 2; #1′ – sample 1 versus sample 2 subtraction; #2′ – sample 2 versus sample 1 subtraction. (B) Typical differential screening results. Randomly picked clones from #1′ subtracted cDNA library (after MOS) were hybridized with radioactively labeled #1′ (filter from the left) and #2′ (filter from the right) subtracted cDNA probes.

Differential screening revealed about 60% and 30% of the strain-specific clones in AB and BA libraries, respectively. Most of the nondifferential DNA sequences were identified as the *mariner* repetitive element, presented by ~7000 copies in each compared genome (Garcia-Fernandez et al. 1993).

17. The recommended number of primary PCR-1 cycles for MOS is the number of SSH primary PCR minus 2. For example, if the SSH primary PCR was visible on agarose/EtBr gel after 31 cycles, you will need 31 – 2 = 29 cycles of primary PCR-1 for MOS.
18. Short PCR primers NP1s and NP2s may be used for insert amplification to reduce hybridization background. However, this is not always necessary.
19. Freshly grown 96-well plates should be used for PCR before bacterial cells precipitate, otherwise 1 µl aliquots will not be equal.
20. It is possible that (5–10% of clones will not yield PCR product as a result of imperfect cloning.
21. The protocol uses 15 ng of ligated tracer cDNA and 450 ng of driver cDNA. The ratio of driver to tracer can be changed if different subtraction efficiency is desired.
22. We highly recommend that you make four identical blots. Two of the blots will be hybridized to forward- and reverse-subtracted cDNAs and the other two can be hybridized to cDNA probes synthesized from tracer and driver mRNAs.
23. The first two probes are the secondary PCR products (Section 6.1.5, step 8 or Section 6.2.1, secondary PCR, step 10) of the subtracted cDNA pool. The last two cDNA probes can be synthesized from the tracer and driver poly(A)+ RNA. They can be used as either single- or double-stranded cDNA probes. Alternatively, unsubtracted tracer and driver cDNA or pre-amplified cDNA from total RNA can be used if enough poly(A)+ RNA is not available. If you have made the MOS-subtracted library, you can still screen it using the same probes.

REFERENCES

Akopyants NS, Fradkov A, Diatchenko L, Hill JE, Siebert PD, Lukyanov SA, Sverdlov ED, Berg DE (1998) PCR-based subtractive hybridization and differences in gene content among strains of Helicobacter pylori. Proc Natl Acad Sci USA 95:13108–13113

Chomczynski P, Sacchi N (1987) Single-step method of RNA isolation by acid guanidinium thiocyanate-phenol-chloroform extraction. Anal Biochem 162:156–159

Diatchenko L, Lau YF, Campbell AP, Chenchik A, Moqadam F, Huang B, Lukyanov S, Lukyanov K, Gurskaya N, Sverdlov ED, Siebert PD (1996) Suppression subtractive hybridization: a method for generating differentially regulated or tissue-specific cDNA probes and libraries. Proc Natl Acad Sci USA 93:6025–6030

Ermolaeva OD, Wagner MC (1995) SUBTRACT: a computer program for modeling the process of subtractive hybridization. Comput Appl Biosci 11:457–462

Ermolaeva OD, Lukyanov SA, Sverdlov ED (1996) The mathematical model of subtractive hybridization and its practical application. Proc Int Conf Intell Syst Mol Biol 4:52–58

Garcia-Fernandez J, Marfany G, Baguna J, Salo E (1993) Infiltration of mariner elements. Nature 364:109–110

Gubler U, Hoffman BJ (1983) A simple and very efficient method for generating cDNA libraries. Gene 25:263–269

Gurskaya NG, Diatchenko L, Chenchik A, Siebert PD, Khaspekov GL, Lukyanov KA, Vagner LL, Ermolaeva OD, Lukyanov SA, Sverdlov ED (1996) Equalizing cDNA subtraction based on selective suppression of polymerase chain reaction: cloning of Jurkat cell transcripts induced by phytohemaglutinin and phorbol 12-myristate 13-acetate. Anal Biochem 240:90–97

Hames BD, Higgins SJ (eds) (1985) Nucleic acid hybridization. A practical approach. IRL Press, Oxford, Washington, DC

Hara E, Kato T, Nakada S, Sekiya S, Oda K (1991) Subtractive cDNA cloning using oligo(dT)30-latex and PCR: isolation of cDNA clones specific to undifferentiated human embryonal carcinoma cells. Nucleic Acids Res 19:7097–7104

Hedrick SM, Cohen DI, Nielsen EA, Davis MM (1984) Isolation of cDNA clones encoding T cell-specific membrane-associated proteins. Nature 308:149–153

Hubank M, Schatz DG (1994) Identifying differences in mRNA expression by representational difference analysis of cDNA. Nucleic Acids Res 22:5640–5648

Jin H, Cheng X, Diatchenko L, Siebert PD, Huang CC (1997) Differential screening of a subtracted cDNA library: a method to search for genes preferentially expressed in multiple tissues. Biotechniques 23:1084–1086

Kellogg DE, Rybalkin I, Chen S, Mukhamedova N, Vlasik T, Siebert PD, Chenchik A (1994) TaqStart Antibody: "hot start" PCR facilitated by a neutralizing monoclonal antibody directed against Taq DNA polymerase. Biotechniques 16:1134–1137

Ko MS (1990) An "equalized cDNA library" by the reassociation of short double-stranded cDNAs. Nucleic Acids Res 18:5705–5711

Luk'ianov KA, Gurskaia NG, Bogdanova EA, Luk'ianov SA (1999) Selective suppression of polymerase chain reaction. Bioorg Khim 25:163–170

Luk'ianov SA, Gurskaia NG, Luk'ianov KA, Tarabykin VS, Sverdlov ED (1994) Highly-effective subtractive hybridization of cDNA. Bioorg Khim 20:701–704

Lukyanov K, Diatchenko L, Chenchik A, Nanisetti A, Siebert P, Usman N, Matz M, Lukyanov S (1997) Construction of cDNA libraries from small amounts of total RNA using the suppression PCR effect. Biochem Biophys Res Commun 230:285–288

Lukyanov KA, Launer GA, Tarabykin VS, Zaraisky AG, Lukyanov SA (1995) Inverted terminal repeats permit the average length of amplified DNA fragments to be regulated during preparation of cDNA libraries by polymerase chain reaction. Anal Biochem 229:198–202

Lukyanov KA, Matz MV, Bogdanova EA, Gurskaya NG, Lukyanov SA (1996) Molecule by molecule PCR amplification of complex DNA mixtures for direct sequencing: an approach to in vitro cloning. Nucleic Acids Res 24:2194–2195

Maruyama K, Sugano S (1994) Oligo-capping: a simple method to replace the cap structure of eukaryotic mRNAs with oligoribonucleotides. Gene 138:171–174

Rebrikov DV, Britanova OV, Gurskaya NG, Lukyanov KA, Tarabykin VS, Lukyanov SA (2000) Mirror orientation selection (MOS): a method for eliminating false positive clones from libraries generated by suppression subtractive hybridization. Nucleic Acids Res 28:e90

Rebrikov DV, Desai SM, Siebert PD, Lukyanov SA (2004) Suppression subtractive hybridization. Methods Mol Biol 258:107–134

Sargent TD, Dawid IB (1983) Differential gene expression in the gastrula of *Xenopus laevis*. Science 222:135–139

Siebert PD, Chenchik A, Kellogg DE, Lukyanov KA, Lukyanov SA (1995) An improved PCR method for walking in uncloned genomic DNA. Nucleic Acids Res 23:1087–1088

Takarada Y (1994) Highly sensitive and specific amplification by primer digestion. Nucleic Acids Res 22:2170–2172

Wang Z, Brown DD (1991) A gene expression screen. Proc Natl Acad Sci USA 88:11505–11509

Zhumabayeva B, Chang C, McKinley J, Diatchenko L, Siebert PD (2001a) Generation of full-length cDNA libraries enriched for differentially expressed genes for functional genomics. Biotechniques 30:512–516, 518–520

Zhumabayeva B, Diatchenko L, Chenchik A, Siebert PD (2001b) Use of SMART-generated cDNA for gene expression studies in multiple human tumors. Biotechniques 30:158–163

CHAPTER 4

STEM-LOOP OLIGONUCLEOTIDES AS HYBRIDIZATION PROBES AND THEIR PRACTICAL USE IN MOLECULAR BIOLOGY AND BIOMEDICINE

ANTON A. BUZDIN*, SERGEY A. LUKYANOV

Shemyakin-Ovchinnikov Institute of Bioorganic Chemistry, Russian Academy of Sciences, 16/10 Miklukho-Maklaya, 117997 Moscow, Russia
*Corresponding author
Phone: +(7495) 3306329; Fax: +(7495) 3306538; E-mail: anton@humgen.siobc.ras.ru

Abstract:	As originally designed, stem-loop (SL) hybridization probe is a single-stranded oligonucleotide containing a sequence complementary to the target that is flanked by inverted repeats forming self-complementary termini, and harbors a fluorophore–quencher pair at the 5′- and 3′-ends. When the target is missing, these molecules form closed pan handle-like structures in which the fluorophore and quencher are located in a close proximity, due to 5′- and 3′-termini spatial neighborhood. Such a quencher proximity represses fluorescence. In contrast, in the presence of the target, probe forms a complex with it, which spatially separates the fluorophore from the quencher. Once the fluorophore and quencher are dissociated, the fluoresce increases, thus enabling quantitative measuring of signals with the threshold value evidencing presence of the target. Alternatively, SL probe can be fluorescence resonance energy transfer (FRET)-labeled, with the fluorescence shift message being monitored in the latter case. SL probes are currently in use in a number of applications, primarily for specific nucleic acid motif detection and in real-time polymerase chain reaction (PCR). Also, SL probes can be utilized in protein–DNA interaction studies, and even for specific inorganic ion detection. This quickly evolving group of methods gave rise to at least 40 international patents and is cited in ~50 peer-reviewed papers annually.
Keywords:	Stem loop, hybridization probe, molecular beacon, mismatched target, secondary structure, linear probe, thermodynamic limitations, fluorophore, quencher, fluorescence resonance energy transfer, FRET, real-time, mismatch, mismatch discrimination, accuracy, sensitivity, nanoparticle, multiplex, scorpion, DNAzyme, catalytic molecular beacon, DNA probe, mispaired base, immobilized.
Abbreviations:	cDNA, complementary DNA; dNTP, deoxyribonucleotidetriphosphate; DABCYL, 4-{[4′-(dimethylamino)phenyl]azo}benzoic acid; FRET, fluorescence

resonance energy transfer; mRNA, messenger RNA; PDGF, platelet-derived growth factor; PCR, polymerase chain reaction; RT, reverse transcription; SL, stem-loop.

TABLE OF CONTENTS

1. Introduction 86
2. Molecular Beacons 87
3. Stem-Loop DNA Probes on Microarrays 92
4. Conclusions 93
References 94

1. INTRODUCTION

Naturally occurring stem-loop (SL)-forming RNA and DNA structures have important regulatory roles in normal cellular functioning (Cech 1993; Pearson and Sinden 1998; Sinden et al. 2002). Apart from recognition by proteins (Varani 1995) and specific gene silencing (Meins et al. 2005), SL structures exhibit some advantageous characteristics in nucleic acids hybridization as compared with other conformations (reviewed by Broude (2002)) and other authors (Fang et al. 2002; Tsourkas and Bao 2003; Drake and Tan 2004; Goel et al. 2005; Santangelo et al. 2006). The use of SL-forming oligonucleotides for the *selective polymerase chain reaction (PCR) suppression* effect, which gave rise to a multitude of useful PCR-based experimental approaches, is reviewed in the Chapter 2 of this book and will not be considered here. In this chapter, we focus on the use of SL oligonucleotides as hybridization probes. The advantage of SL-forming oligonucleotides was clearly demonstrated when analyzing hybridization of linear and SL (also termed *molecular beacon*) DNA probes has revealed significantly higher specificity of the latter (Bonnet et al. 1998, 1999). By analyzing free-energy phase diagrams of molecular beacons in solution with matched and mismatched targets, the authors showed that structurally constrained probes can generally distinguish mismatches over a wider range of temperatures than unstructured probes can do (Bonnet et al. 1998; Broude 2002).

The probable explanation could be a lower free energy of SL probe annealing provided by smaller number of configurations (and, consequently, decreased entropy of hybridization) (Roberts and Crothers 1991). The loop regions, which are generally complementary to the targets, are better exposed for hybridization, as compared with linear probes. It should be mentioned, however, that secondary structures of SL probes dictate kinetic and thermodynamic limitations: pan handle-like structure formation proceeds somewhat slower (but is still fast), and results in products with lower melting temperatures and free energies (Bonnet et al. 1998; Broude 2002). Nevertheless, the beauty of structured probes lies in their ability to better discriminate mismatches in different applications (Riccelli

Stem-loop oligonucleotides as hybridization

et al. 2001). SL structures may be hybridized in solution (and thus used as capture devices in many applications, e.g. as a specific probe in a quantitative real-time PCR (Yesilkaya et al. 2006)), or being attached to a solid support so that the loop is immobilized on a surface and dangling ends are used for hybridization (Riccelli et al. 2001; Broude 2002).

In this chapter, we review the recent developments in using SL probes for DNA microarrays and molecular beacon approaches.

2. MOLECULAR BEACONS

As originally designed, a molecular beacon is a single-stranded oligonucleotide probe containing a sequence complementary to the target that is flanked by inverted repeats forming self-complementary termini, and harbors a fluorophore–quencher pair at the 5'- and 3'-ends (Tyagi and Kramer 1996; Tyagi et al. 1998) (Figure 1). When the target is missing, these molecules form closed pan handle-like structures in which the fluorophore and quencher are located in a close proximity, due to 5'- and 3'-termini spatial neighborhood. Such a quencher proximity represses fluorescence. In contrast, in the presence of the target, the molecular beacon forms a complex with it, which spatially separates the fluorophore from the quencher (like commonly used 4-{[4'-(dimethylamino)phenyl]azo}benzoic acid (DABCYL)). Once the fluorophore and quencher are dissociated, the fluoresce increases, thus enabling quantitative measuring of signals with the threshold value evidencing presence of the target (Figure 1) (Broude 2002).

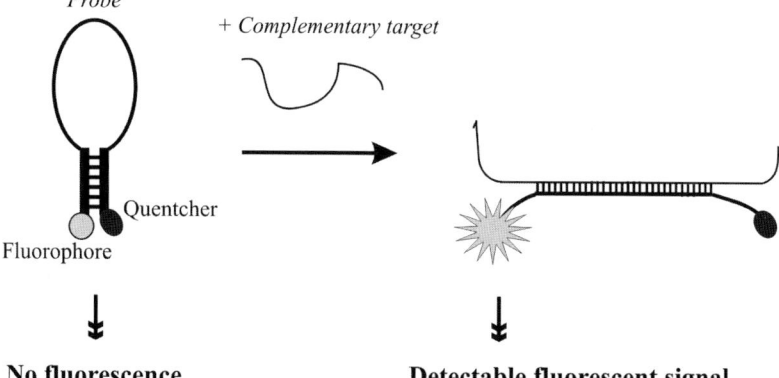

Figure 1. Principle of action of the first generation molecular beacons. The probe, which forms stem-loop structure due to intramolecularly base pairing of the inverted repeats on its termini, is labeled with fluorophore and quencher at the opposite ends. When a complementary target is added, a stronger complementary pairing between target and molecular beacon destroys the initial stem-loop structure and the fluorophore label may produce fluorescence due to a greater distance from quencher.

Alternatively, molecular probes utilizing fluorescence resonance energy transfer (FRET) can be adopted for using in molecular beacon format (Figure 2). Excitation energy is sometimes transferred from one fluorescent dye to another by resonance energy transfer. After excitation by the absorption of a photon, the donor or D dye can transfer its excited state energy to a second chromophore or dye (the acceptor or A) nonradiatively (i.e. without photon emission). The efficiency of this transfer depends on, among other things, the extent of overlap of the D emission and A absorption spectra, the relative orientation of the transition dipoles of D and A and – most importantly – the distance between D and A. FRET can measure distances between 20 and 100 Å (Stryer 1978), as well as detect conformational changes and determine their magnitudes (Johnson 2005). FRET is most widely used for protein interaction studies, but is also of a great usefulness for DNA hybridization probe design (Baker et al. 2006; Kim et al. 2006; Okamura and Watanabe 2006; Wahlroos et al. 2006). In the case of molecular beacons, one attaches donor and acceptor dyes to two different probe termini forming the stem of the pan handle-like structure. When both dyes are colocalized, resonance transfer may occur and the respective fluoresce peak (typical of the acceptor) will be observed; in contrast, when the SL probe conformation is

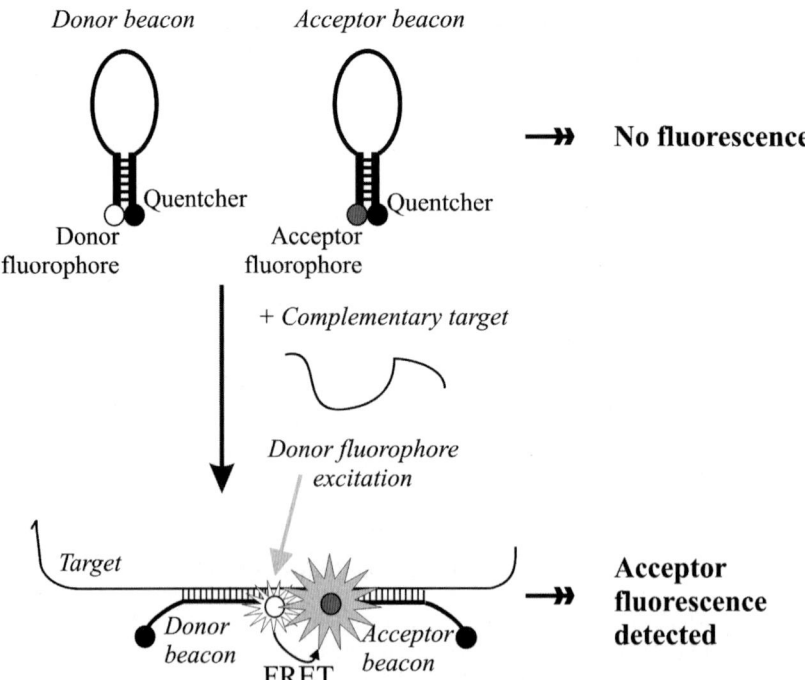

Figure 2. Example of the technique utilizing FRET-labeled molecular beacons.

altered, donor and acceptor fluorophores are too far to enable an efficient energy transfer, and a donor peak will be registered.

Molecular beacons have rapidly found many applications because of the relative simplicity of the assay and the option of real-time process monitoring (Fang et al. 2002). Because of their superiority in mismatch discrimination, DNA molecular beacons were used to detect single nucleotide polymorphisms (Kostrikis et al. 1998; Piatek et al. 1998; Marras et al. 1999), in quantitative PCR (Tyagi et al. 1998; Vet et al. 1999), as DNA microarray-immobilized probes (Liu and Tan 1999; Liu et al. 2000; Steemers et al. 2000), and as complementary probes for detecting RNA targets *in vivo* (Sokol et al. 1998; Broude 2002). Several improvements and further developments of molecular beacon technology have been reported. For example, several targets may be detected simultaneously in one assay by using differently colored molecular beacons (Tyagi et al. 1998; Vet et al. 1999). The method turned to be very sensitive in many applications, including pathogenic viral or bacterial strain detection (Gore et al. 2003; Henry et al. 2006; Patel et al. 2006).

In another approach, the method accuracy and sensitivity was substantially increased by using gold nanoparticles as a quencher due to the superior quenching ability of gold clusters (Dubertret et al. 2001). When fluorescently tagged oligonucleotides are located near metal surfaces, their emission intensity is impacted by both electromagnetic effects (i.e. quenching and/or enhancement of emission) and the structure of the nucleic acids (e.g. random coil, hairpin, or duplex) (Stoermer and Keating 2006). The use of metal nanoparticles in the molecular beacon strategies opens up bright perspectives for creating a variety of quenchers with very different properties, by changing the shape, size, or composition of the metal cluster. A variety of metal-based quenchers with different peculiarities may be used in the future for simultaneous detection of multiple different targets in a single tube (Dubertret et al. 2001). Alternatively, this problem of multiplexing target identifications may be solved by using wavelength-shifting molecular beacons (Tyagi et al. 2000). Nevertheless, highly sensitive multiplex molecular beacon assays using probes labeled with different fluorophores became practice in our days (Sinsimer et al. 2005; Balashov et al. 2006; Gubala and Proll 2006).

Whitcombe and colleagues combined both molecular beacon and a PCR primer in a single SL probe termed "scorpion" (Whitcombe et al. 1999). In this approach, the PCR primer is designed to have a 5′-extension that has all the attributes of a beacon: a loop region complementary to a target flanked by the self-complementary stems, and a fluorophore–quencher pair at the 5′- and 3′-ends of the extension, respectively (Figure 3) (Broude 2002). The SL extension is linked to the PCR primer via a linker, which stops DNA polymerase from replicating the SL. During PCR, when the primer is extended and the target is synthesized, the SL unfolds and the loop sequence hybridizes intramolecularly with its target, which increases fluorescence (Figure 3). Thus, the scorpion–primer approach uses a monomolecular mechanism of probe–target hybridization,

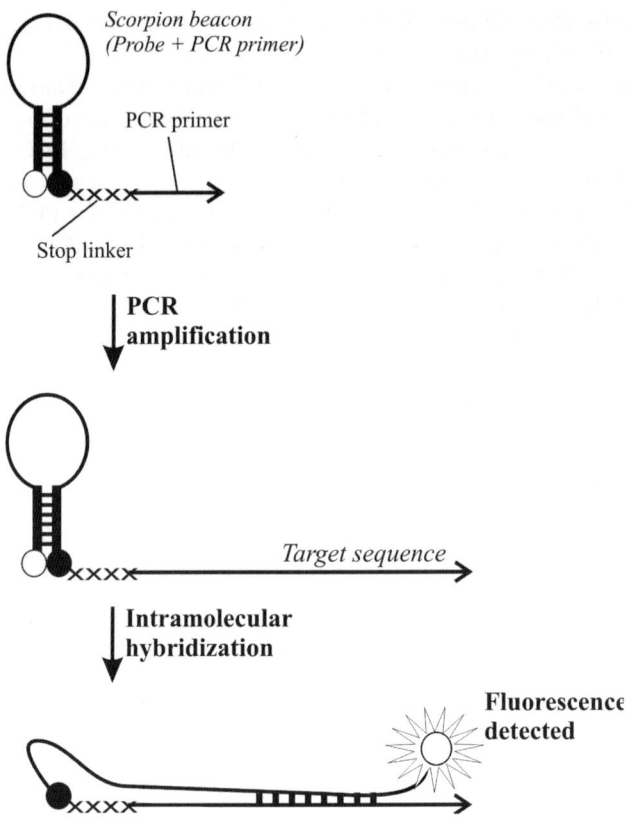

Figure 3. So-called "scorpion" molecular beacon probe. Scorpion probe has both molecular beacon and PCR primer modules. When the target sequence is amplified using scorpion primer, beacon unfolds and fluorophore–quencher pair is separated, thus allowing fluorescence.

compared with the bimolecular recognition in the traditional molecular beacon assay. This provides faster kinetics and greater stability of the probe–target complex (Thelwell et al. 2000; Broude 2002).

Another promising technique utilizes catalytic DNAs (DNAzymes) in combination with the molecular beacon approach. In this case, the PCR primer contains a sequence complementary to a DNAzyme (Todd et al. 2000). During PCR, an active amplicon is synthesized that cleaves the fluorescent beacon-shaped substrate included in the reaction mixture, increasing the fluorescence. This approach can be generalized so that one generic DNAzyme and one corresponding beacon substrate are used to detect different genomic targets (Broude 2002).

Catalytic molecular beacons represent a next generation of molecular probes with the potential to amplify signals and to detect nucleic acid targets without PCR amplification. In this case, a complex double-stranded DNA probe is engineered

Stem-loop oligonucleotides as hybridization

that combines the features of a molecular beacon and a hammerhead-type DNAzyme with RNase activity; these are located on two opposite DNA strands (Stojanovic et al. 2001) (Figure 4). In the absence of a target, the DNAzyme module is blocked being hybridized with the beacon module. When the target is added, the beacon module changes its conformation and allows the DNAzyme to hydrolyze substrate, which is a linear oligonucleotide having fluorophore and quencher at the opposite ends. The DNAzyme cuts the substrate at a cleavable ribonucleotide, which releases fluorophore from quencher and results in increasing fluorescence. The digested substrate dissociates, and the DNAzyme module hybridizes with the beacon module again, and the cycle is repeated. Although the fluorescence response from the substrate turnover was several times smaller than the response caused by the opening of the conventional SL

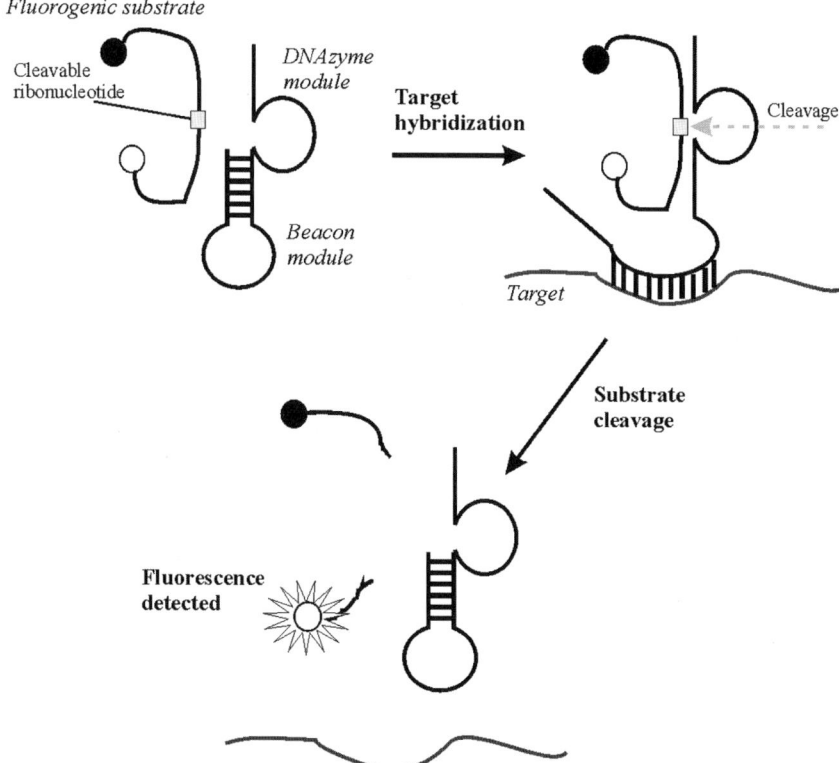

Figure 4. Molecular beacon acting as the DNAzyme. In the absence of a target, the DNAzyme module is blocked being hybridized with the beacon module. When the target is added, the beacon module changes its conformation and allows the DNAzyme to hydrolyze substrate, which is a linear oligonucleotide having fluorophore and quencher at the opposite ends. The DNAzyme cuts the substrate at a cleavable ribonucleotide, which releases fluorophore from quencher and results in increasing fluorescence.

molecular beacon (Stojanovic et al. 2001), this approach initiated catalytic events (Broude 2002). In contrast to the classical molecular beacon method based on binding, these methods utilize catalytic cleavage to release the fluorophore for detection and quantification, making it possible to take advantage of catalytic turnovers for signal amplification.

Unlike classical molecular beacons that detect only nucleic acids, catalytic molecular beacons can be applied to different DNAzymes to detect a broad range of analytes. The methods aimed at the metal ion recovery (e.g. bivalent lead (Chang et al. 2005)) are based on the finding that almost all known transcleaving DNAzymes share a similar structure comprised of a catalytic DNAzyme core flanked by two substrate recognition arms. Using a typical DNAzyme called the "8–17" DNAzyme as an example, the design of highly sensitive and selective Pb^{2+} sensors became practice. The initial design employs a single fluorophore–quencher pair in close proximity, with the fluorophore on the 5'-end of the substrate and the quencher on the 3'-end of the enzyme, with an additional quencher attached to the 3'-end of the substrate (required to improve the efficiency of quenching and to suppress background fluorescence) (Liu and Lu 2006). The dual quencher method allows the sensor to perform at ambient temperatures with a high signal-to-noise ratio.

If the structure of the catalytic beacon is optimized for faster substrate turnover, this approach might be a good alternative to amplification-based detection of nucleic acid targets, or even of inorganic ions. However, currently, this method is still in the development stage and is far from being applied routinely.

Interestingly, molecular beacon can be delivered directly into the living cells, by using standard transfection reagents, or, more efficiently, by attaching a cholesterol unit to a free terminus of the hairpin stem. Such a molecular probe that penetrates cellular membrane may be employed as an excellent biosensor for *in vivo* monitoring in the real time of gene expression at the RNA level and studying further transcript fates (Santangelo et al. 2006; Seo et al. 2006).

3. STEM-LOOP DNA PROBES ON MICROARRAYS

It is not a secret that the accuracy of conventional linear-probe microarrays at detecting mutations is often insufficient (Hacia 1999), and alternative approaches may be advantageous (reviewed in Chapter 9). SL DNA probes, with their enhanced fidelity in mismatch discrimination, have also found applications in DNA microarray technology to improve the array sensitivity by discriminating mispaired bases in a better way. Hairpin probes were attached to surfaces in topologically different manners, depending on the experimental design. For example, capturing probes with single-stranded overhangs were immobilized through the loop (Broude et al. 2001; Riccelli et al. 2001; Zhao et al. 2001), whereas the molecular beacons were tethered through the 5'- or 3'-end (Ortiz et al. 1998; Steemers et al. 2000), or through the linker attached to the stem (Figure 5) (Liu and Tan 1999; Broude 2002). The probes may be immobilized

Stem-loop oligonucleotides as hybridization

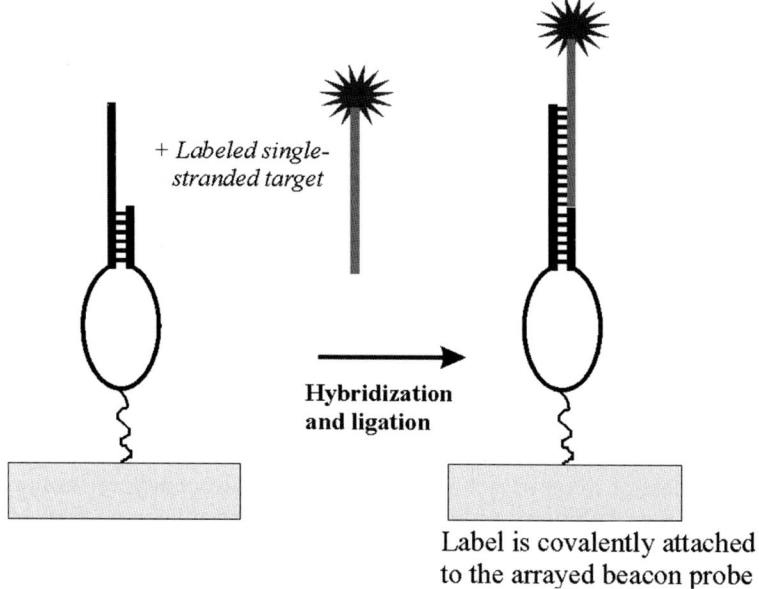

Figure 5. Hybridization–ligation approach for target detection using arrayed molecular beacons.

with either streptavidin–biotin binding (Ortiz et al. 1998; Steemers et al. 2000; Riccelli et al. 2001), or can they be covalently bound to the surfaces with different chemicals (Broude et al. 2001; Zhao et al. 2001; Wang et al. 2005).

Immobilized DNA or DNA–peptide nucleic acid (PNA) chimeric beacons preserve the ability to detect label-free complementary probes (e.g. cRNA, cDNA, or PCR amplicons) (Ortiz et al. 1998; Du et al. 2005). SL DNA probes with dangling overhangs can be used as capture devices, thus showing a twofold increase in hybridization rates and greater stability of the probe–target complex than the corresponding linear probes (Riccelli et al. 2001). An elegant technique combining hybridization and enzymatic ligation using SL DNA probes was recently developed for mutation detection in single-stranded DNA targets (Figure 5) (Broude et al. 2001). It should be noted here that the arrays with SL DNA probes are still at the early development stage.

4. CONCLUSIONS

It is clear now that SL DNA probes already have a wide range of applications in essentially all aspects of molecular biology and biomedicine. Two major factors that are responsible for such broad applications of the SL constructs are: (1) enhanced specificity of the probe–target recognition and (2) the possibility of one-tube formats and real-time reaction monitoring with molecular beacons.

There is no doubt that in the future, new applications will be developed with the sensitivity of future arrays being crucial.

Techniques that combine DNA SL probes with other molecular devices (e.g. ribozymes and DNAzymes, inorganic nanoparticles) and technological developments (e.g. microarrays, advanced optic techniques) promise to become sensitive and robust, high-throughput research and diagnostic methods. Finally, molecular beacon strategies, especially those coupled with aptamer-based bioassays, provide instrument for DNA–protein interaction studies. For example, FRET-labeled molecular-beacon aptamer may bind to a specific protein biomarker, platelet-derived growth factor (PDGF) (Vicens et al. 2005). This bioassay is compatible with pH, temperature, and monovalent cation levels typically encountered in biological samples, and phosphorothioate backbone-modified aptamer is able to exhibit specific FRET. With minimal sample processing and without optimization, the assay is able to detect as little as 10 ng PDGF per microgram of serum proteins from cell culture media (Vicens et al. 2005).

Overall, the success of SL approaches, as well as their universality, versatility, high reproducibility, and ubiquity make us believe in their bright perspectives of being widely useful for many future and current biomedical applications (~50 PubMed citations per year, 40 international patents to November 2006).

REFERENCES

Baker ES, Hong JW, Gaylord BS, Bazan GC, Bowers MT (2006) PNA/dsDNA complexes: site specific binding and dsDNA biosensor applications. J Am Chem Soc 128:8484–8492

Balashov SV, Park S, Perlin DS (2006) Assessing resistance to the echinocandin antifungal drug caspofungin in *Candida albicans* by profiling mutations in FKS1. Antimicrob Agents Chemother 50:2058–2063

Bonnet G, Krichevsky O, Libchaber A (1998) Kinetics of conformational fluctuations in DNA hairpin-loops. Proc Natl Acad Sci USA 95:8602–8606

Bonnet G, Tyagi S, Libchaber A, Kramer FR (1999) Thermodynamic basis of the enhanced specificity of structured DNA probes. Proc Natl Acad Sci USA 96:6171–6176

Broude NE (2002) Stem-loop oligonucleotides: a robust tool for molecular biology and biotechnology. Trends Biotechnol 20:249–256

Broude NE, Woodward K, Cavallo R, Cantor CR, Englert D (2001) DNA microarrays with stem-loop DNA probes: preparation and applications. Nucleic Acids Res 29:e92

Cech TR (1993) Catalytic RNA: structure and mechanism. Biochem Soc Trans 21:229–234

Chang IH, Tulock JJ, Liu J, Kim WS, Cannon DM, Jr., Lu Y, Bohn PW, Sweedler JV, Cropek DM (2005) Miniaturized lead sensor based on lead-specific DNAzyme in a nanocapillary interconnected microfluidic device. Environ Sci Technol 39:3756–3761

Drake TJ, Tan W (2004) Molecular beacon DNA probes and their bioanalytical applications. Appl Spectrosc 58:269A–280A

Du H, Strohsahl CM, Camera J, Miller BL, Krauss TD (2005) Sensitivity and specificity of metal surface-immobilized "molecular beacon" biosensors. J Am Chem Soc 127:7932–7940

Dubertret B, Calame M, Libchaber AJ (2001) Single-mismatch detection using gold-quenched fluorescent oligonucleotides. Nat Biotechnol 19:365–370

Fang X, Mi Y, Li JJ, Beck T, Schuster S, Tan W (2002) Molecular beacons: fluorogenic probes for living cell study. Cell Biochem Biophys 37:71–81

Goel G, Kumar A, Puniya AK, Chen W, Singh K (2005) Molecular beacon: a multitask probe. J Appl Microbiol 99:435–442

Gore HM, Wakeman CA, Hull RM, McKillip JL (2003) Real-time molecular beacon NASBA reveals hblC expression from *Bacillus* spp. in milk. Biochem Biophys Res Commun 311:386–390

Gubala AJ, Proll DF (2006) Molecular-beacon multiplex real-time PCR assay for detection of Vibrio cholerae. Appl Environ Microbiol 72:6424–6428

Hacia JG (1999) Resequencing and mutational analysis using oligonucleotide microarrays. Nat Genet 21:42–47

Henry KM, Jiang J, Rozmajzl PJ, Azad AF, Macaluso KR, Richards AL (2006) Development of quantitative real-time PCR assays to detect *Rickettsia typhi* and *Rickettsia felis*, the causative agents of murine typhus and flea-borne spotted fever. Mol Cell Probes 21:17–23

Johnson AE (2005) Fluorescence approaches for determining protein conformations, interactions and mechanisms at membranes. Traffic 6:1078–1092

Kim H, Kane MD, Kim S, Dominguez W, Applegate BM, Savikhin S (2006) A molecular beacon DNA microarray system for rapid detection of *E. coli* O157:H7 that eliminates the risk of a false negative signal. Biosens Bioelectron 22:1041–1047

Kostrikis LG, Tyagi S, Mhlanga MM, Ho DD, Kramer FR (1998) Spectral genotyping of human alleles. Science 279:1228–1229

Liu J, Lu Y (2006) Fluorescent DNAzyme biosensors for metal ions based on catalytic molecular beacons. Methods Mol Biol 335:275–288

Liu X, Tan W (1999) A fiber-optic evanescent wave DNA biosensor based on novel molecular beacons. Anal Chem 71:5054–5059

Liu X, Farmerie W, Schuster S, Tan W (2000) Molecular beacons for DNA biosensors with micrometer to submicrometer dimensions. Anal Biochem 283:56–63

Marras SA, Kramer FR, Tyagi S (1999) Multiplex detection of single-nucleotide variations using molecular beacons. Genet Anal 14:151–156

Meins F, Jr., Si-Ammour A, Blevins T (2005) RNA silencing systems and their relevance to plant development. Annu Rev Cell Dev Biol 21:297–318

Okamura Y, Watanabe Y (2006) Detecting RNA/DNA hybridization using double-labeled donor probes with enhanced fluorescence resonance energy transfer signals. Methods Mol Biol 335:43–56

Ortiz E, Estrada G, Lizardi PM (1998) PNA molecular beacons for rapid detection of PCR amplicons. Mol Cell Probes 12:219–226

Patel JR, Bhagwat AA, Sanglay GC, Solomon MB (2006) Rapid detection of Salmonella from hydrodynamic pressure-treated poultry using molecular beacon real-time PCR. Food Microbiol 23:39–46

Pearson CE, Sinden RR (1998) Trinucleotide repeat DNA structures: dynamic mutations from dynamic DNA. Curr Opin Struct Biol 8:321–330

Piatek AS, Tyagi S, Pol AC, Telenti A, Miller LP, Kramer FR, Alland D (1998) Molecular beacon sequence analysis for detecting drug resistance in *Mycobacterium tuberculosis*. Nat Biotechnol 16:359–363

Riccelli PV, Merante F, Leung KT, Bortolin S, Zastawny RL, Janeczko R, Benight AS (2001) Hybridization of single-stranded DNA targets to immobilized complementary DNA probes: comparison of hairpin versus linear capture probes. Nucleic Acids Res 29:996–1004

Roberts RW, Crothers DM (1991) Specificity and stringency in DNA triplex formation. Proc Natl Acad Sci USA 88:9397–9401

Santangelo P, Nitin N, Bao G (2006) Nanostructured probes for RNA detection in living cells. Ann Biomed Eng 34:39–50

Seo YJ, Jeong HS, Bang EK, Hwang GT, Jung JH, Jang SK, Kim BH (2006) Cholesterol-linked fluorescent molecular beacons with enhanced cell permeability. Bioconjug Chem 17:1151–1155

Sinden RR, Potaman VN, Oussatcheva EA, Pearson CE, Lyubchenko YL, Shlyakhtenko LS (2002) Triplet repeat DNA structures and human genetic disease: dynamic mutations from dynamic DNA. J Biosci 27:53–65

Sinsimer D, Leekha S, Park S, Marras SA, Koreen L, Willey B, Naidich S, Musser KA, Kreiswirth BN (2005) Use of a multiplex molecular beacon platform for rapid detection of methicillin and vancomycin resistance in *Staphylococcus aureus*. J Clin Microbiol 43:4585–4591

Sokol DL, Zhang X, Lu P, Gewirtz AM (1998) Real time detection of DNA. RNA hybridization in living cells. Proc Natl Acad Sci USA 95:11538–11543

Steemers FJ, Ferguson JA, Walt DR (2000) Screening unlabeled DNA targets with randomly ordered fiber-optic gene arrays. Nat Biotechnol 18:91–94

Stoermer RL, Keating CD (2006) Distance-dependent emission from dye-labeled oligonucleotides on striped au/ag nanowires: effect of secondary structure and hybridization efficiency. J Am Chem Soc 128:13243–13254

Stojanovic MN, de Prada P, Landry DW (2001) Catalytic molecular beacons. Chembiochem 2:411–415

Stryer L (1978) Fluorescence energy transfer as a spectroscopic ruler. Annu Rev Biochem 47:819–846

Thelwell N, Millington S, Solinas A, Booth J, Brown T (2000) Mode of action and application of Scorpion primers to mutation detection. Nucleic Acids Res 28:3752–3761

Todd AV, Fuery CJ, Impey HL, Applegate TL, Haughton MA (2000) DzyNA-PCR: use of DNAzymes to detect and quantify nucleic acid sequences in a real-time fluorescent format. Clin Chem 46:625–630

Tsourkas A, Bao G (2003) Shedding light on health and disease using molecular beacons. Brief Funct Genomic Proteomic 1:372–384

Tyagi S, Kramer FR (1996) Molecular beacons: probes that fluoresce upon hybridization. Nat Biotechnol 14:303–308

Tyagi S, Bratu DP, Kramer FR (1998) Multicolor molecular beacons for allele discrimination. Nat Biotechnol 16:49–53

Tyagi S, Marras SA, Kramer FR (2000) Wavelength-shifting molecular beacons. Nat Biotechnol 18:1191–1196

Varani G (1995) Exceptionally stable nucleic acid hairpins. Annu Rev Biophys Biomol Struct 24:379–404

Vet JA, Majithia AR, Marras SA, Tyagi S, Dube S, Poiesz BJ, Kramer FR (1999) Multiplex detection of four pathogenic retroviruses using molecular beacons. Proc Natl Acad Sci USA 96:6394–6399

Vicens MC, Sen A, Vanderlaan A, Drake TJ, Tan W (2005) Investigation of molecular beacon aptamer-based bioassay for platelet-derived growth factor detection. Chembiochem 6:900–907

Wahlroos R, Toivonen J, Tirri M, Hanninen P (2006) Two-photon excited fluorescence energy transfer: a study based on oligonucleotide rulers. J Fluoresc 16:379–386

Wang Y, Wang H, Gao L, Liu H, Lu Z, He N (2005) Polyacrylamide gel film immobilized molecular beacon array for single nucleotide mismatch detection. J Nanosci Nanotechnol 5:653–658

Whitcombe D, Theaker J, Guy SP, Brown T, Little S (1999) Detection of PCR products using self-probing amplicons and fluorescence. Nat Biotechnol 17:804–807

Yesilkaya H, Meacci F, Niemann S, Hillemann D, Rusch-Gerdes S, Barer MR, Andrew PW, Oggioni MR (2006) Evaluation of molecular-Beacon, TaqMan, and fluorescence resonance energy transfer probes for detection of antibiotic resistance-conferring single nucleotide polymorphisms in mixed Mycobacterium tuberculosis DNA extracts. J Clin Microbiol 44:3826–3829

Zhao X, Nampalli S, Serino AJ, Kumar S (2001) Immobilization of oligodeoxyribonucleotides with multiple anchors to microchips. Nucleic Acids Res 29:955–959

CHAPTER 5

NORMALIZATION OF cDNA LIBRARIES

ALEX S. SHCHEGLOV*, PAVEL A. ZHULIDOV, EKATERINA A. BOGDANOVA, DMITRY A. SHAGIN

Shemyakin-Ovchinnikov Institute of Bioorganic Chemistry, Russian Academy of Sciences, 16/10 Miklukho-Maklaya, 117997 Moscow, Russia
*Corresponding author
Phone: +(7495) 4298020; Fax: +(7495) 3307056; E-mail: jukart@mail.ru

Abstract:	In a cellular transcriptome, the number of mRNA copies per gene may differ by several orders. In cDNA libraries, performed from mRNA, these proportions are the same. Normalization methods allow us to equalize numbers of gene's copies in the library. Normalized cDNA libraries are used to discover new genes transcribed at relatively low levels or for functional screenings. Here, we observed different cDNA libraries normalization methods, which were based on hybridization (renaturation) of cDNA or DNA, or RNA. Also we described duplex-specific nuclease (DSN) normalization protocol – simple and effective cDNA libraries normalization method.
Keywords:	Genetic regulation, differentially expressed mRNA, mRNA study, poly(A) + RNA, full-length cDNA, incomplete cDNA, cDNA representation, transcript concentration, normalized cDNA libraries, poorly transcribed, clone coverage, saturating hybridization, enzymatic removal, physical separation, biotinylation, biotinylation of initial RNA, frequent cutter, duplex-specific nuclease (DSN), hydroxyapatite column, plasmid vector, phagemid vector, first strand cDNA, second strand cDNA, immobilization.
Abbreviations:	cDNA, complementary DNA; dNTP, deoxyribonucleotidetriphosphate; DSN, duplex-specific nuclease; mRNA, messenger RNA; ITR, inverted terminal repeat; PCR, polymerase chain reaction; rRNA, ribosomal RNA; RT, reverse transcription.

TABLE OF CONTENTS

1. Introduction 98
   1.1 Normalized cDNA Libraries: What are they needed for? 98
   1.2 Evaluation of cDNA Library Normalization Efficacy 98
   1.3 Basic Approaches to Generate Normalized cDNA Libraries 100
2. Methods of cDNA Normalization 102
   2.1 cDNA Normalization by Means of Hydroxyapatite
       Column Chromatography 102
   2.2 Generation of Normalized Full-Length-Enriched cDNA
       Libraries Using DNA Immobilization on a Solid Support 104
   2.3 Normalization of Full-Length cDNA with the
       use of Biotinylated RNA 105
   2.4 Normalization of Fragmented cDNA by Means
       of Selective Amplification 107
   2.5 cDNA Normalization Using Frequent-Cutter
       Restriction Enzymes 109
   2.6 Normalization of Full-Length-Enriched cDNA
       with Duplex-Specific Nuclease (DSN Normalization) 109
3. cDNA Preparation for DSN Normalization 111
   3.1 RNA Requirements 111
   3.2 cDNA Synthesis 112
   3.3 cDNA Purification 113
4. DSN Normalization Protocol 113
   4.1 Materials 113
   4.2 cDNA Precipitation 114
   4.3 Hybridization 115
   4.4 DSN Treatment 116
   4.5 First Amplification of Normalized cDNA 117
   4.6 Preliminary Analysis of the Normalization Results 119
   4.7 Second Amplification of Normalized cDNA 121
References 122

1. INTRODUCTION

1.1 Normalized cDNA Libraries: What are they needed for?

Genetic regulation of various biological processes can be effectively studied by means of examining the involved mRNA pool. Modern methods allow to reveal differentially expressed mRNAs, to examine their spatial or temporal localization in an organism and, finally, to analyze the function of a respective protein. Analysis of a complete set of cellular mRNAs (transcriptome) became very popular in the last decade as an effective alternative for genome sequencing because such a procedure requires less resources and makes it possible to identify

Normalization of cDNA libraries

gene structures much more accurately than any kind of bioinformatical predictions based on genome sequence (Guigo et al. 2000).

In eukaryotic cell, mRNA constitutes ~1–5% of total RNA mass and the number of expressing genes varies from several thousands to several tens thousands. In a cellular transcriptome, the number of mRNA copies per gene may differ by several orders. As a rule, 5–10 major housekeeping genes, whose mRNA constitutes totally ~20% of the cellular mRNA mass, 500–2000 genes transcribed at an intermediate level (give 40–60% of the total cellular mRNA mass), and 10,000–20,000 moderately transcribed genes (20–40% of the total cellular mRNA mass) are expressed in an eukaryotic cell (Alberts et al. 1994).

Methods of mRNA study operate mostly not with an mRNA molecule itself, but with its complementary DNA (cDNA). Cloned pool of cDNAs corresponding to mRNA pool of a biological object under study refers to as a cDNA library. Today, several methods providing to generate fitting libraries from limited starting amounts (0.1–1 µg) of total RNA have been developed and are widely in use. The possibility to use total RNA as a template for cDNA synthesis without a stage of labor-intensive and relatively expensive poly(A) + RNA isolation makes these methods advantageous.

cDNA libraries should meet several requirements. First, such libraries must be representative, i.e. they should contain DNA copies for nearly all transcripts that are typical of the biological object under study. In the case of polymerase chain reaction (PCR)-amplified cDNA libraries, this requirement limits the number of the respective PCR cycles. Typically, the maximum of 25 PCR cycles may be used for this purpose; otherwise (is the number of PCR cycles is higher), the library cannot be considered representative (Matz 2002).

Second, the libraries that are enriched in full-length cDNA sequences are preferable for many purposes. For instance, the sequencing of full-length cDNA provides a complete amino acid sequence of the respective protein, whereas in the case of an incomplete cDNA fragment sequencing, special additional procedures may be required to determine the whole structure. Moreover, functional screening of cDNA library becomes impossible if it is not enriched in full-length cDNAs.

Third, for many cDNA library applications (e.g. search for differentially expressed genes), it is essential to minimize the distortion of cDNA representation in a library with respect to initial mRNA. In other words, the content of individual cDNAs in the library must be proportional to the copy number of the initial RNAs. In contrast, for some other tasks the concentrations of different individual cDNAs in a library must be equalized.

Methods to decrease the prevalence of highly abundant transcripts and to equalize transcript concentrations in a cDNA library are called "cDNA normalization" methods. Libraries, in which the disparity in concentrations of cDNAs for various genes is smaller than the initial disparity in concentrations of cDNAs or mRNAs for these genes in the original sample, are referred to as *normalized* ones. In the ideal case (which is never the case in the reality), all cDNA concentrations are equal in the normalized library.

Normalized cDNA libraries are used mostly to discover new genes transcribed at relatively low levels, which are expressed in the biological objects under study. The initial frequency of cDNA abundance for poorly transcribed genes in a general cDNA library may be under 2×10^{-6}, whereas in an ideally normalized library it will be ~25-fold higher (Soares et al. 1994). Therefore, the so-called complete sequencing of normalized libraries (more or less comprehensive library sequencing with at least fivefold clone coverage) requires at least one order less sequenced clones as compared with nonnormalized ones. Also, cDNA normalization significantly reduces the number of analyzed clones in functional screenings.

1.2 Evaluation of cDNA Library Normalization Efficacy

To control the efficacy of cDNA library normalization, it is possible to compare the concentrations of "major" and "minor" transcripts in the respective library before and after normalization step. In such cases one can use quantitative (or semiquantitative) PCR or pseudo Northern blot hybridization. The more efficient is the normalization, the less is the difference between major and minor transcript concentrations after normalization.

Another approach comprises an exhaustive sequencing of a cloned cDNA library and further analysis of the transcripts presented there. Any particular transcript will be presented by n copies, where n means a whole positive number. Then, it is helpful to plot the relationship between the number of individual transcripts (the number of genes) occurring n times, and n itself graphically. The closer is this graph to Poisson standard distribution, the more effective is the normalization.

1.3 Basic Approaches to Generate Normalized cDNA Libraries

Today, two basically distinctive approaches to the generation of normalized cDNA libraries have been proposed.

The first approach is a physical isolation of all transcripts in equal quantities from the original cDNA library. Theoretically, denatured genomic DNA of the same organism immobilized on an inert carrier can be used for this task (Weissman 1987). Saturating hybridization of such absorbent with denatured cDNA, followed by the separation and subsequent elution of bound cDNA will result in generation of normalized cDNA library (since almost all genes are presented in the genome by one or few copies, roughly equal quantities of cDNAs for these genes will hybridize with genomic DNA). Nevertheless, rare transcripts escape an analysis in this system, as it is impossible for them to reach highly enough concentrations, which are required for an efficient hybridization with genomic DNA.

The second group of methods is based on denaturation – hybridization of double-stranded cDNA molecules (or heteroduplexes mRNA or first strand cDNA). For each specific transcript, the hybridization rate will be proportional to the square of its concentration since nucleic acid hybridization kinetics follows a second-order equation. Therefore, the higher the initial concentration of individual cDNA molecules, the more molecules will reassociate after the denaturation (i.e. convert back to double-stranded form). Thus, after denaturation – reassociation step, the cDNA library consists of two pools: (1) the double-stranded fraction, which accumulates the most part of initial cDNA and (2) the single-stranded fraction that includes initial individual cDNAs in substantially equalized rather low concentrations (Young and Anderson 1985; Galau et al. 1977).

For the latter group of methods, the challenging step is the separation of single- stranded pools from double-stranded pools in the cDNA molecules. Until last few years, no simple, efficient, and reliable methods were available to solve this problem, but recently a significant progress in this field was achieved (Zhulidov et al. 2004).

The following approaches are broadly used now for the separation of single- and double-stranded fractions of cDNA libraries: (1) physical separation, (2) selective amplification, and (3) enzymatic removal of the double-stranded pool.

1. Physical separation of single- and double-stranded fractions is mostly realized through column chromatography with hydroxyapatite (Ko 1990; Soares et al. 1994), preliminary immobilization of cDNA first strand on solid carrier beads (Sasaki et al. 1994; Tanaka et al. 1996), and through the biotinylation of initial RNA (Carninci et al. 2000). However, these methods share three substantial disadvantages: they are labor-intensive and complicated, they require large amounts of starting material and, frequently, they are poorly reproducible.
2. Selective amplification of normalized cDNA libraries is based on PCR suppression (Luk'yanov et al. 1996) (see Chapter 2). Being adequate for many research strategies, this method, however, is applicable only for fragmented cDNA libraries (those lacking complete cDNAs) and, moreover, results in a loss of rare transcripts.
3. Theoretically, specific enzymatic removal of the double-stranded fraction is considered the most convenient and robust method of cDNA library normalization. The enzyme (or enzyme mix) used should efficiently hydrolyze the double-stranded cDNA while leaving single-stranded cDNA intact. Such enzyme should be active at high temperature conditions (60–70°C), thus allowing avoiding formation of secondary structures by single-stranded cDNAs (to prevent the background digestion by the enzyme of secondary structure-forming single-stranded cDNAs, which would be the case at lower temperatures). The attempt to use a mix of frequent-cutter restriction enzymes turned out to be rather inefficient (Coche and Dewez 1994). Recently, a new enzyme – duplex-specific nuclease (DSN) with above-mentioned features – was isolated and described (Shagin et al. 2002). It was the basis to develop a simple and effective method of full-length-enriched cDNA normalization (so-called duplex-specific nuclease [DSN] normalization, see Section 2) prior to library cloning (Zhulidov et al. 2004).

2. METHODS OF cDNA NORMALIZATION

2.1 cDNA Normalization by Means of Hydroxyapatite Column Chromatography

2.1.1 Normalization of fragmented cDNA

The first method of normalized cDNA library generation was proposed by Minoru Ko and coauthors in 1990 (Ko 1990). The proposed procedure involves the following stages:
1. A double-stranded cDNA sample synthesized on a poly(A) + RNA basis is sonicated to generate shorter fragments
2. 200–400 bp long cDNA fragments are purified from the agarose gel
3. Purified cDNA fragments are ligated to double-stranded oligonucleotide adapter, and are further PCR amplified using the adapter as a site for primer annealing
4. After cDNA denaturation and reassociation, single- and double-stranded pools are separated with standard chromatography on hydroxyapatite columns (Sambrook et al. 1989)
5. The fraction enriched in single-stranded DNA (that which was not bound by the column) is overprecipitated and amplified

It should be noted that his approach requires several cycles of normalization, i.e. the amplified cDNA obtained after the first round of normalization is again redenatured and rehybridized with subsequent column separation. After three rounds of normalization, the amplified cDNA is cloned. In a model experiment, described by Ko and colleagues, the overall normalization was rather efficient: after three rounds, the mean difference between the frequencies of occurrence of major and minor transcripts decreased from ~20,000-fold to 40-fold value.

Independently and almost at the same time, Patanjali and coauthors proposed a very similar protocol (Patanjali et al. 1991). Their protocol provided a sufficiently normalized cDNA library of 400–1600 bp long fragments in just one round of normalization. The authors succeeded to almost completely equalize the concentrations of different individual cDNAs. However, in other hands this method was poorly reproducible.

2.1.2 Normalization of full-length-enriched cDNA libraries

The next step in the evolution of normalization methods based on the physical separation of fractions by means of hydroxyapatite column chromatography was the development of a technique to normalize full-length-enriched cDNA libraries. The method proposed by Soares et al. (1994) involves the following stages (Figure 1):
1. Generation of a directional full-length-enriched cDNA library in a phagemid vector with subsequent isolation of its single-stranded form
2. Synthesis of short (~200 bp) second strand cDNA fragments by means of limited primer extension while using single-stranded cDNA (generated at the previous step) as a template

Normalization of cDNA libraries

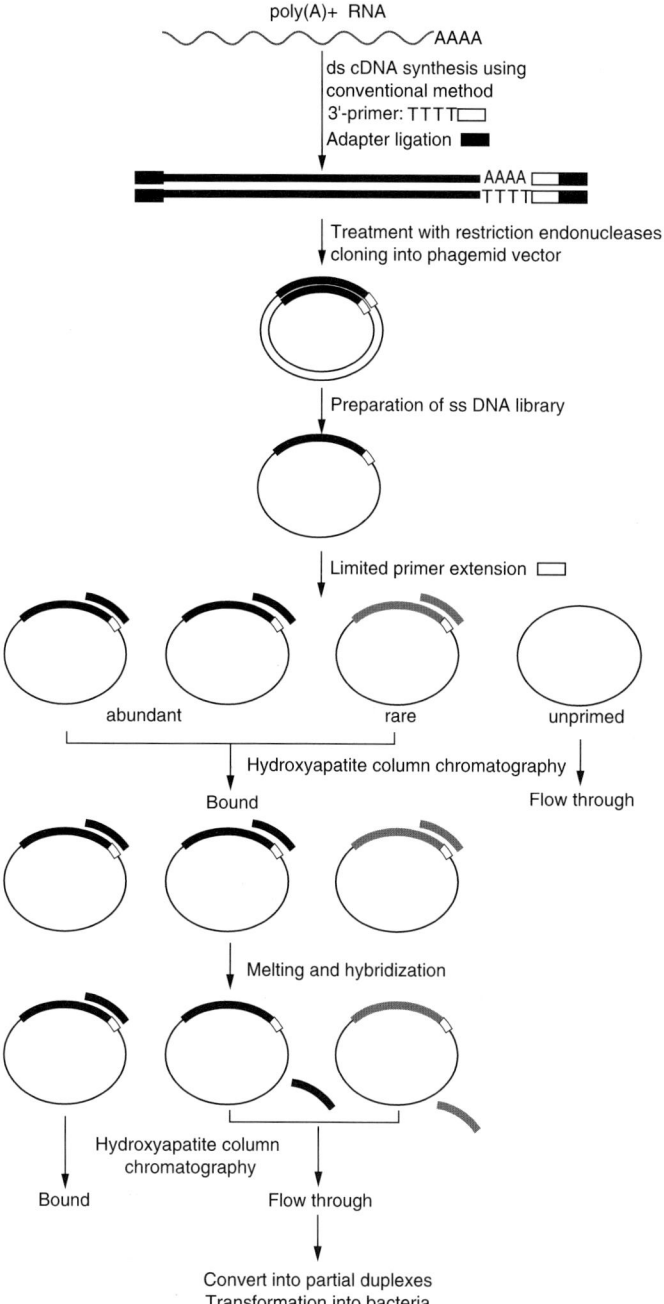

Figure 1. Schematic outline of the normalization of full-length-enriched cDNA libraries using hydroxyapatite column chromatography. (From Soares et al. 1994.) Thick black lines represent abundant transcripts, thick gray line represents rare transcripts. Rectangle represents adapter sequence and its complementary sequence.

3. Elimination of excess circular single-stranded molecules (which were not involved in the synthesis of the second strand cDNA) with hydroxyapatite column
4. Denaturation and rehybridization of double-stranded fragments separated at the previous stage
5. Isolation of a normalized single-stranded DNA pool (from the previous step) by means of hydroxyapatite column chromatography
6. Conversion of single-stranded plasmids isolated at the stage (5) to partial duplexes
7. Electroporation to deliver the plasmids into bacteria to propagate normalized cDNA library

After two rounds of normalization, the authors managed to produce a library with very low (less than tenfold) differences between the frequencies of analyzed transcripts. Marra and colleagues further reported (Marra et al. 1999), that the full-length sequences constitute ~27% of a total cDNA pool in a normalized library generated by means of such protocol.

As compared with other techniques for cDNA normalization, the method proposed by Soares et al. (1994) is the most widely used. For example, it was successfully applied in several expressed sequence tag (EST) large-scale sequencing projects (Urmenyi et al. 1999; Reddy et al. 2002; Caetano et al. 2003).

However, the procedures with hydroxyapatite columns are labor-intensive. To effectively separate fractions on hydroxyapatite columns, 10 µg of initial cDNA are needed. Chromatography-based separation also requires high temperature conditions (60°C) thus complicating manipulations with a sample. Besides, low resolution of hydroxyapatite columns results in contamination of a single-stranded pool with short double-stranded molecules, thus reducing the normalization efficiency. This approach also requires performing several cycles of denaturation – hybridization steps followed by the subsequent chromatography.

Some general shortcomings of this first group of methods should be mentioned. Briefly, the methods proposed by Ko and Pananjali are not applicable to generate normalized full-length cDNA libraries. The method by Soares et al. (1994) includes amplification of the cloned cDNA library before normalization. At this stage, the growth of cDNA-harboring colonies varies depending on the plasmid length; this results in underrepresentation of many long full-length cDNAs after bulk amplification of the library.

2.2 Generation of Normalized Full-Length-Enriched cDNA Libraries Using DNA Immobilization on a Solid Support

Simultaneously with Soares et al. (1994), Sasaki and coauthors proposed an alternative approach to normalize full-length-enriched RNA. The authors hybridized mRNA with short first strand cDNA immobilized on latex beads; this allowed an easy elimination of hybrids by means of sample centrifugation

Normalization of cDNA libraries

after hybridization (Sasaki et al. 1994; Tanaka et al. 1996). In this case, normalization is performed in the following manner:
1. First strand cDNA (the so-called driver) is synthesized through the seeding of poly(T) oligonucleotides covalently bound with solid-phase support (latex beads)
2. After cDNA synthesis, the cDNA/RNA heteroduplexes are denaturated; cDNAs fixed on latex beads are separated by centrifugation
3. cDNAs are prehybridized with poly(A) oligonucleotides (to reduce the background of the further hybridization); poly(A) + RNA is added to the resulting cDNAs fixed on latex beads, and the whole mix is allowed to hybridize
4. After hybridization, latex beads with DNA/RNA heteroduplexes are eliminated by centrifugation; RNAs remaining in solution are reprecipitated and then utilized in the next round of normalization
5. "Normalized" RNAs are used for cDNA synthesis

Effective normalization required 3–4 consistent cycles of the above procedures. After four cycles, ~96–98% RNA molecules was eliminated. Remaining RNA was considered a normalized one; this fraction was used to synthesize cDNA. The authors of this technique succeeded to generate full-length-enriched cDNA library with less than 30-fold difference of transcript abundances. Using similar protocol to generate normalized cDNA library from human heart muscle, Tanaka et al. (1996) managed to create a library with 1395 nonoverlapping (unique) clones from a total of 3040 sequenced ones (~46% of unique clones in the library).

This technique did not become popular due to its multistep protocol, large starting amounts of poly(A) + RNA required (~25 µg), and difficulties in manipulations with RNA. Besides, hybridization of DNA immobilized on a solid-phase support is less effective and slower as compared with hybridization in solution (Young and Anderson 1985).

2.3 Normalization of Full-Length cDNA with the use of Biotinylated RNA

Another approach to generate normalized libraries based on physical separation of fractions has been proposed by Carninci et al. (2000). This procedure involves hybridization of biotinylated RNA and first strand cDNA with subsequent elimination of bio-RNA–DNA hybrids using magnetic beads. The cDNA first strands synthesized via *cap trapper* method (Carninci et al. 1996) are hybridized with biotinylated natural or synthetic RNA. Hybridization in a formamide buffer allows one to reduce the reaction temperature down to 42°C thus preventing the degradation of long single-stranded DNA molecules. To eliminate generated bio-RNA–cDNA hybrids, streptavidine beads are further used. The single-stranded cDNA remaining in solution is used to synthesize cDNA second strand and to further create cDNA library. The basic stages of cDNA synthesis by means of *cap trapper* method and full-length cDNA normalization are illustrated in Figure 2.

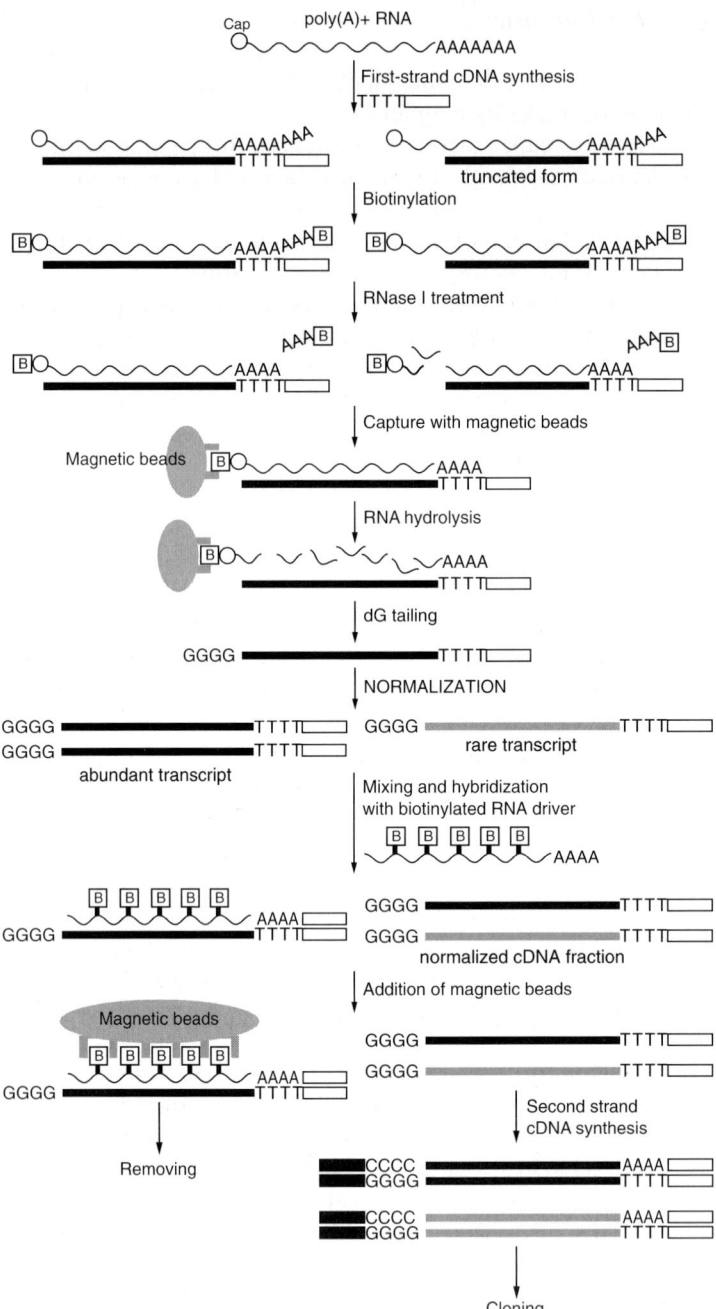

Figure 2. Schematic outline of the normalization of full-length-enriched cDNA with the use of biotinylated RNA. (From Carninci et al. 2000.) Black lines represent abundant transcripts, gray lines represent rare transcripts. Rectangle represents adapter sequence and its complementary sequence.

The method described above provides normalized full-length cDNA libraries. This approach was demonstrated to be highly efficient. Representation of major housekeeping gene transcripts in a normalized library decreased 250-fold as compared with the initial library. The difference between representations of moderately abundant and highly abundant transcripts in a normalized library was less than 30-fold. Full-length sequences constituted ~65–70% of all clones. As to disadvantages of this method, large amounts of poly(A) + RNA required and labor-intensiveness because of its multistep protocol and complexities in manipulations with RNA should be mentioned. Besides, several cycles of biotinylation are required due to rather low efficacy of RNA labeling with biotin. Finally, separation of biotin-modified and nonmodified pools usually involves several rounds.

2.4 Normalization of Fragmented cDNA by Means of Selective Amplification

cDNA normalization method free from any kind of physical separation of fractions has been proposed by Luk'yanov et al. (1996). This method is based on a selective amplification of normalized cDNA pool due to selective PCR suppression. As it has been shown, amplification of DNA flanked with inverted terminal repeats (ITR) in PCR with the primer that corresponds to ITR external segment is inhibited (the primer length should be well below ITR length). The principle of PCR suppression is described in detail in Chapter 2.

For normalization, double-stranded cDNA is treated with frequent-cutter restriction enzyme(s) and is separated into two samples. Each sample is ligated with specific suppression adapter oligonucleotide to generate ITRs at the ends of cDNA molecules. After ligation, samples are denatured and allowed to reanneal in different tubes (first hybridization). The double-stranded pool resulting from the first hybridization consists of homoduplexes harboring identical adapter sequences on their both 5′-termini. The two samples are then mixed (without denaturation) and left to hybridize together (second hybridization). Only normalized single-stranded fraction (which did not form homoduplexes at the previous stage) is exposed to the second hybridization. Together with homoduplexes, heteroduplexes flanked with different adapter sequences are generated in the second hybridization as well. Thus, following second hybridization, normalized single-stranded DNA partially converts to heteroduplex fraction. Subsequent to complementary extension of 3′-termini, heteroduplexes may be selectively PCR amplified with primers corresponding to external segments of suppressive adapters. Amplification of hybrids flanked with ITRs is suppressed by stem-loop structures produced in the course of PCR (*selective PCR suppression*; see Chapter 2). Single-stranded molecules are not amplified since they contain only one primer annealing site (Figure 3).

The procedure described above provides substantially normalized cDNA libraries within just one round of normalization. This technique is rather efficient, easy, and very well-reproducible. However, it is not free from certain imperfections

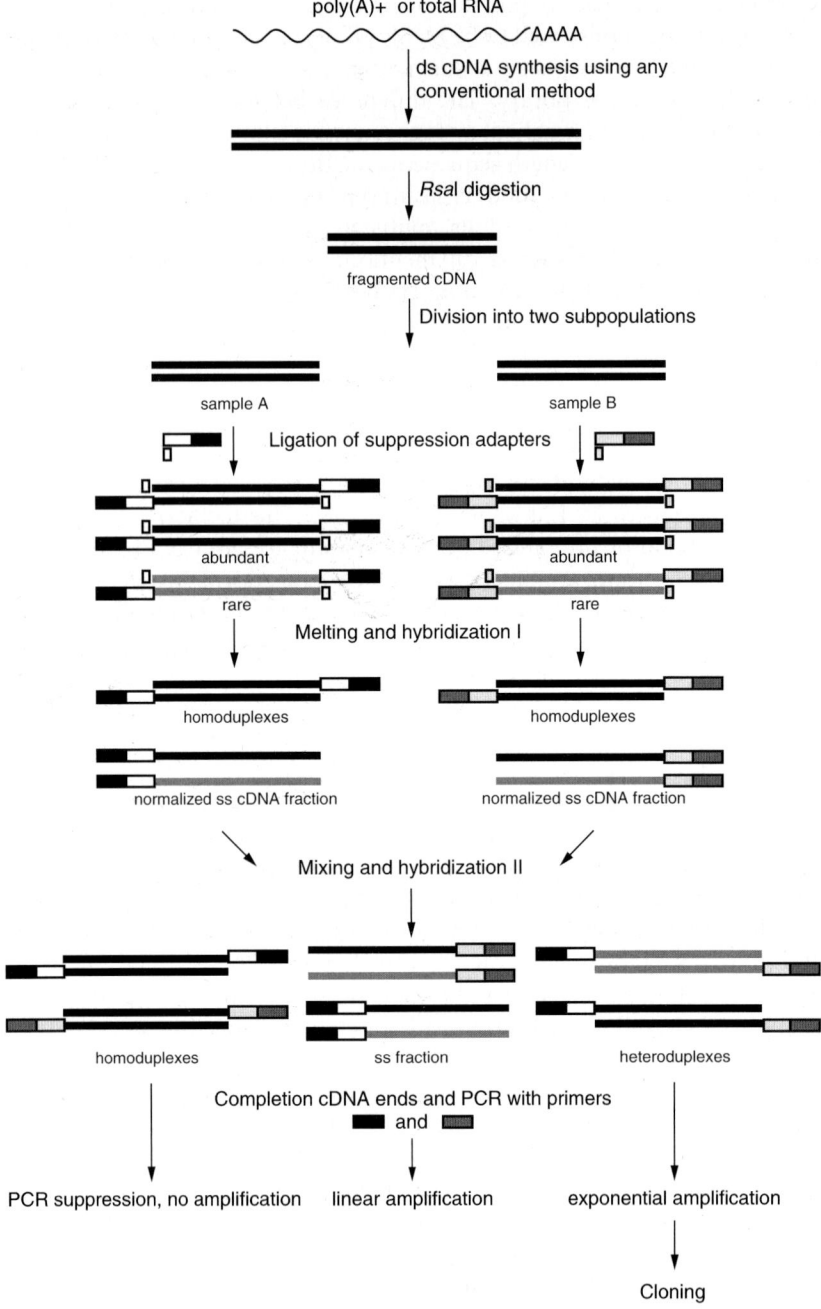

Figure 3. Schematic outline of the normalization of fragmented cDNA libraries using selective amplification. (From Lukyanov et al. 1996.) Black lines represent abundant transcripts, gray lines represent rare transcripts. Rectangle represents adapter sequence and its complementary sequence.

Normalization of cDNA libraries

as well: it does not permit to normalize full-length cDNA since the efficiency of PCR suppression is greatly reduced when long DNA molecules are amplified (see Chapter 2). Thus, this method of normalization is applicable only to fragmented cDNA libraries. Moreover, the concentration of low abundant transcripts may be less than it is required for an efficient hybridization, especially during the second hybridization (see above); this frequently results in their underrepresentation in the normalized library.

2.5 cDNA Normalization Using Frequent-Cutter Restriction Enzymes

While seeking to overcome the complexity and low reproducibility of cDNA normalization techniques based on physical separation of single- and double-stranded pools, Coche and Dewez (1994) proposed to use a mixture of frequent-cutter restriction endonucleases to eliminate double-stranded DNA fraction. According to the proposed procedure, first strand cDNA is synthesized through the seeding of oligo(T)-containing oligonucleotides covalently bound with magnetic beads. After second strand synthesis and the subsequent ligation of the cDNA with oligonucleotide adapter, cDNA is further denatured, reannealed, and then exposed to restriction endonucleases. Hydrolyzed fragments of double-stranded cDNA and single-stranded cDNA appear to be immobilized on magnetic beads. Then, second strand cDNAs are synthesized using adapter-specific primer extension and are further PCR amplified. At this stage, restriction endonuclease-digested cDNA fragments will not be amplified as they do not contain two primer annealing sites necessary for exponential PCR. The resulting PCR products were used for the next round of normalization.

Contrary to the expectations, this method was found to be rather inefficient. Even after three normalization steps, concentrations of major transcripts were greatly higher as compared with minor ones. This most probably could be due to a low efficiency of hybridization of cDNA immobilized on a solid-phase carrier. Moreover, treatment with restriction endonucleases is performed at a low temperature. In this case, significant loss of cDNAs tending to secondary structure formation occurs (as the resulting intermolecular duplex may contain a site recognized by the endonuclease).

2.6 Normalization of Full-Length-Enriched cDNA with Duplex-Specific Nuclease (DSN Normalization)

This approach is schematized in the Figure 4. The method involves denaturation of double-stranded cDNA flanked by adapter sequences and its subsequent renaturation, followed by enzymatic degradation of double-stranded DNA fraction (mostly formed by reannealed abundant transcripts) using DSN. Single-stranded fraction containing transcripts in more or less equalized concentrations is further PCR amplified (Zhulidov et al. 2004, 2005).

Similar to previously described techniques, DSN normalization is based on second-order in-solution hybridization kinetics. Standard hybridization conditions

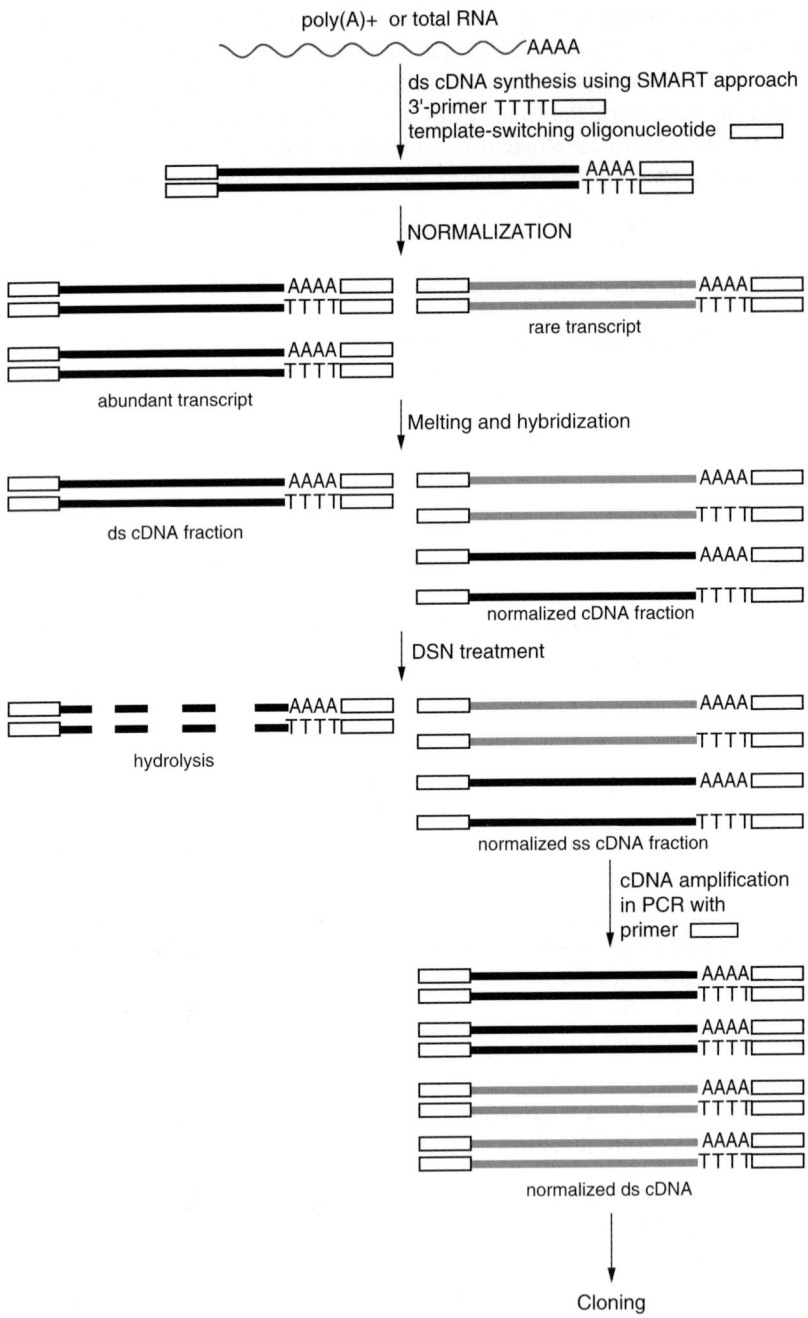

Figure 4. DSN normalization scheme. Black lines represent abundant transcripts, gray lines represent rare transcripts. Rectangle represents adapter sequence and its complementary sequence.

Normalization of cDNA libraries

identical to those routinely used for the suppression subtractive hybridization (SSH) (Diatchenko et al. 1996) (Chapter 3) were utilized. Formamide buffer employed by Carninci et al. (2000), which efficiently prevents secondary structures formation could not be used here because it inhibits DSN necessary for double-stranded fraction digestion.

The rationale of this method is degradation of double-stranded fraction using DSN. A number of specific DSN features make it ideal for removing double-stranded DNA from complex mixtures of nucleic acids. DSN displays a strong preference for cleaving double-stranded DNA in both DNA–DNA and DNA–RNA hybrids, as compared to single-stranded DNA and RNA, irrespective of the sequence length. Moreover, the enzyme remains stable over a wide range of temperatures and displays optimal activity at 55–65°C (Shagin et al. 2002). Consequently, degradation of double-stranded DNA-containing fraction by this enzyme may occur at elevated temperatures, thereby avoiding loss of transcripts due to formation of secondary structures and nonspecific hybridization involving adapter sequences.

DSN normalization includes a step of PCR amplification of normalized single stranded cDNA fraction. To overcome PCR tendency to amplify shorter DNA fragments more efficiently than longer ones two basic approaches were applied. The first is the use of well-known long-distance PCR system (Barnes 1994) and the second is regulation of average length of complex PCR product by partial PCR suppression (Shagin et al. 1999). The last approach requires that cDNAs to be normalized have identical inverted adapter sequences at both termini (see Chapter 2 for the detailed description of the PCR suppression effect).

Adapter sequences can be introduced at the cDNA termini using various approaches, for instance, by adapter ligation or during cDNA synthesis utilizing the *SMART* approach (Zhu et al. 2001) or its modifications. SMART-prepared cDNA is enriched with full-length sequences and could be obtained both from poly(A) + and total RNA (even if only small amount of starting material is available). Synthetic adapter sequences are introduced to both 5'- and 3'-ends of cDNA during cDNA synthesis.

3. cDNA PREPARATION FOR DSN NORMALIZATION

3.1 RNA Requirements

The sequence complexity and the average length of the normalized cDNA library noticeably depend on the quality of experimental RNA starting material. There are several methods to isolate RNA providing stable RNA preparation from a majority of biological objects, for example, Trizol method (GIBCO/Life Technologies), Chomczynski and Sacchi method (Chomczynski and Sacchi 1987), and RNeasy kits (QIAGEN). After RNA isolation, RNA quality should be evaluated by means of denaturing formaldehyde or agarose gel electrophoresis (Sambrook et al. 1989).

The following characteristics indicate successful RNA isolation:
- For mammalian total RNA, two intensive bands at ~4.5 and 1.9 kb should be observed against a light smear. These bands represent 28S and 18S ribosomal RNAs (rRNAs). The ratio of intensities of these bands should be ~1.5–2.5:1. Intact mammalian poly A + RNA appears as a smear sized from 0.1 to 4–7 kb (or more) with faint 28S and 18S rRNA bands.
- With RNA from other sources (plants, insects, yeast, amphibians), the normal mRNA smear may not exceed 2–3 kb. Moreover, the overwhelming majority of invertebrates have 28s rRNA with a "hidden break" (Ishikawa 1977). In some organisms, interaction between the parts of 28s rRNA is rather weak, so the total RNA preparation exhibits a single 18s-like rRNA band. In other species, the 28s rRNA is more robust, so it is still visible as the second band.

If experimental RNA is shorter than expected and/or degraded according to electrophoresis data, preparation of fresh RNA is recommended after checking the quality of RNA purification reagents. In some cases, partially degraded RNA is only available (e.g. tumor samples or hard treated tissues). This RNA can be used to synthesize cDNA; however, the cDNA sample will contain reduced number of full-length molecules.

As representation of the resulting amplified cDNA depends on the initial amount of RNA used for the first strand cDNA synthesis, we recommend using higher starting amounts of RNA indicated in the selected cDNA synthesis protocol. In our experiments, the best results were achieved when cDNA synthesis was started with 0.5–1.5 µg of poly(A) + or total RNA.

3.2 cDNA Synthesis

The DSN normalization protocol provided below is optimized for cDNA prepared by the SMART approach (Zhu et al. 2001) or its modifications. However, some other full-length cDNA synthesis systems might be used for this purpose as well, e.g. RLM FirstChoice cDNA synthesis system produced by Ambion, which is based on the specific enzymatic modification of RNA 5′-cap by the tobacco acid pyrophosphatase and further ligation of an RNA adapter to it; this ensures full-length double-stranded cDNA production, if oligo d(T)-containing primer is used for the first strand cDNA synthesis. Also, an appropriate double-stranded cDNA can be prepared using the following commercially available cDNA synthesis kits:
- Mint cDNA synthesis kit (Evrogen Catalog #: SK001)
- SMART PCR cDNA synthesis kit (Clontech Catalog #: 634902)
- Super SMART PCR cDNA synthesis kit (Clontech Catalog #: 635000)

cDNA obtained is flanked by the identical adapter sequences (e.g. 5′-AAG CAG TGG TAT CAA CGC AGA GTA CT-3′) at both termini and allows nondirectional cloning of cDNA library only. Do not perform cDNA ends polishing before normalization procedure.

Normalization of cDNA libraries

If a directionally cloned cDNA library is required, we recommend using commercially available kits SMART cDNA library construction kit (Clontech Catalog # 634901) or Creator SMART cDNA library construction kit (Clontech Catalog # 634903) except for CDS III/3′ PCR primer. During first strand cDNA synthesis, the CDS III/3′ PCR primer must be replaced by CDS-3M adapter (5′-AAG CAG TGG TAT CAA CGC AGA GTG GCC GAG GCG GCC $(T)_{20}$VN-3′, where N = A, C, G, or T; V = A, G, or C). This adapter has the same exterior sequence as SMART IV oligonucleotide (provided in the kits listed above) and allows following amplification of the first strand cDNA in PCR with only one PCR primer (5′ PCR primer: 5′-AAG CAG TGG TAT CAA CGC AGA GT-3′).

First strand cDNA synthesis should be performed according to the manufacturer protocol (SMART cDNA synthesis by LD PCR, section A, first-strand cDNA synthesis) with foregoing modification. For PCR amplification of double-stranded cDNA the kit manufacturer protocol should be slightly modified: the PCR mixture should contain doubled amount of 5′ PCR primer and should not contain CDS III/3′ PCR primer.

Synthetized cDNAs are flanked by adapter sequences with a common exterior part and different interior sequences with *SfiIA* or *SfiIB* restriction sites for cloning.

The use of other cDNA synthesis kits may require additional optimization of the DSN normalization protocol provided.

It is important to realize that adapters used to synthesize starting cDNA for directional cloning are longer than those used for nondirectional cloning. Using longer adapters leads to a reasonable decrease in the cDNA average length during PCR and often causes the appearance of a low-molecular weight fraction in the resulting normalized cDNA (which, in turn, makes it necessary to include a size-separation procedure to remove short cDNA fragments before cloning). Therefore, if directional cloning of cDNA library is not crucial for research purposes, we recommend preparing cDNA with shorter adapters (that are suitable for nondirectional cloning only).

3.3 cDNA Purification

Before normalization, double-stranded cDNA must be purified from primer excess, deoxyribonucleotidetriphosphate (dNTPs), and salts. Several commercial kits are suitable for cDNA purification like QIAquick PCR purification kit (Catalog # 28104, 28106, QIAGEN Inc.) or equivalent. It is necessary that the kit used effectively removes the primer excess.

4. DSN NORMALIZATION PROTOCOL

4.1 Materials

– Double-stranded cDNA flanked by the sequence 5′-AAG CAG TGG TAT CAA CGC AGA GT-3′ at both ends (see Section 5.3 for details)
– 4X Hybridization buffer (200 mM HEPES, pH 7.5; 2 M NaCl)*

- DSN enzyme (Evrogen)* in 50 mM Tris-HCl, pH 8.0 with 50% glycerol (1 U/µl)
- DSN storage buffer (50 mM Tris-HCl, pH 8.0)*
- 2X DSN master buffer (100 mM Tris-HCl, pH 8.0; 10 mM $MgCl_2$; 2 mM DTT)*
- DSN stop solution (5 mM EDTA)*
- Thermostable single-stranded DNA binding (SSB) protein (1.5–5 µg/µl; optional, see Section 5.4.4, step 3)
- PCR primer M1 (10 µM, 5′-AAG CAG TGG TAT CAA CGC AGA GT-3′)*
- PCR primer M2 (10 µM, 5′-AAG CAG TGG TAT CAA CGC AG-3′)*
- Long-distance (LD) PCR enzyme mix with 10X PCR buffer (hot start must be used to reduce nonspecific DNA synthesis during the PCR setup. PCR kits that are adapted for long-distance PCR and also include automatic hot start are recommended, for example, Advantage 2 PCR kit, Clontech Catalog #: 639206, 639207)
- 50X dNTP Mix (10 mM each nucleotide)
- Sterile nuclease free (millipore-filtered) water
- Mineral oil
- Blue ice
- Sterile 0.5 ml PCR tubes (thin-wall PCR tubes recommended. These tubes are optimized to ensure more efficient heat transfer and to maximize thermal-cycling performance)
- Sterile microcentrifuge 1.5 ml tubes
- Pipettors (P20, P200)
- Pipette tips
- Base and tray or retainer for holding tubes
- Vortex mixer
- Microcentrifuge
- Agarose gel electrophoresis equipment and reagents (1X TAE buffer, 1.2–1.5% agarose with ethidium bromide, 1 kb DNA size markers)
- PCR thermal cycler
- 98% ethanol
- 80% ethanol
- 3M NaAc (sodium acetate), pH 4.8

*These reagents are available from Evrogen Trimmer kits (Evrogen Catalog # NK001, NK002)

4.2 cDNA Precipitation

Note: Do not use any coprecipitants in the precipitation procedure.
1. Aliquot cDNA solution containing ~700–1300 ng of purified cDNA (see Section 5.3 for details) into a separate sterile tube. Store the remaining cDNA solution at −20°C.
2. Add 0.1 volumes of 3 M NaAc, pH 4.8, to the reaction tube.

Normalization of cDNA libraries

3. Add 2.5 volumes of 98% ethanol to the reaction tube.
4. Vortex the mixture thoroughly.
5. Centrifuge the tube for 15 min at 12,000–14,000 rpm at room temperature.
6. Remove the supernatant carefully.
7. Gently overlay the pellets with 100 μl of 80% ethanol.
8. Centrifuge the tubes for 5 min at 12,000–14,000 rpm at room temperature.
9. Carefully remove the supernatant.
10. Repeat steps 4.2.6–4.2.8.
11. Air dry the pellet for 10–15 min at room temperature. Be sure the pellet has dried completely.
12. Dissolve the pellet in sterile water to the final cDNA concentration of ~100–150 ng/μl.
13. To check the cDNA quality and concentration, analyze 1 μl of cDNA solution using gel electrophoresis alongside 0.1 μg of 1 kb DNA size markers on a 1.5% agarose/EtBr gel in 1X TAE buffer.

4.3 Hybridization

Note: Before starting hybridization, make sure 4X hybridization buffer has been allowed to stay at room temperature for at least 15–20 min. The buffer used should have no visible pellet or precipitate. If necessary, warm the buffer at 37°C for about 10 min to dissolve any precipitate.

1. For each sample to be normalized combine the following reagents in a sterile 1.5 ml tube: 4–12 μl double-stranded cDNA solution from step 4.2.12 (~600–1200 ng of cDNA); 4 μl 4X hybridization buffer; and sterile water to the total volume of 16 μl.
2. Mix the contents and spin the tube briefly in a microcentrifuge.
3. Aliquot 4 μl of the reaction mixture into each of the four appropriately labeled (e.g. see Table 1) sterile PCR tubes.

Table 1. Setting up DSN treatment

Component\tube	Experimental			Control
	Tube 1	Tube 2	Tube 3	Tube 4
	(S1 DSN1) (μl)	(S1 DSN1/2) (μl)	(S1 DSN1/4) (μl)	(S1 control) (μl)
DSN enzyme in storage buffer	1	–	–	–
1/2 DSN dilution	–	1	–	–
1/4 DSN dilution	–	–	1	–
DSN storage buffer	–	–	–	1

4. Overlay the reaction mixture in each tube with a drop of mineral oil and centrifuge the tubes at 14,000 rpm for 2 min.
5. Incubate the tubes in a thermal cycler at 98°C for 2 min.
6. Incubate the tubes at 68°C for 5 h, and then proceed immediately to DSN treatment. Samples may be hybridized for as little as 4 h, or as long as 7 h. Do not allow the incubation to proceed for more than 7 h. Do not remove the samples from the thermal cycler before DSN treatment.

4.4 DSN Treatment

1. Shortly before the end of the hybridization procedure, prepare the following dilutions of the DSN enzyme in two sterile tubes:
   (a) Add 1 µl of DSN storage buffer and 1 µl of DSN solution to the first tube. Mix by gently pipetting the reaction mixture up and down. Label the tube as 1/2 DSN.
   (b) Add 3 µl of DSN storage buffer and 1 µl of DSN solution to the second tube. Mix by gently pipetting the reaction mixture up and down. Label the tube as 1/4 DSN.
   (c) Place the tubes on ice.
2. Preheat the DSN master buffer at 68°C.
3. Add 5 µl of the preheated DSN master buffer to each tube containing hybridized cDNA (see Section 5.4.3, step 6), spin the tube briefly in a microcentrifuge and return it to the thermal cycler. Do not remove the tubes from the thermal cycler except for the time necessary to add preheated DSN master buffer.

Note: When integrity of very long transcripts (more than 5 kb) is crucial, 1 µl of the thermostable SSB protein (with concentration 1–5 µM) can also be added to the reaction mixture. In this case, less PCR cycles are required (for 3–4 cycles) to amplify normalized cDNA (see Section 5.4.5, step 11).

4. Incubate the tubes at 68°C for 10 min.
5. Add DSN enzyme as specified in Table 1. After DSN adding return the tubes immediately to the thermal cycler. Do not remove the tubes from the thermal cycler except for the time necessary to add DSN enzyme. When the tube is left at room temperature after DSN adding, nonspecific digestion of secondary structures formed by ss DNA may occur to decrease the efficacy of normalization.
6. Incubate the tubes in the thermal cycler at 68°C for 25 min.
7. Add 10 µl of DSN stop solution, mix the contents, and spin the tubes briefly in a microcentrifuge.
8. Incubate the tubes in the thermal cycler at 68°C for 5 min.
9. Take the tubes off the thermal cycler and place them on ice.
10. Add 20 µl of sterile water to each tube. Mix the contents and spin the tubes briefly in a microcentrifuge. Place the tubes on ice. The samples obtained can be stored at −20°C for up to 2 weeks and used afterwards to prepare more normalized cDNA.

4.5 First Amplification of Normalized cDNA

1. Prepare a PCR master mix for all reaction tubes. In a sterile 1.5 ml tube, combine the following reagents in the order shown (per rxn): 40.5 µl sterile water, 5 µl 10X PCR buffer, 1 µl 50X dNTP mix, 1.5 µl PCR primer M1, 1 µl 50X polymerase mix (total volume is 49 µl).
2. Mix well by vortexing and spin the tube briefly in a microcentrifuge.
3. Aliquot 1 µl of each diluted cDNA (from Step 4.4.10) into an appropriately labeled sterile PCR tube.
4. Aliquot 49 µl of the PCR master mix into each of these reaction tubes.
5. Mix the contents by gently flicking the tubes. Spin the tubes briefly in a microcentrifuge.
6. If the thermal cycler used is not equipped with a heated cover, overlay each reaction with a drop of mineral oil. Close the tubes, and place them into the preheated thermal cycler.
7. Subject the tubes to 7 PCR cycles using the following program: 95°C for 7 s, 66°C for 30 s, 72°C for 4 min. Please note than this program is optimized for MJ Research PTC-200 thermal cycler and polymerase mixtures noted in Section 5.4.1. Optimal PCR parameters may vary with different thermal cyclers, polymerase mixes, and templates.
8. Use the control tube (see Table 1) to determine the optimal number of PCR cycles as described in following steps 9–10 (below). Store other tubes on ice.
9. For each control tube, determine the optimal number of PCR cycles as shown in Figure 5):
   (a) Transfer 12 µl from the 7-cycle PCR tube to a clean microcentrifuge tube (for agarose/EtBr gel analysis).
   (b) Run two additional cycles (for a total of 9) with the remaining 38 µl of the PCR mixture.
   (c) Transfer 12 µl from the 9-cycle PCR tube to a clean microcentrifuge tube (for agarose/EtBr gel analysis).
   (d) Run two additional cycles (for a total of 11) with the remaining 26 µl of the PCR mixture.
   (e) Transfer 12 µl from the 11-cycle PCR tube to a clean microcentrifuge tube (for agarose/EtBr gel analysis).
   (f) Run two additional cycles (for a total of 13) with the remaining 14 µl of the PCR mixture.
10. Use 5 µl of each aliquot of each PCR reaction (from Step 9) for gel electrophoresis alongside 0.1 µg of 1 kb DNA size markers on a 1.5% agarose/EtBr gel in 1X TAE buffer. Store the remaining material on ice. Use Figure 7 to determine optimal number of PCR cycles required for amplification of each of the control tubes ("X"). Choosing the optimal number of PCR cycles ensures that the double-stranded cDNA will remain in the exponential phase of amplification. When the yield of PCR products stops increasing with every additional cycle, the reaction has

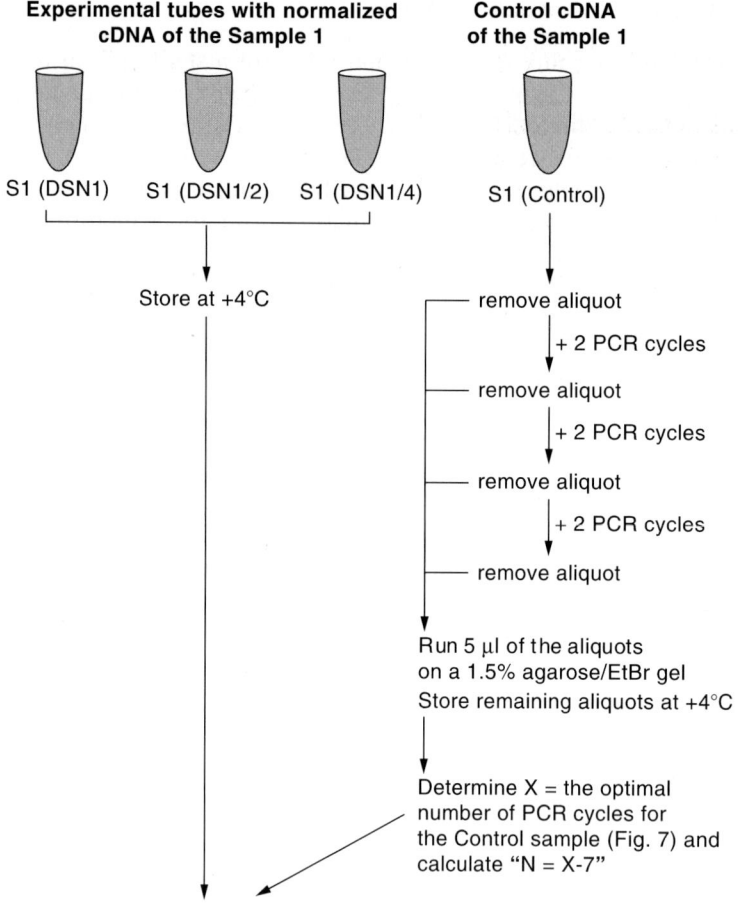

Figure 5. Optimizing PCR parameters for normalized cDNA amplification. Scheme of the experimental procedures.

reached its plateau. The optimal number of cycles for your experiment should be one or two cycles less than that needed to reach the plateau. Figure 6 provides an example of how the analysis should proceed. Be conservative: when in doubt, it is better to use fewer cycles than too many.

11. Retrieve the 7-PCR tubes from ice, return them to the thermal cycler, and if necessary, subject them to additional N cycles (where $N = X - 7$). Then, immediately, subject the tubes to additional 9 cycles. Altogether, control tube should be subjected to X PCR cycles, whereas experimental tubes should be subjected to $X + 9$ PCR cycles, where X is the optimal number of PCR cycles determined for the control tube. In the example shown in Figure 6, the optimal number of

Normalization of cDNA libraries

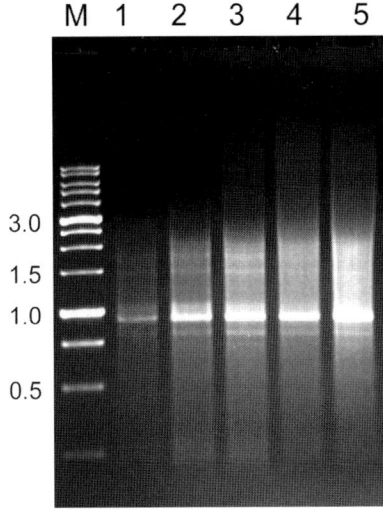

Figure 6. Analysis for optimizing PCR parameters. 5 μl of the aliquots from the control tube were analyzed using gel electrophoresis on a 1.5% agarose/EtBr gel in 1X TAE buffer: 1 – after 7 PCR cycles; 2 – after 9 cycles; 3 – after 11 cycles; 4 – after 13 cycles; 5 – after 15 PCR cycles. Lane M: 1 kb DNA ladder size markers, 0.1 μg loaded. In this experiment, after 11 cycles, a smear appears in the high-molecular weight region of the gel, indicating that the reaction is overcycled. Because the plateau is reached after 11 cycles, the optimal number of cycles for this experiment is 9.

PCR cycles determined for control cDNA in the control tube was 9. Thus, in this example $X = 9$, and $N = 9–7 = 2$. Hence, in this example, 7-PCR experimental tubes should be subjected to $2 + 9$ additional PCR cycles.

Note: If SSB protein is used during DSN treatment (see Section 5.4.4, step 3), less PCR cycles are required to amplify normalized cDNA, e.g. in this case, experimental tubes should be subjected to $N + 6$ additional PCR cycles.

12. When the cycling is completed, analyze 5 μl from each tube using gel electrophoresis alongside 5 μl aliquot from control PCR tube (with the optimal PCR cycle number) and 0.1 μg of 1 kb DNA size markers on a 1.5% agarose/EtBr gel in 1X TAE buffer.
13. Select the tube(s) with efficient normalization as described in Section 5.4.6. For comparison, Figure 7 shows a characteristic gel profile of normalized human placenta cDNA. If cDNA from two or more tubes seems well normalized, combine the contents of these tubes in one sterile 1.5 ml tube, mix well by vortexing and spin the tube briefly in a microcentrifuge. Amplified normalized cDNA obtained can be stored at –20°C for up to 1 month and used afterwards to prepare more normalized cDNA.

4.6 Preliminary Analysis of the Normalization Results

1. Compare the intensity of the banding pattern of your PCR products from experimental tubes with that from the control tube and with the 1 kb DNA ladder size markers (0.1 μg run on the same gel). Use Figure 7 as an example.
   – If the smear from the experimental tubes is much fainter than that shown for the control, PCR undercycling could be the problem. Subject experimental

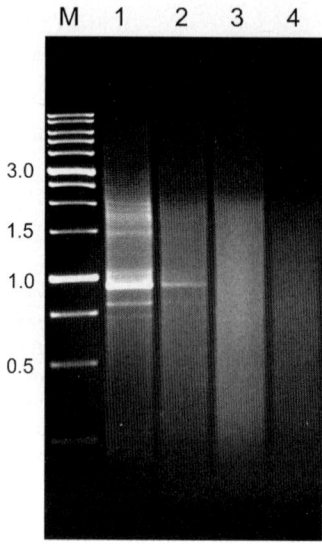

Figure 7. Analysis of cDNA normalization results. 5 μl aliquots of the PCR products were analyzed using gel electrophoresis on a 1.5% agarose/EtBr gel. Lane M: 1 kb DNA ladder size markers, 0.1 μg loaded. Lane 1: cDNA from control tube Lane 2: cDNA from S1_DSN1/4 Tube. Lane 3: cDNA from S1_DSN1/2 tube. Lane 4: cDNA from S1_DSN1 tube. In this experiment efficient normalization was achieved in the S1_DSN1/2 tube (lane 3). In the S1_DSN1/4 tube (lane 1) normalization was not completed, in the S1_DSN1 tube (lane 4) DSN treatment was excessive, resulting in partial cDNA degradation.

tubes to two or three additional PCR cycles and repeat electrophoresis. If there is still a strong difference between the overall signal intensity of PCR products from all experimental tubes and from the control tube, it may indicate that normalization process was superfluous.
  – If the overall signal intensity of PCR products from the experimental tubes is much stronger than that shown for the Control, especially if the bright bands are distinguishable, it may indicate that normalization process was not successful. Test DSN activity and repeat normalization with doublet amounts of DSN.
  – If the overall signal intensity of PCR products from the experimental tubes is similar to that in the control tube, select the tube(s) with efficient normalization using the instruction below.
2. A typical result, indicative of efficient normalization, should have the following characteristics:
  – The pattern of PCR products from the experimental tube(s) containing efficiently normalized cDNA looks like smears without clear bands, whereas a number of distinct bands are usually present in the pattern of PCR products from the nonnormalized control tube.
  – The average length of PCR products from the experimental tube(s) containing efficiently normalized cDNA is congruous with the average length of the PCR products from the nonnormalized control tube.

4.7 Second Amplification of Normalized cDNA

Reamplification of normalized cDNA allows to avoid cDNA degradation due to residual DSN activity and to prepare more cDNA for library cloning. If you plan to estimate normalization efficiency before cloning, it is essential to amplify control nonnormalized cDNA simultaneously.

1. Aliquot 2 µl of normalized cDNA (see Section 5.4.5, step 13) into a sterile 1.5 µl tube; add 20 µl of sterile water to the tube, mix well by vortexing and spin the tubes briefly in a microcentrifuge.
2. Aliquot 2 µl of control cDNA (from aliquot with optimal PCR cycling, see Section 5.4.5, steps 9 and 10) into another sterile 1.5 µl tube; add 20 µl of sterile water to the tube, mix well by vortexing and spin the tubes briefly in a microcentrifuge.
3. Aliquot 2 µl of diluted normalized cDNA from step 1 into an appropriately labeled sterile PCR tube.
4. Aliquot 2 µl of diluted control cDNA from step 2 into another appropriately labeled sterile PCR tube.
5. Preheat a thermal cycler to 95°C.
6. Prepare a PCR master mix for all reaction tubes. In a sterile 1.5 ml tube, combine the following reagents in the order shown (per rxn): 80 µl sterile water; 10 µl 10X PCR buffer; 2 µl 50X dNTP mix; 4 µl PCR primer M2; 2 µl 50X polymerase mix (total reaction volume 98 µl).
7. Aliquot 98 µl of the PCR master mix into each of the reaction tubes (from steps 3 and 4).
8. Mix contents by gently flicking the tubes. Spin the tubes briefly in a microcentrifuge.
9. If your thermal cycler is not equipped with a heated cover, overlay each reaction with two drops of mineral oil. Close the tubes, and place them into the preheated thermal cycler.
10. Commence thermal cycling using the following program: 95°C for 7 s, 66°C for 30 s, 72°C for 4 min. Subject the tubes to 12 cycles. Please note than this program is optimized for MJ Research PTC-200 thermal cycler and polymerase mixtures noted in Section 5.4.1. Optimal PCR parameters may vary with different thermal cyclers, polymerase mixes, and templates.
11. When the cycling is completed, analyze 5 µl of the PCR product using gel electrophoresis alongside 0.1 µg of 1 kb DNA size markers on a 1.5% agarose/EtBr gel in 1X TAE buffer to check the PCR quality and concentration. If necessary, subject the PCR tubes to 1–2 additional PCR cycles.
12. Now, you have normalized double-stranded cDNA that can be used for cloning into a vector of your choice. This cDNA can be stored at −20°C for up to 3 months. To estimate normalization efficiency after cDNA library cloning, sequence 100 randomly picked clones from the library. In a well-normalized library, redundancy of the first 100 sequences should not exceed 5%. You can also estimate normalization efficiency before cloning using

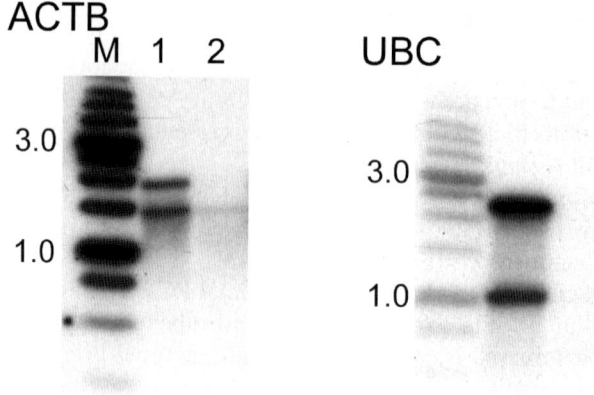

Figure 8. Virtual Northern blot analysis of abundant (ACTB, UBC) transcripts in nonnormalized (lane 1) and normalized (lane 2) cDNAs.

quantitative PCR or Virtual Northern blotting (Franz et al. 1999) with marker genes of known abundance. In both cases, it is done by comparing the abundance of known cDNAs before and after normalization. A typical result of Virtual Northern blot of nonnormalized and normalized cDNA with *ACTB*- and *UBC*-derived probes is shown in Figure 8.

Note: ACTB and *UBC* genes are expressed at high levels in most human tissues and cell lines; however, there could be some exceptions. If in the samples of your particular interest *ACTB* and *UBC* transcripts belong to intermediate or low abundance groups, unchanged or slightly increased concentration of these transcripts in normalized cDNA could be observed. In this case, select other marker genes that are highly abundant your samples to test normalization efficiency.

REFERENCES

Alberts B, Bray D, Lewis J, Raff M, Roberts K, Watson JD (1994) Molecular Biology of the Cell. Garland Publishing, New York

Barnes WM (1994) PCR amplification of up to 35-kb DNA with high fidelity and high yield from lambda bacteriophage templates. Proc Natl Acad Sci USA 91:2216–2220

Caetano AR, Johnson RK, Pomp D (2003) Generation and sequence characterization of a normalized cDNA library from swine ovarian follicles. Mamm Genome 14:65–70

Carninci P, Kvam C, Kitamura A, Ohsumi T, Okazaki Y, Itoh M, Kamiya M, Shibata K, Sasaki N, Izawa M, Muramatsu M, Hayashizaki Y, Schneider C (1996) High-efficiency full-length cDNA cloning by biotinylated CAP trapper. Genomics 37:327–336

Carninci P, Shibata Y, Hayatsu N, Sugahara Y, Shibata K, Itoh M, Konno H, Okazaki Y, Muramatsu M, Hayashizaki Y (2000) Normalization and subtraction of cap-trapper-selected cDNAs to prepare full-length cDNA libraries for rapid discovery of new genes. Genome Res 10:1617–1630

Chomczynski P, Sacchi N (1987) Single-step method of RNA isolation by acid guanidinium thiocyanate-phenol-chloroform extraction. Anal Biochem 162:156–159

Coche Th, Dewez M (1994) Reducing bias in cDNA sequence representation by molecular selection. Nucleic Acids Res 22:4545–4546

Diatchenko L, Lau Y-FC, Campbell AP, Chenchik A, Mogadam F, Huang B, Lukyanov S, Lukyanov K, Gurskaya N, Sverdlov ED, Siebert PD (1996) Suppression subtractive hybridization, a method for generating differentially regulated or tissue-specific cDNA probes and libraries. Proc Natl Acad Sci USA 93:6025–6030

Franz O, Bruchhaus II, Roeder T (1999) Verification of differential gene transcription using virtual northern blotting. Nucleic Acids Res 27:e3

Galau GA, Klein WH, Britten RJ, Davidson EH (1977) Significance of rare mRNA sequences in liver. Arch Biochem Biophys 179:584–599

Guigo R, Agarwal P, Abril JF, Burset M, Fickett JW (2000) An assessment of gene prediction accuracy in large DNA sequences. Genome Res 10:1631–1642

Ishikawa H (1977) Evolution of ribosomal RNA. Comp Biochem Physiol B 58:1–7

Ko MS (1990) An "equalized cDNA library" by the reassociation of short double-stranded cDNAs. Nucleic Acids Res 18:5705–5711

Luk'yanov KA, Gurskaya NG, Matts MV, Khaspekov GL, D'vachenko LB, Chenchik AA, D'evich-Stuchkov SG, Luk'yanov SA (1996) A method for obtaining equalized cDNA libraries based on polymerase chain reaction suppression. Rus J Bioorg Chem 22:587–591

Marra M, Hillier L, Kucaba T, Allen M, Barstead R, Beck C, Blistain A, Bonaldo M, Bowers Y, Bowles L, Cardenas M, Chamberlain A, Chappell J, Clifton S, Favello A, Geisel S, Gibbons M, Harvey N, Hill F, Jackson Y, Kohn S, Lennon G, Mardis E, Martin J, Mila L, McCann R, Morales R, Pape D, Person B, Prange C, Ritter E, Soares M, Schurk R, Shin T, Steptoe M, Swaller T, Theising B, Underwood K, Wylie T, Yount T, Wilson R, Waterston R (1999) An encyclopedia of mouse genes. Nat Genet 21:191–194

Matz MV (2002) Amplification of representative cDNA samples from microscopic amounts of invertebrate tissue to search for new genes. Methods Mol Biol 183:3–18

Patanjali SR, Parimoo S, Weissman SM (1991) Construction of a uniform-abundance (normalized) cDNA library. Proc Natl Acad Sci USA 88:1943–1947

Reddy AR, Ramakrishna W, Sekhar AC, Ithal N, Babu PR, Bonaldo MF, Soares MB, Bennetzen JL (2002) Novel genes are enriched in normalized cDNA libraries from drought-stressed seedlings of rice (*Oryza sativa* L. subsp. *indica* cv. Nagina 22). Genome 45:204–211

Sambrook J, Fritsch EF, Maniatis T (1989). Molecular Cloning: A Laboratory Manual, 2nd edn. Cold Spring Harbor Laboratory Press, Cold Spring Harbor, NY

Sasaki YF, Ayusawa D, Oishi M (1994) Construction of a normalized cDNA library by introduction of a semi-solid mRNA-cDNA hybridization system. Nucleic Acids Res 22:987–992

Shagin DA, Lukyanov KA, Vagner LL, Matz MV (1999) Regulation of average length of complex PCR product. Nucleic Acids Res 27(18):e23

Shagin DA, Rebrikov DV, Kozhemyako VB, Altshuler IM, Shcheglov AS, Zhulidov PA, Bogdanova EA, Staroverov DB, Rasskazov VA, Lukyanov S (2002) A novel method for SNP detection using a new duplex-specific nuclease from crab hepatopancreas. Genome Res 12:1935–1942

Soares M, Bonaldo M, Jelene P, Su L, Lawton L, Efstratiadis A (1994) Construction and characterization of a normalized cDNA library. Proc Natl Acad Sci USA 91:9228–9232

Tanaka T, Ogiwara A, Uchiyama I, Takagi T, Yazaki Y, Nakamura Y (1996) Construction of a normalized directionally cloned cDNA library from adult heart and analysis of 3040 clones by partial sequencing. Genomics 35:231–235

Urmenyi TP, Bonaldo MF, Soares MB, Rondinelli E (1999) Construction of a normalized cDNA library for the *Trypanosoma cruzi* genome project. J Eukaryot Microbiol 46:542–544

Weissman SM (1987) Molecular genetic techniques for mapping the human genome. Mol Biol Med 4:133–143

Young BD, Anderson MLM (1985) Quantitative analysis of solution hybridisation. In: Hames BD, Higgins SJ (eds) Nucleic Acid Hybridisation. IRL Press, Oxford-Washington DC, pp 47–71

Zhu YY, Machleder EM, Chenchik A, Li R, Siebert PD (2001) Reverse transcriptase template switching: a SMART approach for full-length cDNA library construction. Biotechniques 30:892–897

Zhulidov PA, Bogdanova EA, Shcheglov AS, Vagner LL, Khaspekov GL, Kozhemyako VB, Matz MV, Meleshkevitch E, Moroz LL, Lukyanov SA, Shagin DA (2004) Simple cDNA normalization using kamchatka crab duplex-specific nuclease. Nucleic Acid Res 32:e37

Zhulidov PA, Bogdanova EA, Shcheglov AS, Shagina IA, Wagner LL, Khazpekov GL, Kozhemyako VV, Lukyanov SA, Shagin DA (2005) A method for the preparation of normalized cDNA libraries enriched with full-length sequences. Rus J Bioorg Chem 31:170–177

CHAPTER 6

PRIMER EXTENSION ENRICHMENT REACTION (PEER) AND OTHER METHODS FOR DIFFERENCE SCREENING

LILIA M. GANOVA-RAEVA

Centers for Disease Control and Prevention, Division of Viral Hepatitis, 1600 Clifton Rd. NE, MS A-33, Atlanta, Georgia 30329, USA
E-mail: lkg7@cdc.gov

Abstract:	Our knowledge on the subject of genetic divergence is ever expanding and encapsulates a wide spectrum of research areas from the study of single nucleotide polymorphism to examination of the complex differences of host–pathogen interactions. The understanding of genetic differences is essential to our ability to address adequately human health issues caused by mono allelic genetic disorders, altered gene expression in cancers, development of drug resistance and the variety of ways organisms respond to infections and the environment. Large number of hybridization-based applications has been developed to query and find such differences. This review introduces a new method for genetic difference screening — the primer extension enrichment reaction (PEER) — presented in the context of similar subtraction-based hybridization methods. PEER is a novel approach to difference screening and is not intended to replace the existing elegant hybridization methods but to expand their scope. PEER is tailored to find unknown targets present in very low copy numbers and in the context of an imperfect genomic match. The PEER method takes advantage of the greater hybridization specificity of shorter oligonucleotides coupled with enzymatic extension specificity.
Keywords:	Primary structure, genetic differences, gene expression, sequencing-by-hybridization, primer extension enrichment reaction (PEER), class II restriction endonuclease, class IIS restriction endonuclease, IIS enzyme, spot hybridization, enrichment, enrichment value, cost efficient, subtractive cloning, arbitrary primed PCR, amplification fragment length polymorphism (AFLP), serial analysis of gene expression (SAGE), genomic signature tag (GST), serial analysis of binding elements (SABE), differential analysis of restriction fragments amplification (DARFA), ligation mediated enrichment, selectively primed adaptive driver RDA, enzymatic degrading subtraction (EDS), phenol, linker capture subtraction (LCS), differential subtraction chain (DSC), cloning of deleted sequences (CODE), negative subtraction chain (NSC), SABRE, DNA enrichment by allele-specific hybridization (DEASH).

Abbreviations: AFLP, amplification fragment length polymorphism; cDNA, complementary DNA; CODE, cloning of deleted sequences; DARFA, differential analysis of restriction fragments amplification; DD, differential display; DEASH, DNA enrichment by allele-specific hybridization; dNTP, deoxyribonucleotidetriphosphate; DSC, differential subtraction chain; ds, double stranded; EDS, enzymatic degrading subtraction; GST, genomic signature tag; LCS, linker capture subtraction; MDR, mispaired DNA rejection; mRNA, messenger RNA; NA, nucleic acid; NSC, negative subtraction chain; PCR, polymerase chain reaction; PEER, primer extension enrichment reaction; RaSH, Rapid subtraction hybridization; RDA, representational differences analysis; RFLP, restriction fragment length polymorphism; RT, reverse transcription; SABE, serial analysis of binding elements; SABRE, selective amplification via biotin- and restriction-mediated enrichment; SAGE, serial analysis of gene expression; SH, subtractive hybridization; SNP, single nucleotide polymorphism; SPAD, Selectively primed adaptive driver; ss, single stranded; SSH, suppression subtractive hybridization; UDG, uracil deglycosilase.

TABLE OF CONTENTS

1. Introduction 127
2. Primer Extension Enrichment Reaction 128
   2.1 Method Outline 128
   2.2 Discussion 132
   2.3 PEER Protocol 135
3. Other Subtraction and Hybridization Based Methods
   for Difference Screening 137
   3.1 Differential Screening 137
   3.2 Subtractive Hybridization 139
   3.3 Subtractive Cloning 140
   3.4 Differential Display 140
   3.5 AFLP, SAGE/CAGE, GSTs, and DARFA 142
   3.6 Representational Differences Analysis 143
   3.7 SPAD–RDA 146
   3.8 Enzymatic Degradation Subtractions (EDS, LCS,
       DSC, NSC, UDG/USA, and CODE) 146
   3.9 Suppression Subtraction Hybridization 150
   3.10 Selective Amplification Via Biotin and
        Restriction-Mediated Enrichment 153
   3.11 DNA Enrichment by Allele-Specific
        Hybridization 154
   3.12 Methods Combining the use of SSH
        and Microarrays 155
   3.13 Conclusions 156
   References 157

1. INTRODUCTION

Identifying and isolating genetic differences without *a priori* knowledge of the primary structure, i.e. sequence, of the genetic material, is technically demanding and laborious. Genetic differences, which can be found at any level of genetic organization or gene expression, can be caused by gene rearrangements, deletions, insertions, or by the presence of genomes from extraneous organisms and can lead to disparate disease outcomes. They can be represented by a single nucleotide polymorphism (SNP) (e.g. hepatitis B virus vaccine escape mutants), single allele differences (e.g. cystic fibrosis, sickle cell anemia, Tay Sachs disease and over 4000 other genetic diseases), gene expression differences in the various cells of one organ that occur in order to execute the specific function of the tissue or organ, gene expression at different stages of an organism's development that provide for cell differentiation related to growth and development, genetic differences between organisms, sex-related differences, complex genetic disorders (e.g. Alzheimer, diabetes), and gene expression prompted by the interaction of different organisms, such as pathogen–host interactions, or in response to different environmental stimuli.

Most genetic differences can be queried by using RNA, DNA, cDNA, or total nucleic acids (NAs), depending on the experimental goal. Direct screening of large plasmid or phage libraries, the earliest approach used for identifying target NAs of unknown sequence, can be inefficient and labor-intensive [1]. Over the last two decades, several new approaches have been developed to improve the efficiency of this task. They can be divided into two broad categories: (1) subtractive approaches, such as representational differences analysis (RDA) [2] and its variants [3,4], differential subtraction chain (DSC) [5], selective amplification via biotin- and restriction-mediated enrichment (SABRE) [6], suppression subtractive hybridization (SSH) [7], differential display (DD) [8], and others [9,10]; and (2) high-throughput [11,12,13,14,15,16,17] and microarray-based [18,19] methods.

All subtractive approaches are based on molecular comparison of two specimens: (1) a "tester" – a specimen that is presumed to contain the unknown target of interest, and (2) a "driver" – a specimen that is a perfect genetic match for the tester but is not believed to contain the target [20]. The comparison is usually done by hybridizing the two specimens to each other, eliminating the hybrids they have in common and screening the final product for a subset of molecules that reflect the differences between the two. Subtractive methods are often used in molecular studies because of their relative simplicity and high efficiency.

High-throughput methods approach gene expression *in toto* and are based on miniaturization, artificial nucleic acid synthesis, and hybridization and often require the heavy use of bioinformatics for data management. In recent years, the integration of subtractive approaches into high-throughput methods [21,22,23,24] has generated efficient methodology for identifying differentially expressed sequences.

Among the subtractive techniques, RDA and the closely related SSH are the most popular and have been used successfully to recover unknown sequences and differentially expressed genes. SSH has been used to find a number of differentially expressed RNA messages [25,26,27,28,29,23] and to identify a new calicivirus in walrus [30]. The GBV–A and –B viruses [31] as well as TT virus (TTV) [32] (all suspect viral hepatitis agents) were discovered by RDA. In practice, SSH can enrich a target gene by approximately 3×10^3 times [7,1]. Although impressive, this performance is nonetheless insufficient for the detection of an infectious agent whose genomes might be present at only a few copies in the specimen of interest [33], for example, in chronic carriers the levels of hepatitis C virus might be as low as 10^3 or less, yet the disease persists and the individual is still a potential source of infection. Such levels are below the molecular resolution of the current enrichment methods. A common limitation of all subtractive approaches is their requirement for perfect hybridization, which goes hand in hand with the need for an abundance of ideally matched drivers. Since such conditions can rarely be achieved, most subtraction methods are intrinsically biased against single-stranded, low-copy-number molecular species [31,33]. The enrichment of the target of interest is usually achieved by hybridization between long DNA fragments and, in the case of cDNA generated by random priming, a heterogeneous population of DNA fragments. If present in low numbers, these molecules have little chance to form complete hybridization products after denaturing. Another limitation of most subtraction methods is that they use the target of interest as a potential PCR template only. If present at very few copies or if it has remained single stranded, the target might fail to amplify efficiently. In addition, many subtraction approaches rely on the presence of a poly-A tail to generate starting material and consequently are not suitable when working with DNA or RNA that is not polyadenylated.

Microarray approaches tend to be costly and require some prior knowledge of the sequence of interest. The more recent sequencing-by-hybridization approaches and massive parallel sequencing methods [13,14,16], which have been proposed as alternative solutions to uncovering differences, are also expensive and might not be suitable for small sample volumes.

The primer extension enrichment reaction (PEER) method belongs to the category of subtraction techniques. It is based on two new strategies: (1) the use of tester DNA to generate both PCR primer and template, and (2) the selective inactivation of primers containing sequences common to the tester and driver to ensure preferential amplification of templates that contain sequences unique to the tester. PEER improves the sensitivity of current subtraction methods and takes direct advantage of the unknown target's unique specificity.

2. PRIMER EXTENSION ENRICHMENT REACTION

2.1 Method Outline

A general outline of PEER is presented in Figure 1. Total NA is extracted from a tester and a driver specimen and used in a modified SMART cDNA protocol (Clontech, Palo Alto, CA) to generate dsDNA with two different sets of

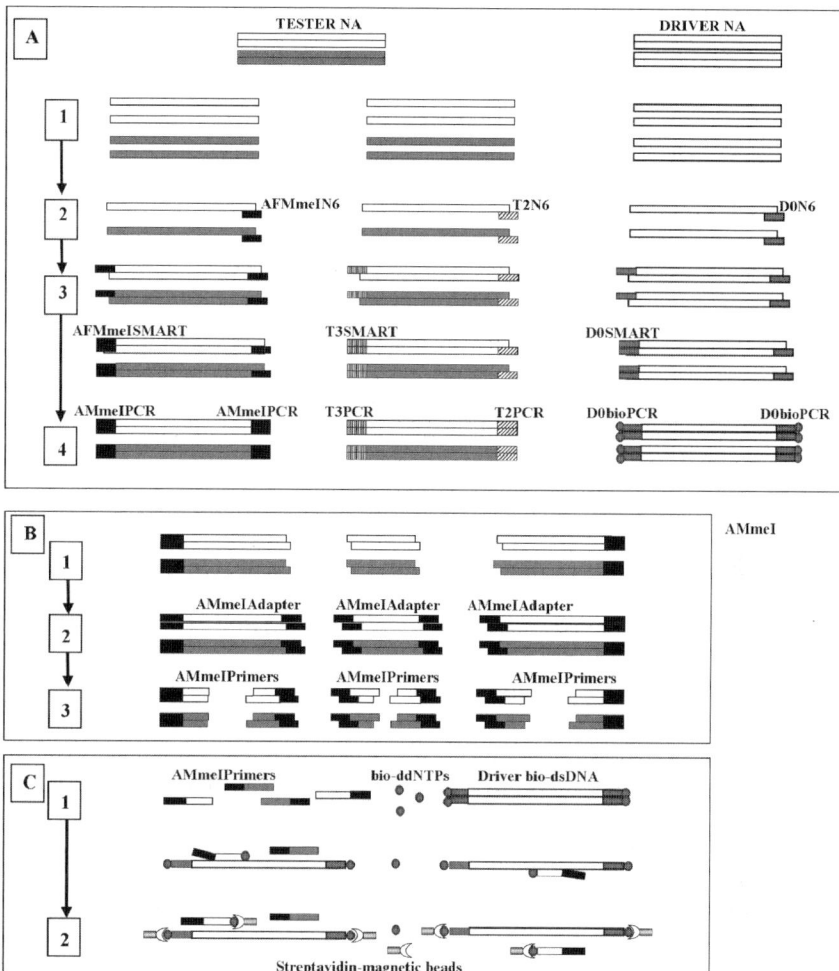

Figure 1. Primer extension enrichment reaction (PEER). Panel A: Generation of dsDNA from total nucleic acid (NA). 1. Tester NA (white and gray rectangle) is split in two aliquots and denatured; driver NA (white rectangle) is denatured as well; 2. Single strands are reverse transcribed (RT) by SuperScript RT with three different primers – AFMmeIN6* for the first Tester aliquot, T2N6 (diagonal fill rectangle) for the second aliquot, and D0N6 (shadowed rectangle) for the driver; 3. Reverse transcription switches templates and copies annealed SMART primers (SMART technology, Clontech); 4. RT products are amplified with Advantage2 Polymerase to yield Tester1 dsDNA with primers AMmeIPCR (black rectangle), Tester23 dsDNA with T3PCR (vertical fill rectangle) and T2PCR (diagonal fill rectangle) and driver bio-dsDNA with D0bioPCR biotinylated at the 5'-end (shadowed rectangle with circle) Panel B: Processing of Tester1 dsDNA. 1. DNA is cleaved by a cocktail of restriction enzymes that leave 3'-GC protruding ends; 2. Ends are treated with the Klenow fragment of DNA Polymerase I in the presence of dCTP only and then ligated to AMmeIAdapter; 3. Tagged fragments are cut to uniform size by MmeI to create multiple AMmeIPrimers. Panel C: Blocking reaction. 1. AMmeIPrimers generated from Tester1 dsDNA are extended on Driver bio-dsDNA template in the presence of biotinylated ddNTPs (red circles) and ThermoSequenase; 2. Biotinylated molecules are captured with streptavidin-coated magnetic beads (white crescent with gray bar) and removed from the reaction.

*black rectangles – primers AFMmeIN6, AFMmeISMART, AMmeIPCR, AMmeIAdapter

(*Continued*)

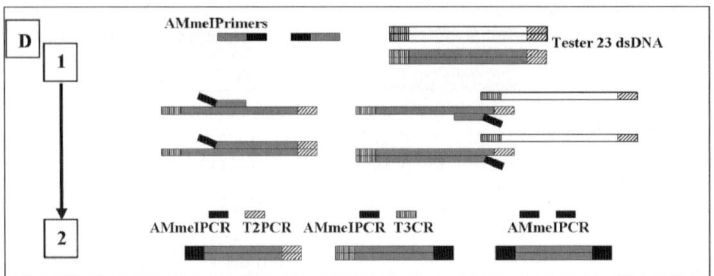

Figure 1. cont'd. Panel D: Retrieval of targets of interest from the Tester23 dsDNA. 1. Capture PCR – AMmeIPrimers that were not blocked and removed in the preceding steps are added to Tester23 dsDNA and in the presence of regular dNTP are annealed and extended to capture the targets of interest; 2. Regular PCR amplification of the capture products with different primer combinations.

primers for the tester (called Tester 1 cDNA and Tester 2 cDNA primers in Table 1) and one set for the driver (Driver cDNA primers in Table 1). The product is referred to as dsDNA to distinguish it from cDNA since it is generated from total NA instead of from RNA alone (Figure 1A). Tester 1 dsDNA material is converted into small fragments by extensive endonuclease cleavage and then tagged by ligation to a specially designed adapter. The 3' end of the adapter incorporates a recognition site for a class IIS restriction endonuclease [34, 35]. After ligation, the fragments are cleaved with the appropriate IIS enzyme to create oligonucleotides with unique sequence at the 3'-end derived from the tester and a 5'-end derived from the adapter (Figure 1B). These adapter-tagged oligonucleotides are annealed to the driver dsDNA template and extended in the presence of biotinylated ddNTPs. All oligonucleotides that prime a reaction from the driver template can acquire biotinylated ddNTP. This event blocks any further extension and allows the removal of the biotinylated molecules from the reaction by use of streptavidin-coated magnetic beads. Primers that share driver sequences are blocked and removed leaving only primers with unique sequences that can only be found in the tester (Figure 1C). In the presence of Tester23 dsDNA and dNTPs, these oligonucleotides can prime an extension reaction from the fragments unique to the tester (target capture). This step converts the tagged primers into DNA templates suitable for PCR amplification by oligonucleotides containing only the adapter sequences or in combination with T2PCR or T3PCR oligonucleotides. The last step in PEER is a standard PCR amplification with primers containing only adapter and T2PCR/T3PCR sequences that can be used without any molarity restrictions. The final step is expected to generate collection of fragments of different size (Figure 1D).

Table 1. Primers used in the PEER study

Primer sequence 5'–3'	Name	Function
MmeI experiments		
AATGCAGACACAGAAGGTCCATCCGAC	AFMmeI	TESTER MmeI adapter forward
P-GGTCGGATGGACCTTCTGTGTCTGC	ARMmeIP[1]	TESTER MmeI adapter reverse
GCTGCAGACACAGAAGGTCCATCCGACNNNNNN	AFMmeIN6	TESTER 1 cDNA
GCTGCAGACACAGAAGGTCCATCCGACGGG	AFMmeISMART	TESTER 1 cDNA
CAGACACAGAAGGTCCATCCGAC	AMmeIPCR	TESTER 1 cDNA PCR
ACACTAGAGCATGCGTCAAGAGAANNNNNN	T2N6	TESTER 23 cDNA
ACACTCCAGGAGGTCAGAAACAACGGG	T3SMART	TESTER 23 cDNA
ACACTAGAGCATGCGTCAAGAGAA	T2PCR	TESTER 23 cDNA PCR
ACACTCCAGGAGGTCAGAAACAAC	T3PCR	TESTER 23 cDNA PCR
AAGCAGTGGTATCAACGCAGAGTANNNNNN	D0N6	DRIVER cDNA
AAGCAGTGGTATCAACGCAGAGTACGCGGG	D0SMART[1]	DRIVER cDNA
Bio-AAGCAGTGGTATCAACGCAGAGTA	D0bioPCR	DRIVER cDNA PCR
*Bpm*I experiments		
ACACTCGAGGAGGTCTGGAGIIIIIII	PEER1*Bpm*N6	TESTER 1 cDNA
ACACTCGAGGAGGTCTGGAGGG	PEER1*Bpm*G	TESTER 1 cDNA
AACACTCGAGGAGGTCTGGAG	PEER1*Bpm*AF	TESTER *Bpm*I adapter forward
CTCCAGACCTCCTCGAGTGTG	PEER1*Bpm*AR	TESTER *Bpm*I adapter reverse
GAGCTGTGGTGAGTTGGTTGGAAIIIIIII	PEERT7N7	TESTER 78 cDNA
AAGCAGAGGCAGCATTGGAGGG	PEERT8G	TESTER 78 cDNA
AGCTGTGGTGAGTTGGTTGG	PEERT7	TESTER 78 cDNA RCR
AGCAGAGGCAGCATTGGAGG	PEERT8	TESTER 78 cDNA RCR
AAGCAGTGGTATCAACGCAGAGTAIIIIIII	D0N6	DRIVER cDNA
AAGCAGTGGTATCAACGCAGAGTACGCGGG	D0SMART[2]	DRIVER cDNA
AAGCAGTGGTATCAACGCAGAGTA	D0PCRbio	DRIVER cDNA PCR
Control primers		
AATGCAGACACAGAAGGTCCATCCGAC*TAATACGACTCACTATAGGG*	AT7[3]	PEER control primer
AATGCAGACACAGAAGGTCCATCCGACG*AAACAGCTATGACCATGAT*	ASK[3]	PEER control primer

I = 5-nitro indol; N = random base
[1]P indicates that the oligo was phosphorylated to improve ligation
[2]According to the SMART cDNA technology (Clontech, Palo Alto, CA)
[3]These primers are not part of PEER but were used to monitor the success of the protocols' steps using a "control" template

2.2 Discussion

The proof of the PEER concept was tested in preliminary experiments. PEER is intended to find unknown targets at unknown and potentially very low concentrations. This goal was challenged in series of experiments designed to identify the minimum amount of a target DNA present in the background of a complex mixture that could be found and "captured" using oligonucleotides that match the target sequence but were used in very low concentrations. If we could create unique primers from the unknown target DNA itself, we would have the ideally matched oligonucleotides for its amplification. However, their molarity may not exceed that of the template. Since such unfavorable primer concentrations would not promote a PCR reaction, we needed to be able to tag them with a different 5′-sequence that we knew of and that could later be supplied in excess. We synthesized oligonucleotides so that the 20 nucleotides at the 3′-terminal match the template and the remaining 5′-nucleotides cannot be found in the template; these were named "capture primers". We then conducted experiments to determine whether the target template could be amplified after "capture" of the target by only the mismatched portion of the capture oligonucleotides. The results indicated that adapter-primed reactions (i.e. PCR with primers whose sequences did not exist in the original template) yielded amplification products from as little as 0.063 amol of template (1360 copies/ml) and with as little as 4 fmol of capture primers.

To test whether a large number of primers could be successfully, specifically, and completely blocked by di-deoxytermination, we tested a variety of polymerases and a range of nucleotide concentrations using an artificial template (pB6, a fragment of WCV cloned in pTAdvantage vector) and 50 pmol each of SK and T7 generic primers. After multiple PCR rounds of extension and blocking in the presence of ddNTPs, an aliquot of the product was used in regular PCR in the presence of a fresh master mix containing conventional Taq Polymerase (Roche) and regular dNTPs (Roche). The best results, as measured by the absence of product in the reactions to which ddNTPs were added initially, were achieved with Thermo Sequenase [36]. We also observed blocking by Vent (exo-) polymerase and Taq polymerase with a ddNTP: dNTP ratio of 10:1, but Thermo Sequenase remained the enzyme of choice because it generated consistent results under all experimental conditions.

The new PEER method exploits unique target sequences by creating primers from the double-stranded material of interest and then using an intact aliquot of the material as a template for amplification. The first strand of double-stranded DNA (dsDNA) is created using total NA as a template for reverse transcription with primers that have a random hexamer at the 3′ end and the appropriate adapter sequences at the 5′ end, so that these can be used in a subsequent PCR step. The use of an enzyme such as SuperScript II Reverse Transcriptase, a derivative of M-MLV with DNA polymerase activity [37], ensures that single-stranded RNA (ssRNA), DNA, or RNA: DNA hybrids will be copied into cDNA and enter the enrichment process. Once priming sites are generated on both ends of the fragments the product can be exponentially amplified by SMART PCR to generate dsDNA. This approach maintains the correct representation [38] of the NAs entering the protocol and, via the PCR step, supplies a renewable source of the target material. We found that SuperScript II Reverse Transcriptase generated good results for both templates tested, one from a DNA virus (hepatitis B virus – HBV) and one from an RNA virus (WCV). In both cases, we had previous knowledge of the viral titer, and the process of cDNA generation that includes a Smart PCR step did not alter it. The initial primer design (Table 1) included 5-nitro-indol instead of random bases at the 3′-end of the RT primers. The random bases were eventually favored because they provided a higher efficiency of the PCR step; additional experiments, not discussed here, showed that the 5-nitro-indol's higher affinity to itself hinders the reaction performance. The PEER protocol can be modified for use with other pre-dsDNA/cDNA procedures. DNase/RNase treatment, filtration, ultracentrifugation, gradient separation, and other procedures, may be incorporated, depending on the application.

Once generated, the double-stranded material is converted into unique primers by extensive endonuclease cleavage (Figure 1B). This strategy ensures that the primers perfectly match the unknown template. To digest the dsDNA into small fragments (i.e. create multiple "primers" from the unknown target), we tested two approaches. First, we used the unique 2.5-cutter CviJI* (cuts RGCY, and RGCR/YGCY, but not YGCR) to generate a maximum number of small fragments with one treatment [39]. CviJI has proven to be a useful tool for generating probes from low-copy-number DNA sequences by thermal-cycle labeling [40]. A drawback of using this enzyme is the formation of blunt ends, which interfere with the efficient ligation of adapters because large numbers of fragments are available to religate to themselves. This inefficiency was confirmed experimentally. As an alternative to CviJI*, we used a cocktail of four-cutter enzymes (AciI, HpaII, HinP1I, MaeII, and TaqI) that have different recognition sites (CCGC, CCGG, GCGC, ACGT, and TCGA, respectively) but leave GC-5′ overhangs. After the digestion, the DNA fragments were treated with Klenow DNA polymerase in the presence of dCTP to fill in the 5′-overhangs with one nucleotide. This step converts the self-complementary 5′-GC protrusions into 5′-C overhangs that can ligate only to the synthetic adapters as specifically designed. To convert these short DNA fragments into primers that can be recovered and used in the enrichment protocol, they were "tagged" by ligation to adapter sequences (Figure 1B).

In the context of the human genome (3.2 Gb) [41], 18 nt is the absolute minimum length (x) required for the creation of a specific oligonucleotide, calculated by the formula $Nx/4^x <1$, where N is the size of the target. For a large viral genome (e.g. N = 100,000 nt), this minimum length is reduced to 10 bp. However, if the aim is to distinguish a viral genome of that size within the context of the human genome as a background, a minimum size of 18 bp is needed to ensure unique sequence specificity. For viral discovery, the PEER protocol uses MmeI [42] adapters to generate primers from the double-stranded cDNA tester with a 5′-end sequence that is artificially introduced by the adapter and a 3′-end that is derived from the target cDNA sequence (Figure 1B). With minor modifications, the adapter can alternatively incorporate sites for other IIS restriction enzymes, such as BpmI, BsgI, and Eco57I, that cleave 16 nt downstream of their recognition sites, a feature that might be useful for other applications (e.g. SNP discovery). Class IIS endonucleases that leave 5′-protruding ends [34] are not suitable for PEER because they will generate self-blocking primers. The IIS cleavage allows all cDNA fragments of various lengths that have acquired adapters to be trimmed to a uniform length. This generates a population of molecules with 3′-ends that are derived directly from the target of interest and that are suitable for extension reactions at a reasonably narrow temperature range. Such approach represents a novel use of these class IIS enzymes.

The enrichment efficiency of PEER was tested on serum containing HBV with a titer 3×10^8 IU/ml and Vero monkey kidney (VMK) cell culture infected with WCV at an inoculums' size of 10^4 copies. The HBV-containing serum was diluted tenfold four times with normal human serum pool to create testers with different viral loads, i.e. 3×10^7, 3×10^6, 3×10^5, and 3×10^3 IU/ml. dsDNA fragments before the enrichment protocol and the corresponding products after the enrichment were cloned in *Escherichia.coli* libraries. Such paired libraries were generated for all tested serum dilutions. As previously established [1], to find a high titer virus (e.g. 10^8 copies/ml) within a library representing the entire human genome one needs only to search through about 100 of the clones since ~2–3% of this library should contain viral sequences; however, if the viral titer is 10^3, one needs to screen 10^7 clones. To circumvent exhaustive screening by hybridization of libraries generated from serum containing low viral titer we resorted to evaluating the copy number of targets of interest in the dsDNA by PCR and hybridization. The PCR approach, although very reliable when used on the dsDNA material prior to enrichment, cannot be applied to PEER products because they might not consist of fragments that will contain both priming sites. To assess the presence of the targets of interest in the PEER product we did spot hybridizations and, from the corresponding libraries, isolated colonies at random and sequenced them. After PEER, the serum with initial titer of 3×10^6 IU/ml was enriched 5.3×10^2 times for HBV sequences and the one with initial titer 3×10^3 IU/ml was enriched 1.3×10^4 times. The observed higher enrichment values that were obtained for the lower titer library may be explained by the fact that the tester material was being diluted with the driver pool, thus creating a population almost perfectly matched to the driver and hence achieving a greater blocking efficiency. In fact, when enriching the low titer WCV virus with a perfect driver (noninfected VMK cells) the achieved enrichment value was 4.45×10^4.

PEER is a conceptually new approach for the subtractive enrichment of complex nucleic acid mixtures and represents a novel use for both class IIS restriction enzymes [43] and di-deoxytermination. Unlike the other techniques that are based on subtractive hybridization of long DNA molecules that are eventually used as PCR templates, PEER centers on selective blocking of short DNA fragments through hybridization and highly specific enzymatic extension, and the use of these fragments as PCR primers. In addition, because PEER was designed to create normalized starting material that is double stranded, the method does not have preference for DNA or any particular type of RNA. PEER is suitable for use with samples of limited volume and is very cost efficient, especially when compared to new high-throughput sequencing methods. We did not observe loss of integrity of the background DNA, i.e. no recombination or insertion/deletion events. We did find some primer multimers among the clones, but did not quantify them since we filtered the sequence data for background vector or primer noise prior to analysis. PEER should also allow for several rounds of enrichment, as do RDA and SSH, i.e. the final PEER product can be digested again with the GC cutter cocktail, adapter-tagged, cut with *Mme* I and blocked on the same driver or even on an alternative driver, depending on the experimental goals. We have not yet attempted such experiments.

Finally, the PEER method is flexible and can be modified for the discovery of single-nucleotide polymorphisms (SNPs) or minor differences in allele states as well as for other subtraction applications such as pathogen discovery and differential expression of genes. Although descriptions of these other uses of PEER are beyond the scope of this chapter, a review of the PEER protocol should identify steps that may be modified to increase the versatility of the technique. For example, one may use different restriction enzymes separately or in various combinations to fragment the dsDNA. The primers can be created with the use of DNAses or exonucleases after the tagging step to generate fragments with randomly distributed ends suitable for a total comparative SNP analysis of the target NAs.

Our findings demonstrate that PEER is very robust, can be applied to different targets, and can detect NAs of unknown sequence at very low concentrations. In our experiments, PEER outperformed the commercially available SSH technique [44]. The method was recently successfully applied to a variety of viruses representing different genome structures – human herpesvirus and Ectromelia virus (dsDNA), human echovirus, West Nile virus, and human respiratory syncytial virus (ssRNA), Orthoreovirus (polysegmented ds RNA), porcine circovirus (circular ssDNA) and was able to recover viral genetic material from as little as 10^3 pfu [45].

2.3 PEER Protocol

2.3.1 Nucleic acid extraction

Total NA is extracted from 100–200 µl of serum or cell culture using Masterpure complete kit (Epicenter Biotechnologies, Madison, WI) or High Pure viral nucleic acid extraction kit (Roche) and resuspended in 10 µl 10 mM Tris (pH 8–8.5).

2.3.2 Modified SMART protocol

About 5 µl of the extracted NA is reverse transcribed (RT) with SuperScript II (Invitrogen, Carlsbad, CA). Two RT reactions are performed for the tester, one using 10 pmol each primer AFMmeIN6 and AFMmeISMART, and the other using 10 pmol primers T2N6 and T3SMART. Primers D0SMART and D0N6 are used for the driver reaction. Reaction volumes and conditions are described in the SMART cDNA synthesis protocol (Clontech). After synthesis, the enzyme is heat-inactivated and the product diluted with 40 µl of TE.

2.3.3 First PCR amplification

About 10 µl of the RT product is amplified with Advantage 2 Polymerase (Clontech) as recommended in the Smart cDNA protocol and using the corresponding PCR primers (AMmeIPCR for Tester 1, T3PCR, and T2PCR for Tester 23 and D0bioPCR for the driver) in triplicate reactions under the conditions suggested by the manufacturer. The amplification parameters are 95°C/1 min, and (95°C/3 s, 68°C/3 min) × 28 cycles. The dsDNA is purified on a Qiagen PCR purification column (Qiagen, Inc., Valencia, CA) and eluted in 75 µl 10mM Tris (pH 8).

2.3.4 Digestion with restriction endonucleases

About 70 µl of the Tester 1 dsDNA are digested overnight with HpaII (MBI Fermentas Amherst, NY), HinP1I, AciI (NEB Ipswich, MA), MaeII (Roche Molecular Biochemicals, Germany), and TaqI (NEB) using 1 µl of each enzyme and TaqI buffer (NEB) at 37°C. After digestion, the enzymes are heat-inactivated, with the fragments purified through a QIAquick PCR purification kit and eluted in 55 µl 10 mM Tris (pH 8).

2.3.5 Klenow treatment

The ends of the fragments are filled in with Klenow polymerase (Roche) in the presence of dCTP for 1 h at 37°C. The enzyme is then heat-inactivated; the reaction mixture purified with QIAquick nucleotide removal column (Qiagen), and the product eluted in 50µl 10 mM Tris (pH 8).

2.3.6 Adapter ligation

Double-stranded adapters are prepared by mixing the forward (AFMmeI) and reverse (ARMmeIP) adapter primers (Table 1) at equimolar ratio (200 pmol each), heating to 96°C for 5 min and slowly cooling to room temperature. 200 pmol of the adapter is ligated overnight to 45 µl of dCTP-filled-in Tester 1 fragments. The ligation products are purified to remove the T4 ligase and buffer with QIAquick nucleotide removal column and eluted in 55 µl 10 mM Tris (pH 8).

2.3.7 MmeI digestion

The ligation products are digested with 5U MmeI (NEB) for 2 hrs. The cleaved DNA is resolved in 10% polyacrylamide gel, the resulting 50 bp fragment is cut out, isolated from the gel with QIAquick gel extraction kit (Qiagen) and resuspended in 50 µl 10mM Tris (pH 8).

2.3.8 Blocking of MmeI-tagged primers

About 25 μl of the fragment is used as primer with 10 μl Driver bio-dsDNA template in the presence of 2.5 μl each ddNTPs-bio (Biotin-11-ddNTPs, NEN Life Science Products Inc., Boston, MA), 0.025 mM each dNTPs (Roche) and Thermo Sequenase (Amersham Pharmacia Biotech, Inc., Piscataway, NJ). The blocking reaction is carried out as follows: 96°C/3 min; and (95°C/2 s, 55°C/20 s, and 68°C/20 s) × 55 cycles. The product is purified with QIAquick nucleotide removal kit to remove the excess ddNTPs and eluted in 100 μl 10 mM Tris (pH 8).

2.3.9 Removal of biotinylated products

The cleaned product is heated to 95°C, and 50 μl of streptavidin-coated magnetic beads are added (SPHERO Streptavidin Magnetic Particles from Spherotech, Inc., Libertyville, IL). After 10 min incubation at >60°C, the beads are captured on a magnet rack (Qiagen) and the supernatant removed to a fresh tube, ensuring that the temperature remains above 55°C.

2.3.10 Capture reaction

About 50 μl of the supernatant (purified nonblocked primers) are used in a 100 μl capture reaction with 5 μl of the Tester 23 cDNA as template under the following conditions: 95°C/2 min; (95°C/20 s, 45°C/30 s, 72°C/2 min) × 10 cycles; (95°C/20 s, 52°C/30 s, and 72°C/2 min) × 30 cycles; and 72°C/7 min.

2.3.11 Final PCR

About 5 μl of the capture product is amplified in a 100 μl final reaction volume with primers AMmeIPCR and T2PCR, AMmeIPCR, and T3PCR or AMmeIPCR alone under the following conditions: 95°C/2 min, and (94°C/10 s, 60°C/20 s, and 72°C/90 s) × 30 cycles. The product is quantified, cloned, and sequenced.

3. OTHER SUBTRACTION AND HYBRIDIZATION BASED METHODS FOR DIFFERENCE SCREENING

Subtractive hybridization is the core approach behind most techniques described here. Differentiating and excluding/subtracting nucleic acid species by hybridization or by physical comparison can be accomplished in a creative variety of ways.

3.1 Differential Screening

Differential screening [46] enables comparison of two complete mRNA populations by probing a cDNA library generated from one of the samples with labeled fragments representing one of the RNA populations. A summary schematic representation of the approach, also known as plus/minus screening, is shown on Figure 2. This protocol can be very thorough and informative because it theoretically allows the acquisition of a complete picture of all mRNAs involved. In actual experimental settings, however, the method has always proved to be very laborious [47]. It requires the building of a good cDNA library that has an adequate number of clones plated at manageable density and the generation of several lifts from said library, that need to have adequate and approximately equal amount of the clone/plaque material on them, so that the positive–negative hybridization calls after the screening are not biased by clone copy numbers. The hybridization itself needs to be done with limiting concentration of the probe as to distinguish differences between mRNA copy numbers in the compared populations; such conditions can be difficult to achieve. Even though many of the technical drawbacks that come with cDNA library construction are no longer a great technical challenge today, there still are obstacles that cannot be easily overcome. The major disadvantages are: the intensive labor

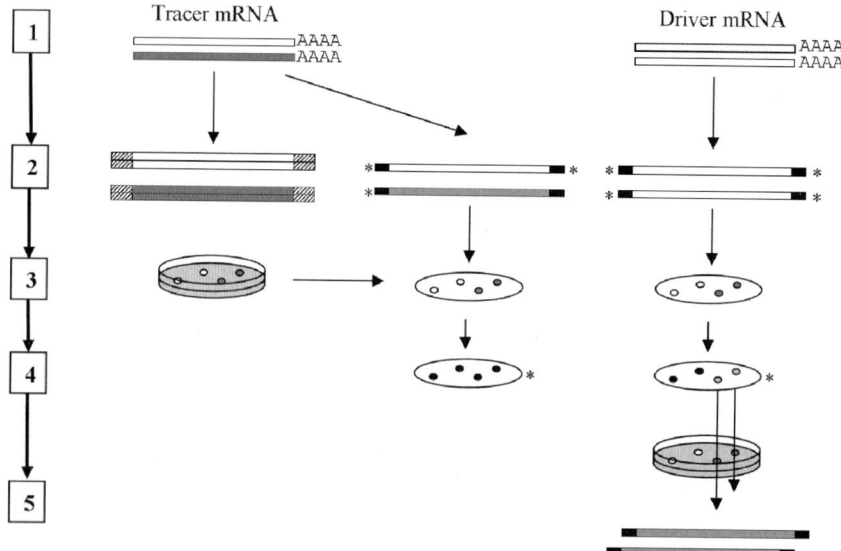

Figure 2. Differential screening or plus/minus screening (DS). 1. Total mRNA from tracer/cancer material (white and gray rectangle) and mRNA from driver/healthy material (white rectangle) are reverse transcribed in the presence of random hexamers and oligo(dT) primer; 2. cDNA is ligated to restriction site linkers (vertical diagonal fill rectangle) and the product is cloned in appropriate λ-expression vector. Aliquots of the mRNA are labeled with P^{32}-ATP (black fill rectangles*) and polynucleotide kinase, or alternatively the corresponding cDNA is synthesized in the presence of P^{32}-dNTPs; 3. Duplicate plaque lifts from the cDNA library are made on nylon membranes; 4. One set of the membranes is probed with the labeled tracer mRNA and the duplicate set is probed with the labeled driver mRNA; 5. Plaques that do not give hybridization signal when probed with the driver mRNA are isolated form the corresponding original plate used to create the lift and subject to secondary screening and further analysis.

Primer extension enrichment reaction

required, poor reproducibility, bias against less abundant mRNA species, and the requirement for multiple controls and secondary screening. Successful uses of this approach include: differential expression studies of polycystic kidney disease [47], TNF-induced genes [48], and early embryonic differentiation [49].

3.2 Subtractive Hybridization

The search for differentially expressed genes [50,46] and the development of subtractive hybridization have gone hand in hand [51]. Subtractive hybridization is any hybridization applied to remove common material that can interfere with the desired outcome of a screen and may be applied to the test sample of interest, to the probe designed to look for targets of interest [52,53,54,55], or to both [56]; it is depicted here in Figure 3. The best enrichment obtained by subtractive

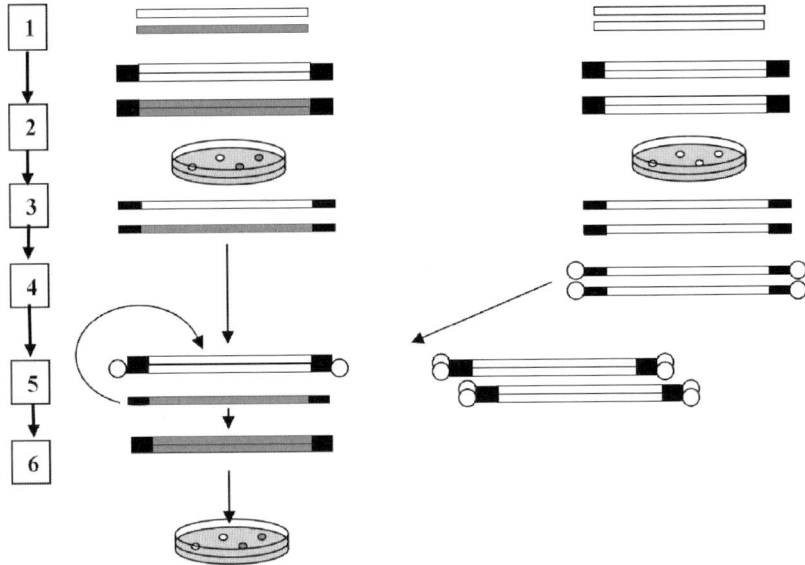

Figure 3. Subtractive Hybridization (SH). 1. Total mRNA from tracer/cancer material (white and gray rectangle) and mRNA from driver/healthy material (white rectangle) are reverse transcribed in the presence of random hexamers and oligo(dT) primer; 2. cDNA is ligated to cloning site linkers and the product is cloned in a λ-bacteriophage vector; 3. Single-stranded phage DNA is recovered by transfection of *Escherichia coli* with in parallel with a helper phage; 4. Driver phage is biotinylated; 5. Tenfold excess of the driver is mixed with the single-stranded tester material and allowed to hybridize; 6. Biotinylated homo and heteroduplex molecules are removed by avidin agarose or streptavidin and the remaining subtracted ssDNA is taken into another round of subtraction with fresh driver; or converted to dsDNA with Klenow enzyme and cloned (subtractive cloning) or labeled and used as positive hybridization probe on libraries containing the target of interest; or amplified by PCR and subsequently used in the same applications.

hybridization followed by amplification and cloning has been between 100- and 1000-fold, depending on the target [55,20]; the kinetics of the process have been described previously [57,58]. An interesting example of recent work that uses straightforward subtractive hybridization is the isolation of all messages involved in the utilization of complex polysaccharides from *Aspergillus nidulans* [59]. The approach used was to collect mRNA from fungi grown on variety of polysaccharide sources different from glucose, and to pool and clone those together without any other selective enrichment or PCR. The resulting library was then probed by negative subtractive hybridization with labeled mRNA from *Aspergillus* grown in the presence of glucose only, which resulted in the isolation of over 3000 negative clones that generated more than 2000 unique contigs. To confirm their function, the clones were arrayed and probed with cDNAs from fungi grown under different sugar conditions. Subtractive hybridization has been applied with success to uncover tumor suppressor genes [60]; to find moderately induced sequences from humans [61] and mycobacteria [62], especially when coupled with PCR; and to study monoallelic imprinted genes [63] and organ-specific gene expression [64].

3.3 Subtractive Cloning

Subtractive cloning takes the differential screening approach a step further by constructing cDNA libraries that are already enriched in differentially expressed sequences [50,46,65]. It is achieved by generating single-stranded cDNA from the material of interest and hybridizing it to excess mRNA from another cell type. The double-stranded hybrids are the genes expressed in common and the single-stranded mRNAs represent the ones unique to the cell of interest that can be further cloned and validated. This approach has the revolutionary advantage of removing the background of common genes that otherwise interferes with the desired outcome. The main disadvantages are the requirement for large quantities of mRNA (only 5% of the general RNA population) and inefficient recovery of the single-stranded material, which thus limits the availability of cDNA for cloning. The approach has been used successfully [47] to isolate a new member of the Ras super family, the T-cell antigen receptor, the murine IL-4 receptor, etc. The subtractive cloning has also seen a great benefit from coupling with methods that take advantage of poly-A tailing of most mRNAs [66], PCR [67,68], secondary hybridization [69] or a combination of these [70,71,72].

3.4 Differential Display

DD methods are not true subtraction methods but can nonetheless play a great role when choosing an approach to find unique sequence characteristics. Many of the methods described below can be used in combination with subtraction technique.

Primer extension enrichment reaction

The mRNA DD [8] or arbitrarily primed PCR [73] came about as an alternative to the differential screening. It takes advantage of the power and specificity of PCR and does not require *a priori* construction of libraries. In its original design, the method was intended to display only a subset (1/12) of the original mRNA population (Figure 4). However, it retains great flexibility as to the kind of arbitrary or specific oligonucleotides used (gene-specific, gene-family specific, anchor-polyA-tail, etc.), as these can be employed in variety of combinations to create subsets of fragments that are unbiased and suitable for screening. Its advantages are the use of amplification, the ability to query with different sets of oligonucleotides, and greater reproducibility and opportunity to compare more than two specimens of interest at a time. However, regardless of the size of the subpopulation displayed, the background can mask true differences; rare transcripts can be missed in the PCR step; multiple primer sets may be required to cover adequately the entire mRNA population, and the results usually reflect uniquely induced rather than up-regulated or down-regulated

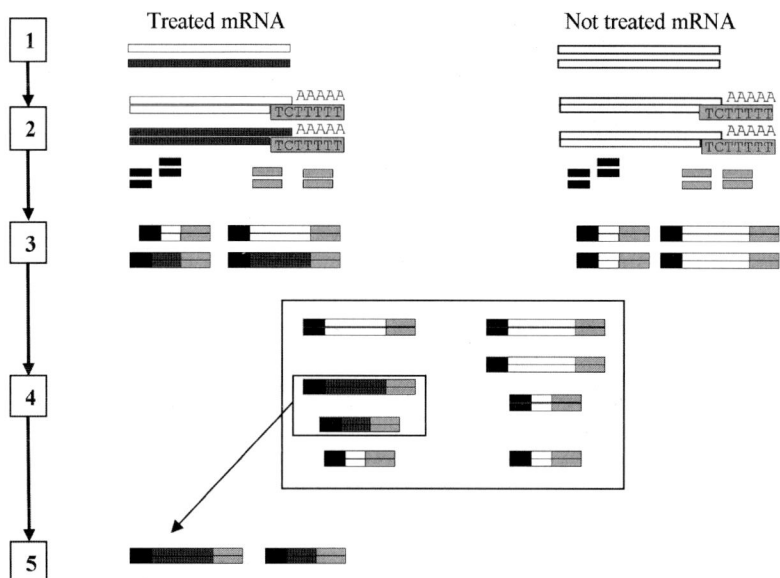

Figure 4. Differential display (DD). Generation of representations from mRNA. 1. mRNA from treated material (white and gray rectangle) and mRNA from untreated material (white rectangle) are reverse transcribed with 5'-T_{11}CA to allow anchored annealing to approximately one twelfth of the entire mRNA population; 2. cDNA is the PCR amplified with primers T_{11}CA and a 10 mer Ltk3 (black rectangles) and labeled with radionuclide (α-^{35}S) to allow visualization on polyacrylamide gel by autoradiography; 3. Fragments bellow 500 bp in size are resolved on a DNA sequencing gel; 4. Resolved fragments display the differences between the compared (could be more than two) specimens and after thorough screening the different bands are isolated form the gel; 5. Isolated fragments are reamplified, cloned, and analyzed.

gene expression. Attempts to circumvent these disadvantages include primer redesign, increasing the cDNA concentration, increasing the 5′-primer:3′-anchor ratio, and devising strategies for systematic rather than arbitrary display [74,75,76,77,9,78]. An interesting attempt to improve this method was the coapplication of the subtraction approach, generating the differential subtraction display [79] from which cDNA eluted from down-regulated gel band is amplified, biotinylated, and used in excess as "driver cDNA", and cDNA form an up-regulated band is amplified and used as "tester cDNA" without biotinylation. The two amplicons were allowed to hybridize, after which the hybrids were removed by streptavidin and the residual cDNA cloned.

3.5 AFLP, SAGE/CAGE, GSTs, and DARFA

Multiple approaches to the display of genetic differences have been described in addition to DD. All approaches are intended to provide complete and unbiased fingerprints of specific transcriptomes or developmental stages. Highlighted here are the more widely used techniques that have become the methods of choice for various targets [80,81].

A novel DNA fingerprinting technique displaying the amplification fragment length polymorphism (AFLP) of total digests of genomic DNAs [82] was based on the restriction fragment length polymorphism technique (RFLP) [83], supplemented and enhanced with PCR with specific primers that provided the ability to amplify only selected subsets of the genome. AFLP has proven indispensable for microbial typing [84,85].

A display method that directly addresses the specific fingerprint of gene expression at the sequence level is serial analysis of gene expression (SAGE) [86]. It is founded on two principles: (1) a short sequence tag contains adequate information to uniquely identify a transcript if it is isolated from a defined position of this transcript, and (2) concatenation of such tags in a single clone will provide instant, rich, and manageable sequence information about the particular transcriptome (Figure 5). The method has shown very good reproducibility ($R^2 = 0.96$) [81] and has found wide application [87,88,89]. A drawback of this method, however, is the fact that frequently (in about 25–30% of the instances) the generated tags are not long enough to be unambiguously assigned to a transcript. SAGE has also been modified to query the 5′-end of transcripts along with the corresponding promoter elements to generate a technique named CAGE [90] and to provide another method for identification and quantitative analysis of genomic DNAs called genomic signature tags (GSTs) [91]. The principle of SAGE has been combined with subtractive hybridization and immunoprecipitation in another recently developed technique called serial analysis of binding elements (SABE), targeted to study gene regulation in humans [92].

One of the newest additions to the DD technologies, called differential analysis of restriction fragments amplification (DARFA) [85], is amenable to both complete transcriptome analysis and DNA fingerprinting. The technique is open,

Primer extension enrichment reaction

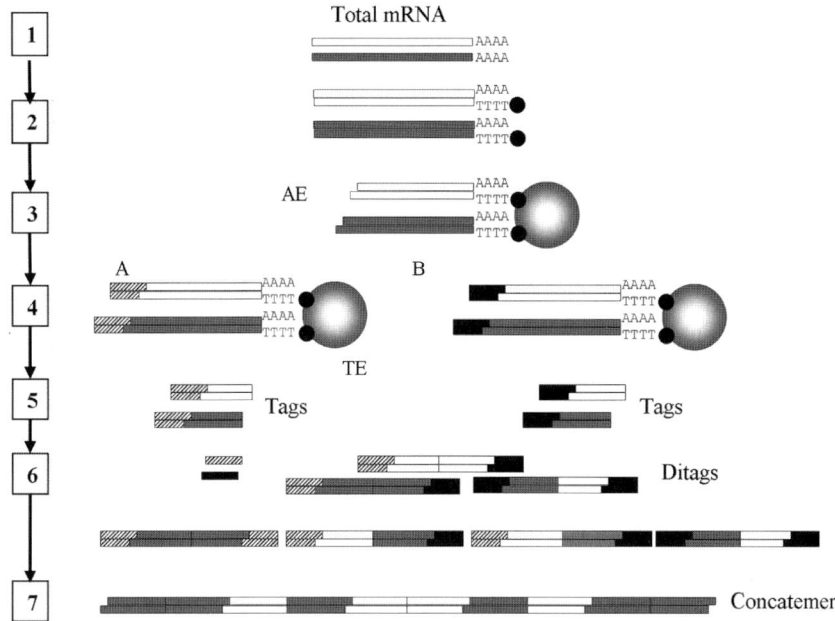

Figure 5. Serial analysis of gene expression (SAGE). 1. Total mRNA from the target material (white and gray rectangle) is reverse transcribed with 5′-bio-oligo(dT) primer (TTTT black-fill circle); 2. cDNA is cleaved with a four-cutter restriction endonuclease-anchoring enzyme (AE), which is expected to cut most transcripts more than once; 3. Fragments are bound to streptavidin beads and divided in two aliquots; 4. Fragments are ligated to different adapters (diagonal stripe rectangles – A, black rectangles – B) containing a site for an IIS endonuclease and then digested with this IIS-tagging enzyme (TE); 5. Released tags are ligated to create ditags; 6. Ditags are PCR amplified with primers A and B and the primers are removed by cleavage with the AE enzyme; 7. Fragments are ligated into concatemers, cloned, and analyzed.

i.e. has no requirement for prior sequence information and claims to be able to display the entire transcriptome due to the characteristic of Hpy188III to generate 120 subpopulations based on the 2 nt 5′-overhang sequence combinations (Figure 6). This approach is intended to ensure that every one restriction fragment will acquire adapter and thus PCR sites and a 4 bp combination identifier of the subpopulation, but the method is no less cumbersome than other display approaches and requires large amounts of starting material. The starting material may also be subject to a preliminary PCR step to provide adequate amount for the downstream procedures, which may diminish the accuracy of the method's representation.

3.6 Representational Differences Analysis

Representation differences analysis (RDA) builds successfully on the subtractive hybridization approach and was designed to find small differences between the sequences of two DNA populations [2,93]. The approach employs a combination

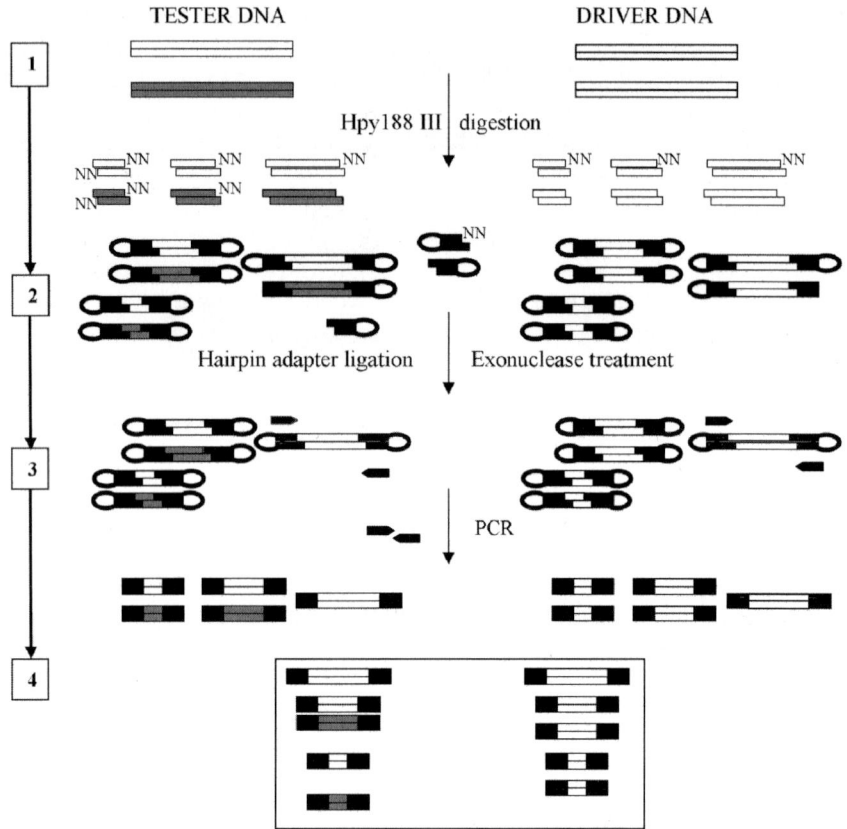

Figure 6. Differential analysis of restriction fragments amplification (DARFA). 1. Generation representations from total DNA. Tester (white and gray rectangle) and driver (white rectangle) DNA or cDNA are digested with Hpy188III restriction endonuclease (TCNNGA) to reduce the complexity of the starting material and create a pool of 120 subpopulations differing by their two nucleotide 5'-overhang sequences; 2. Fragments are ligated to hairpin adapters (black staggered rectangles with loop); 3. Products are treated with exonucleases to remove the unligated fragments and hairpin adapters and amplified by PCR with primers derived from the first 18–3'-nucleotides of the adapters (black-filled directional rectangle); 4. Amplicons are separated on 5% denaturing acrylamide gel and the difference bands isolated for further analysis.

of subtractive and kinetic enrichment of PCR amplicons. DNA "representations" are created by cleavage of DNA by a fairly infrequent restriction endonuclease, ligation to oligonucleotide adapters, and PCR amplification. This generates a representative subset of the genome whose complexity is reduced by 10–50 times depending on the endonuclease of choice. The adapters of the tester representation product are removed by cleavage and replaced by a different new set of adapters. The tester then is allowed to hybridize to the driver (subtraction step, where the tester/driver heterodimers will represent the identical sequences),

Primer extension enrichment reaction

and the ends are filled in. The resulting material is not subject to physical separation but amplified with the second-adapter oligo alone (Figure 7). This represents the kinetic enrichment step where tester sequences gain amplification advantage because they alone have two priming sites and thus will be amplified exponentially. The approach has great merit and its best applications are the generation of restriction fragment polymorphism probes and difference cloning. Many variations to improve the method's range have been suggested, including

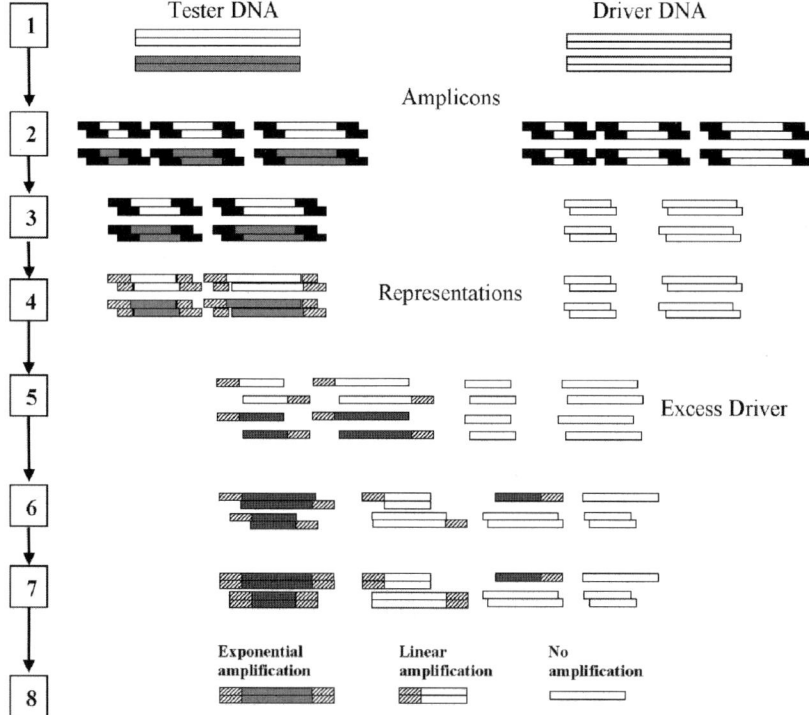

Figure 7. Representational differences analysis (RDA). Generation representations from total DNA: 1. Tester DNA (white and gray rectangle) in two and driver DNA (white rectangle) are digested with restriction endonuclease (Bam HI, Bgl II and Hind III) to reduce the complexity of the starting material; 2. Fragments are ligated to Adapters (black rectangles); 3. Products are amplified for 20 cycles by PCR with adapter primers thus creating representation of molecules smaller than 1 kb; 4. All PCR adapters are removed by cleavage and the tester is ligated to a different set of dephosphorylated adapters (upward diagonal fill rectangles). Subtraction and Kinetic Enrichment; 5. The tester representation is mixed with excess driver and denatured; 6. Mixture is allowed to rehybridize and if the amount of driver is adequate the only hybrid molecules with adapters on both strands will be the ones that are specific to the tester (subtraction); 7. Hybrids are treated with Taq polymerase in the presence of all dNTPs to fill in the 3'-ends; 8. PCR amplification with the second set adapter primers generates exponential amplification only form templates with priming sites on both strands (kinetic enrichment). Products may be additionally treated with mung bean nuclease to remove the single-stranded material and reamplified.

a variety of restriction endonucleases and different adapters to make it usable for the comparison of more that two genomes from environmental samples [94]. Rapid subtraction hybridization (RaSH) was introduced [95] to provide a streamlined RDA-based cloning. RDA has been used successfully not only on DNA but also on cDNA representations [96,97,98], and applications such as the discovery of GBV-A and GBV-B [99,100], and TTV [32], cloning of apoptosis-related genes [101], iron-regulated gene expression in bacteria [3,102], identification of new tumor suppressor genes [103], changes of expression in malignant formations [104] and identification of developmentally regulated genes [105]. RDA served as a base concept for a creative alternative PCR subtraction technique termed ligation mediated enrichment (Limes) [106]. The later was specifically designed to address the problem of genetic backgrounds that have high repeats content and uses Taq DNA ligase to join only perfectly matched ends, thus creating amplifiable templates only from perfect hybrids.

3.7 SPAD–RDA

Selectively primed adaptive driver (SPAD)–RDA is a fairly novel adaptation of the subtractive hybridization and differs from RDA by an alternative approach to the generation of the tester and the driver and by the optimization of the driver material in the course of subtraction [4] (Figure 8).This improvement potentially circumvents the recognized drawbacks of any PCR-based subtraction, that is if the complexity of the starting material is at too high or the target at too low a concentration, the enrichment will be ineffective. In essence, SPAD–RDA combines RDA [2] with the use of selective primers, as for AFLP [82], and driver-control subtraction, as for SABRE [6]. The approach leads to improved recovery of viral sequences on a greatly reduced background but does not address the low-copy-number problem.

3.8 Enzymatic Degradation Subtractions (EDS, LCS, DSC, NSC, UDG/USA, and CODE)

Many adaptations of subtraction hybridization (SH) PCR approaches have been proposed in attempt to circumvent one of SH's most problematic areas – the background created by incomplete driver/tester hybridization. One large category of adaptations consists of methods that employ one or multiple enzymatic degradation steps to disable the unwanted templates.

The enzymatic degrading subtraction (EDS) [107] is one of the earliest proposed alternatives for the construction of subtractive libraries from PCR-amplified cDNA. With EDS, the tester DNA is blocked by thionucleotide incorporation that renders it resistant to exonucleases III treatment; the rate of tester/driver hybridization is accelerated by phenol–emulsion reassociation [108], and the driver cDNA and hybrid molecules are enzymaticaly removed by digestion with exonucleases III and VII rather than by physical partitioning.

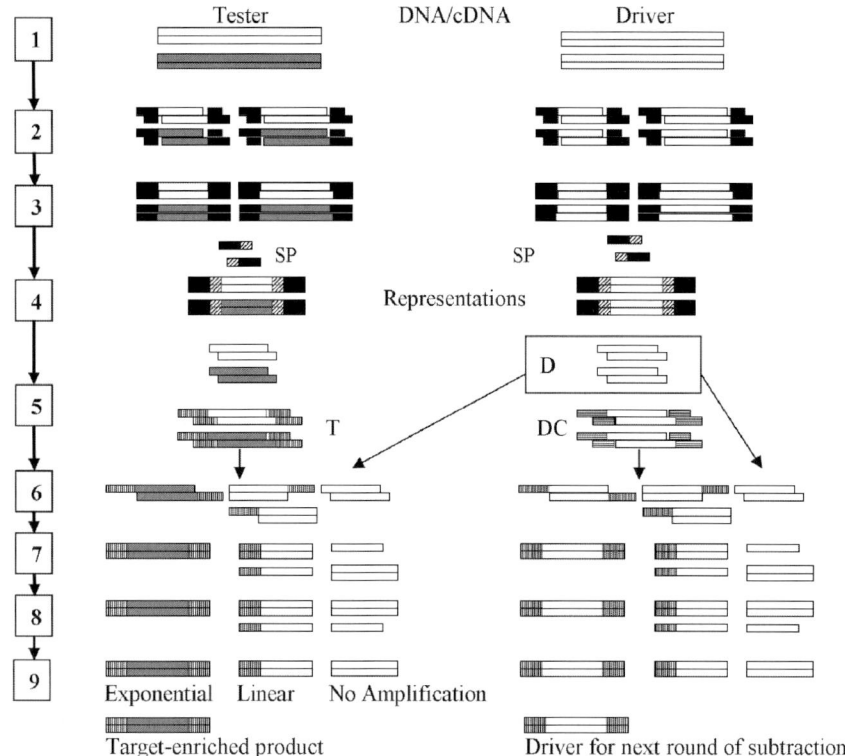

Figure 8. Selectively primed adaptive driver–RDA (SPAD–RDA). 1. Tester DNA (white and gray rectangle) and driver DNA (white rectangle) are digested with restriction endonuclease SauIIIA to reduce the complexity of the starting material; 2. Fragments are ligated to 5′-dephosphorylated adapters (black rectangles) and after column purification the ends are filled in; 3. Products are amplified by PCR with selective primers (SP) to create representations; 4. SP sequences are removed by cleavage; 5. Tester (T) is ligated to a different set of dephosphorylated adapters (vertical fill rectangles) and so is a portion of the driver to create a driver control (DC); 6. T and excess adapter-free driver (D) are mixed, denatured, and reannealed (subtraction); the same is done in parallel with a D/DC mix (driver control subtraction); 7. 3′-ends of hybrids are filled in; 8. PCR amplification with the second set adapter primers generates exponential amplification only from templates with priming sites on both strands (kinetic enrichment); 9. Product is treated with mung bean nuclease to remove the single-stranded material and reamplified.

The utility of EDS has been demonstrated by constructing a subtractive library enriched for cDNAs differentially expressed in adult rat brains [107].

Shortly after EDS, a new modification of SH was introduced [109], which could potentially achieve the cloning of differentially expressed genes by linker capture subtraction (LCS) where the tester and driver are digested, ligated to a linker, and amplified. The linker is removed from the driver and after hybridization the products are subjected to mung bean nuclease action to remove ssDNA (Figure 9).

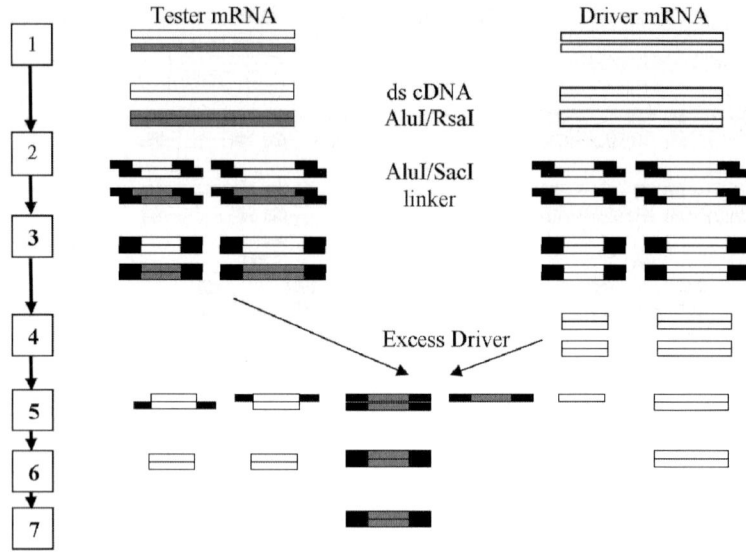

Figure 9. Linker capture subtraction (LSC). Panel A. 1. Tester mRNA (white and gray rectangle) and driver mRNA (white rectangle) are converted to double-stranded cDNA and digested with restriction endonucleases AluI and RsaII; 2. Fragments are ligated to linkers that contain AluI/SacI restriction site (black staggered rectangles); 3. Products are amplified for 20 cycles by PCR with adapter primers derived from the top strand of the linkers; 4. Linkers removed from the driver PCR products by cleavage; 5. Tester product is mixed with excess driver and denatured. Mixture is allowed to rehybridize and if the amount of driver is adequate the only hybrid molecules with adapters on both strands will be the ones that are specific to the tester (subtraction); 6. Hybrids are treated with mung bean nuclease to remove all ssDNA; 7. PCR amplification with the same linker primers generates exponential amplification only form templates with priming sites on both strands. Product may reenter the process at step 5 for additional rounds of enrichment.

The subtraction step can be repeated several times and, unlike RDA, where the enrichment is mostly kinetic at the final PCR step, the target is selected by the specifically preserved priming sites. This idea is appealing as it seems to simplify the RDA approach. Like RDA, LCS takes advantage of the improved hybridization kinetics of nucleic acid mixtures with reduced complexity with each round of subtraction providing better levels of subtraction.

DSC is another approach that is very similar to LSC [5] and is also based on the alternative "negative" amplification strategy. The main principle is that if tester sequences have counterparts in the driver, these can be rendered unamplifiable to leave the desired unique sequences available for amplification. This availability is achieved by digestion of separate driver and tester pools with a restriction endonuclease followed by ligation of different adapters to provide unique PCR priming sites. The adapter-tagged fragments are then amplified and the adapter sequences are removed from the driver by digestion. The products

Primer extension enrichment reaction

are then left to hybridize and the resulting single-stranded ends are digested away with mung bean nuclease from the tester molecules that have found a homologue in the driver (Figure 10). The result is the conversion of the tester sequences that have counterparts in the driver to new driver molecules with the tester population depleted and the driver population enriched and available for further rounds of subtraction with an exponentially increasing amount of seemingly appropriate driver.

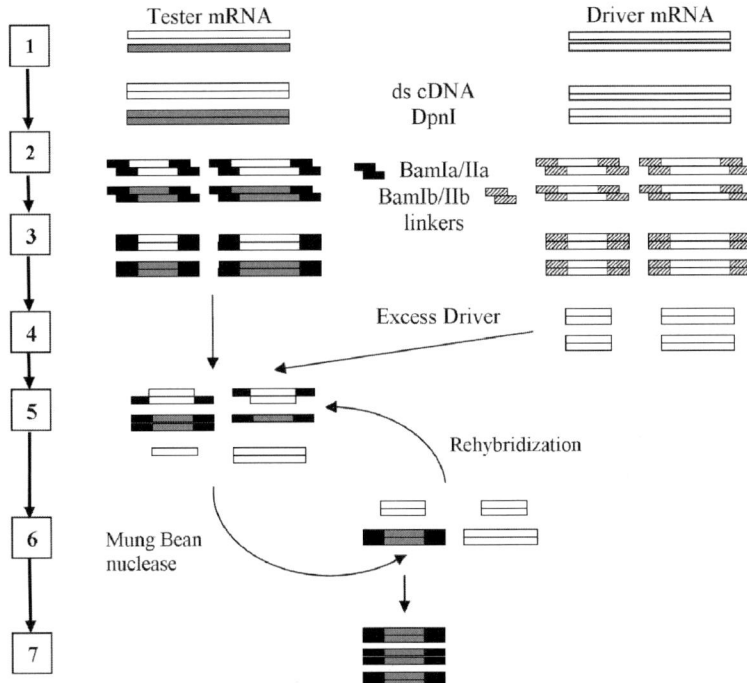

Figure 10. Differential subtraction chain (DSC). Panel A: 1. Tester mRNA (white and gray rectangle) and driver mRNA (white rectangle) are converted to double-stranded cDNA and digested with restriction endonuclease DpnI; 2. Fragments are ligated to two separate sets of primer/adapters (black staggered rectangles and diagonal fill staggered rectangles) that contain BamHI restriction site; 3. Products are amplified by PCR with adapter primers derived from the top strand of the adapters; 4. linkers removed from the driver PCR products by cleavage; 5. Tester product is mixed with excess driver and denatured. Mixture is allowed to hybridize and if the amount of driver is adequate the only hybrid molecules with adapters on both strands will be the ones that are specific to the tester (subtraction); 6. Hybrids are treated with mung bean nuclease to remove all ss DNA; this also removes all single-stranded adapter sequences found in tested/driver heterohybrids homologous molecules, thus creating additional driver fragments that can reenter the enrichment step (rehybridization); 7. PCR amplification with the same linker primers generates exponential amplification only form templates with priming sites on both strands. Product could re-enter the process at step 5 for additional rounds of enrichment.

Potential drawbacks of this approach is the possibility that the sequence of interest might contain insertions or deletions that will allow the formation of a hybrid and from this point on such sequence can be lost for the enrichment process. Another problem might be the "destructive" nature of the approach itself, since rare products may not reanneal easily and thus if they remain in single-stranded form, will be removed from further selection. The bias against single-stranded species is a common shortfall to all subtraction approaches. Still, the method is quite insightful and has been used successfully in cancer studies for the identification of a several differentially expressed sequences [110,111].

Cloning of deleted sequences (CODE) is yet another alternative of the classic subtraction method, influenced by DSC and designed to address cloning of deleted sequences, i.e. of sequences present in the tester but no longer in the driver [112]. This procedure combines (1) the use of restriction endonucleases to reduce the complexity of the sample; (2) utilization of dUTPs and Uracil deglycosilase (UDG) to remove unwanted driver DNA after the hybridization step; and (3) a biotinylated primer to rescue the fragments of interest.

A recent review on SH [113] that briefly highlights all major current subtraction strategies also gives a good account of how a combination of UDG and single-strand-specific nuclease treatment can eliminate unwanted tester/driver and driver/driver hybrids (Figure 11).

Another addition to the approaches aimed at reducing or eliminating subtraction background created by cross-hybridization is a technique called mispaired DNA rejection (MDR) [114]. MDR also takes advantage of the abilities of common mismatch repair enzymes (i.e. mung bean nuclease is able to attack single-stranded loops in double-stranded structures and the Surveyor nuclease recognizes and cleaves mispaired structures within DNA duplexes). The method is elegantly designed to test the ability of the treatment to remove from the final subtraction products such as repetitive clones and chimera clones. A very useful application of MDR is also the recovery of highly conserved sequences between different genomes.

The DSC method has evolved into negative subtraction chain NSC [115], another SH-based approach that takes advantage of DSC idea and employs improved adapters to remove the background after a round of classic SH. Another recent modification of SH, coupled with template switching universal long PCR [116] and single-strand deletion has been proposed as a quick method to identify unknown viral agents [117].

3.9 Suppression Subtraction Hybridization

Subtractive hybridization methods have proven valuable for the identification of significant genes from multiple systems representing the growth and differentiation of cells [118,119,46,120]. Most of the methods are quite tedious and involve complicated protocols that require great accuracy, attention to detail, and often

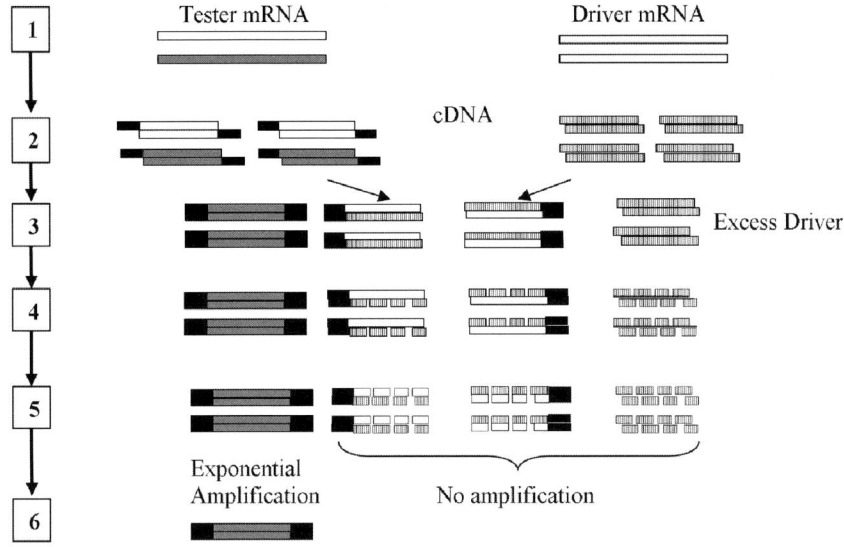

Figure 11. Uracil–DNA subtraction assay (USA). 1. Tester cDNA (white and gray rectangle) is ligated to 5′-adapters (black rectangles) and driver cDNA (vertical stripe rectangle) is synthesized from mRNA (white rectangles) in the presence of Uracil; 2. Tester is mixed with excess of the driver, the mixture melted, and allowed to reanneal; 3. Hybrids are treated with Klenow and then with uracil–DNA glycosidase to generate nicks in all molecules containing driver fragments; 4. Hybrid mixture is then treated by singe-strand-specific nuclease digestion that generates short unamplifiable fragments from the driver homohybrids and the tester–driver heterohybrids; 5. Product is then amplified by the tester adapter primers to generate tester-specific amplicons; 6. The first difference product could be used in the same scheme for another round of subtraction, back in step 3 to generate a second difference product.

substantial amounts of starting material to ensure that no rare molecular species of importance will be omitted. One of SSH's central features is the utilization of PCR suppression [121]. PCR suppression is based on the premise that if long inverted terminal repeats are attached to the ends of DNA fragments, they will form stable panhandle structures at the end of each denaturation and annealing cycle; in the presence of PCR primers that have the sequence of these repeats, no amplification will occur. SSH has another feature of the method is that it normalizes or equalizes the sequence abundance in the target cDNA population by splitting the target tester material in two aliquots and ligating them to different sets of adapters with inverted repeats. The normalization occurs as the more abundant species readily reassociate with tester or homologous driver sequences; the rare species remain single stranded. The two testers are then hybridized to excess driver and then mixed together and extra denatured driver added in a second hybridization step. This creates tester heterohybrids from the difference material that are the only molecules available for PCR (Figure 12). SSH is capable of enriching the target ~1000-fold after

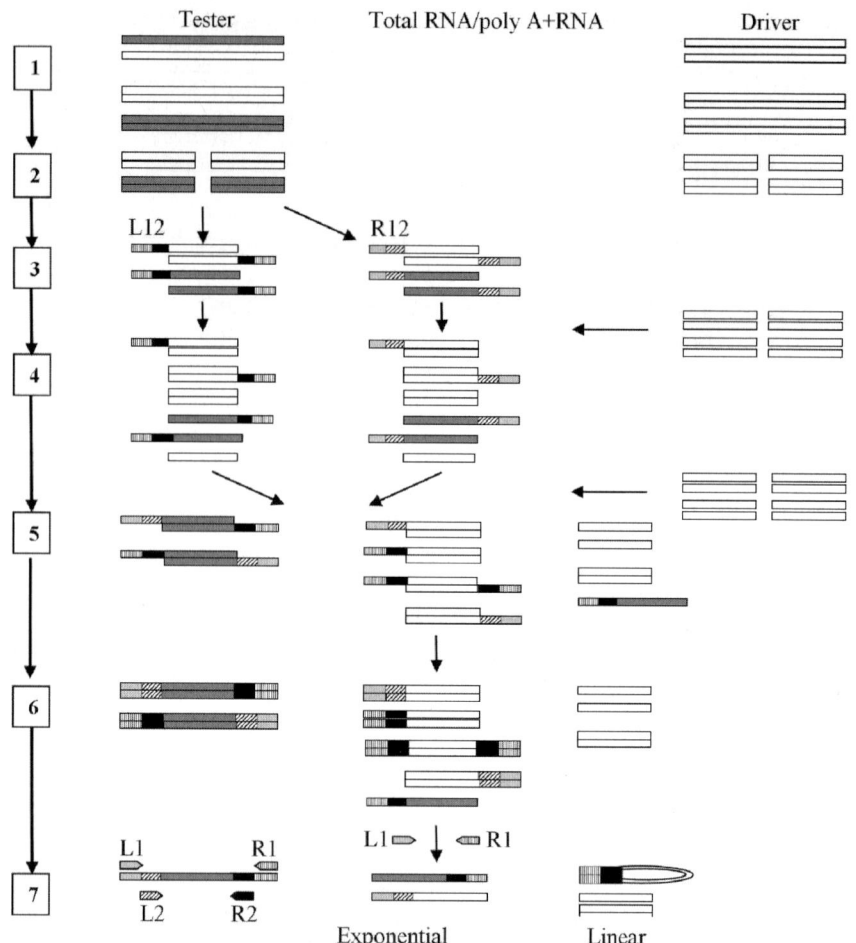

Figure 12. Suppression subtraction hybridization (SSH). 1. Total RNA or PolyA + RNA from the tester (white and gray rectangle) and the driver (white rectangle) are converted independently to double-stranded cDNA by conventional or SMART RT–PCR; 2. All cDNAs are digested with RsaI; 3. Tester is divided in two aliquots and each is ligated to different set of nonphosphorylated adapters (L12 and R12) 4. First hybridization step. Both tester aliquots are subtracted independently with excess of the driver; 5. Second hybridization step. Products from the first step are mixed and without further denaturation are supplemented with fresh denatured driver and the mixture is allowed to rehybridize; 6. Ends of the hybrids are filled in the 3′-ends; 7. Product is amplified by two rounds of PCR with nested primers L1/R1 followed by L2/R2.

only one round of subtraction [122,123,124,125,126]. SSH is an improvement on the subtraction approaches [7] and has found a wide application [127,128,129,130]. SSH is a clean technique with truly good performance; however, SSH-generated libraries contain some background clones that do not represent differentially expressed transcripts. SSH is not ideally suited when dealing with transcripts that

Primer extension enrichment reaction 153

are only moderately (2–4-folds) enhanced, and generally requires a perfect driver match for optimal performance. SSH could miss unique sequences that are present at very low copy numbers and have remained single stranded and thus had not acquired adapters. The problem posed by false positive clones has been tackled by the mirror orientation selection (MOS) approach [131], which deals with redundant molecules that have evaded the hybridization subtraction step and have remained in the process. The rationale behind MOS is that such molecules have only one orientation relative to the adapters, while each genuine product of the enrichment process has been generated by both orientations of the molecule; hence, if one of the adapters is removed enzymatically, the subtracted product is allowed to denature and reanneal again, so the background molecules will remain single stranded with only one priming site which can allow only their linear amplification. SSH has served as a basis for ligation-mediated suppression PCR [132,133,134] that has been used successfully for genome walking into unknown sequence areas. SSH or the combination of SSH and MOS have been used successfully for detection of differentially expressed genes in many systems [135,136], including work identifying diversity in an environmental genome [112], early mammalian embryonic development studies where the starting material can be very scarce and some transcripts extremely rare [25,33]; disease studies [26,137,138,139,140,29,23,141,142,143]; pharmacogenomics [144]; infectious agents studies [84,122]; and virus–host interaction studies [145,146,147,22,148].

3.10 Selective Amplification Via Biotin and Restriction-Mediated Enrichment

SABRE is an approach that uses the selective enrichment principle of RDA, combines it with the use of biotin–streptavidin affinity and restriction enzyme site reconstitution to achieve purification of the desired tester homohybrid population [6]. It was designed to take advantage of restriction-mediated reduction of the complexity of the starting material and biotin- and restriction-mediated recovery of tester homohybrids. The protocol demonstrates the ability to detect moderately rare (representing ~0.03% of the total) mRNA species and 2–10-fold elevation in their expression levels (Figure 13). In addition to the tester/driver subtraction, the driver cDNA is subtracted in a parallel control experiment with another batch of the driver DNA amplified with tested adapters. The product of this subtraction control reaction is then used for a second round of subtraction of the tester material to ensure that any differences that may arise form PCR irregularities due to the adapter do not contribute to a background accumulation of false positives. This thoughtful approach uses virtually identical primers for the generation of the tester and the driver, thereby ensuring that both representations will be comparable, allows for multiple subtraction rounds, and, using two separate elements (both streptavidin capture and restriction], ensures the selection of tester-derived molecules [149]. A possible drawback of this method is bias against low-copy tester material that might remain single

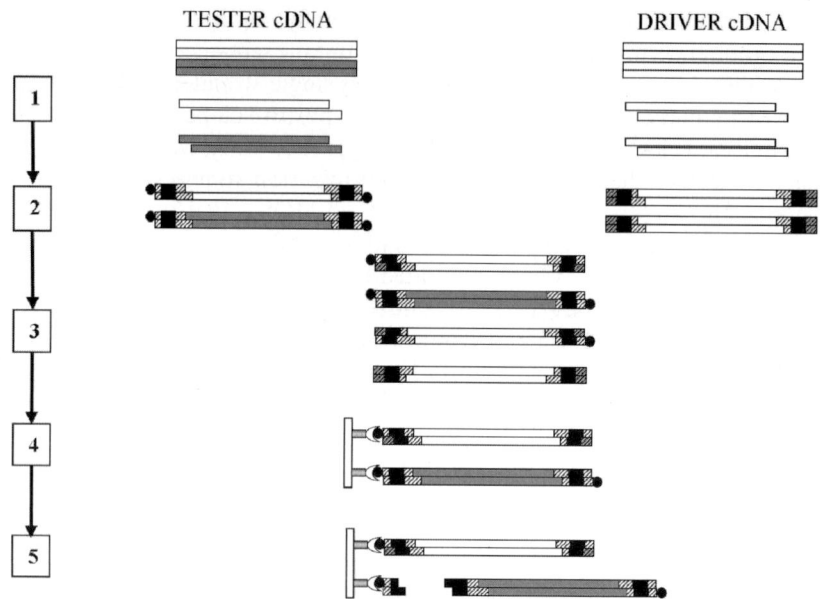

Figure 13. Selective amplification via biotin and restriction-mediated enrichment (SABRE). 1. Tester (white and gray rectangle) and Driver (white rectangle) double-stranded cDNA are digested exhaustively with restriction endonuclease MboI; 2. MboI-linker-adapters are ligated to both tester and driver. Tester adapters are biotinylated and have a functional BamHI site; the driver adapters are not biotinylated and lack the correct BamHI site. Tester and driver may be are amplified with virtually identical PCR oligonucleotides designed to anneal; 3. Products are mixed, denatured, and hybridized in the presence of 30-fold excess driver by phenol–salt emulsion, followed by digestion with S1 nuclease to remove single-stranded moieties; 4. Resulting products are captured by streptavidin-coated paramagnetic beads and the tester homohybrids are release specifically by BamHI digestion; Steps 2–5 could be repeated to amplify the differences between the compared molecules.

stranded due to incomplete hybridization. Moreover, some species might be lost due to preferential amplification in the PCR step that supplies the starting material.

3.11 DNA Enrichment by Allele-Specific Hybridization

DNA enrichment by allele-specific hybridization (DEASH) is an interesting and important approach to subtraction application that detects small differences that are not the result of differential gene or the presence of extraneous genetic material is [150]. Such small differences might arise from low-frequency base substitutions, haplotypes, SNPs, sequence variants, and recombinant molecules that represent rare mutations or pathological recombination events. dsDNA that contains different alleles is mixed with biotinylated allele-specific oligonucleotide directed to a chosen variant and a nonbiotinylated competitor complementary to the other allele (Figure 14). The specificity of the hybridization

Primer extension enrichment reaction   155

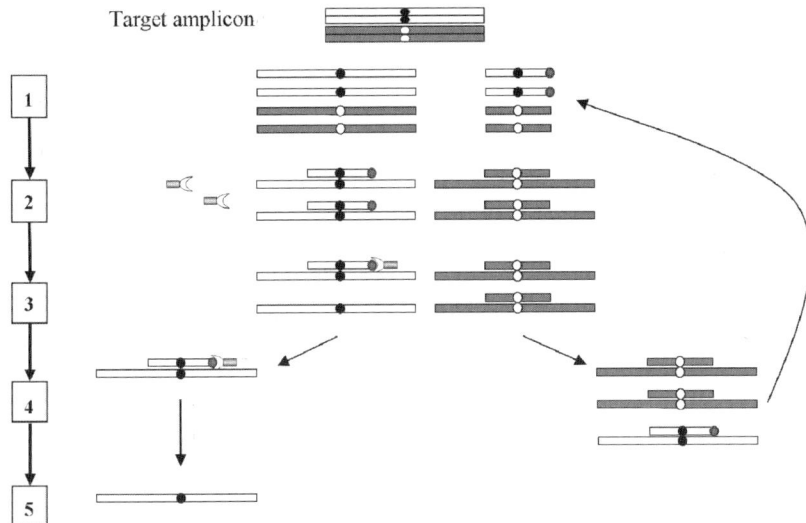

Figure 14. DNA Enrichment by allele-specific hybridization (DEASH). 1. Target amplicon (DNA, cDNA, or PCR representation in white and gray rectangle) that contains mixture of alleles differing by a chosen base substitution (white circle and black circle) is mixed with biotinylated allele-specific oligonucleotide (bio-ASO, white rectangle with black and grey circle) and the corresponding competitor ASO (gray rectangle with white circle) and denatured; 2. Streptavidin-coated magnetic beads (white crescent with gray bar) are added to the mix to retrieve the target amplicons; 3. Magnetic beads are recovered form the mixture with a magnet; 4. The biotin-captured amplicons/alleles are eluted from the beads; 5. Remaining nucleic acid mix can enter another round of enrichment with bio-ASO.

might in so doing, be improved. The hybrids are captured on streptavidin-coated paramagnetic beads and then thermally eluted in low salt buffer. The protocol can be repeated for several rounds and has the potential to detect rare variants. Its accuracy is unbiased by amplification since the separation occurs prior to PCR. It is, however, not suitable for detection of unknown changes since it requires preliminary information about the sequence of interest in order to design the specific sequence enrichment probes.

3.12 Methods Combining the use of SSH and Microarrays

The microarray approach was introduced more than 10 years ago [151]. In essence the microarray approach is a visualized subtractive hybridization method with an added internal control that does not physically remove common molecular species and can query any desired set of messages depending on the experimental goal. This method became particularly powerful with the improvement and standardization of array printing and of microhybridization methods, and with the acquisition of complete genome sequences of many medically and industrially important organisms, including the human genome [152,41]. The method still requires substantial investment, and even though it was quickly commercialized, it has generated consistency and reproducibility issues. Another important factor that restricted the use of such data acquisition method relates to data analysis, because the ability to acquire data is exceeding the computational power required to extract comprehensive information [153]. Fortunately, there have been rapid improvements in the data analysis fields [139,154,155,156,157,158]. A combination of SSH and microarray approach [24] is certainly a streamlined way to differential gene expression profiling that can ensure that the data could be complete and manageable at the same time. This has already been done in breast cancer studies [21], virus–host interaction studies [22], and plant genetics [159]. The microarray approach to gene discovery has been compared to SSH [160] and DD [161], and the assessment was that these methods can be used as an alternative and/or complimentary transcript profiling tool, especially when the targets are new genes and transcripts of low abundance.

Finally, several methods have been developed to address the issue of quality control of the subtractive process, and the integrity and quality of the starting material. Regardless of the downstream protocol (SH, RDA, SSH, DEASH, etc.,) an essential requirement for the success of any method designed to look at differences is to provide an unbiased representation of the starting material. A number of control approaches [162,163,2,164,165,116,1,166,167,168,134] have been developed to ensure the quality of the starting material are found in references, and they may be applied discriminately depending on the application and the target of interest.

3.13 Conclusions

PEER is a new member of the subtraction hybridization-based methods designed to query genetic differences. The major advantages of PEER are that it takes direct advantage of the unique specificity of the target's sequence by using it both as a template and a primer. The blocking of common primers is an extra step that adds hybridization and enzymatic specificity to the selection of oligonucleotides that enter the final amplification process. PEER is versatile and not restricted to DNA or RNA starting material and does not require excess of driver material since the hybridization/exclusion step is done by PCR. The sensitivity and applicability of the method to both RNA and DNA targets was recently demonstrated with a variety of viruses representing different genome structures – human herpesvirus and Ectromelia virus (dsDNA), human echovirus, West Nile virus, and human respiratory syncytial virus (ssRNA), Orthoreovirus (polysegmented ds RNA), porcine circovirus (circular ssDNA) [45].

Most existing subtraction methods suffer from the inability to completely remove the background of common or highly repetitive sequences, are limited in the recovery of rare molecular species, and may omit single-stranded molecules.

Generally more advantageous are the hybridization methods that have their sensitivity and specificity enhanced by enzymatic steps and that could be coupled with good controls for the background. Standardized "chip" assays that combine subtraction and microarray approaches will clearly dominate in the near future, however, methods that are based on RDA, SSH, or DD have become "classical" and are widely used, have good versatility and general reproducibility, have been well standardizes and cannot be dismissed when screening for genetic differences at any level of complexity.

REFERENCES

1. Muerhoff AS, Leary TP, Desai SM, Mushahwar IK (1997) Amplification and subtraction methods and their application to the discovery of novel human viruses. J Med Virol 53:96–103
2. Lisitsyn N, Lisitsyn N, Wigler M (1993) Cloning the differences between two complex genomes. Science 259:946–951
3. Allander T, Emerson SU, Engle RE, Purcell RH, Bukh J (2001) A virus discovery method incorporating DNase treatment and its application to the identification of two bovine parvovirus species. Proc Natl Acad Sci USA 98:11609–11614
4. Birkenmeyer LG, Leary TP, Muerhoff AS, Dawson GJ, Mushahwar IK, Desai SM (2003) Selectively primed adaptive driver RDA (SPAD-RDA): an improved method for subtractive hybridization. J Med Virol 71:150–159
5. Luo J-H, Puc JA, Slosberg ED, Yao Y, Bruce JN, WrightJr TC, Becich MJ, Parsons R (1999) Differntial subtraction chain a method for identifying differences in genomic DNA and mRNA. Nucleic Acids Res 27:e24
6. Lavery DJ, Lopez-Molina L, Fleury-Olela F, Schibler U (1997) Selective amplification via biotin- and restriction-mediated enrichment (SABRE) a novel selective amplification procedure for detection of differentially expressed mRNAs. Proc Natl Acad Sci USA 94:6831–6836
7. Diatchenko L, Lau YF, Campbell AP, Chenchik A, Moqadam F, Huang B, Lukyanov S, Lukyanov K, Gurskaya N, Sverdlov ED, Siebert PD (1996) Suppression subtractive hybridization: a method for generating differentially regulated or tissue-specific cDNA probes and libraries. Proc Natl Acad Sci USA 93:6025–6030
8. Liang P, Pardee AB (1992) Differential display of eucaryotic messenger RNA by means of the polymerase chain reaction. Science 257:967–971
9. Matz MV, Lukyanov SA (1998) Different strategies of differential display: areas of application. Nucleic Acids Res 26:5537–5543
10. Shaw-Smith CJ, Coffey AJ, Huckle E, Durham J, Campbell EA, Freeman TC, Walters JR, Bentley DR (2000) Improved method for detecting differentially expressed genes using cDNA indexing. Biotechniques 28:958–964
11. Agaton C, Unneberg P, Sievertzon M, Holmberg A, Ehn M, Larsson M, Odeberg J, Uhlen M, Lundeberg J (2002) Gene expression analysis by signature pyrosequencing. Gene 289:31–39
12. Bishop R, Shah T, Pelle R, Hoyle D, Pearson T, Haines L, Brass A, Hulme H, Graham SP, Taracha EL, Kanga S, Lu C, Hass B, Wortman J, White O, Gardner MJ, Nene V, de Villiers EP (2005) Analysis of the transcriptome of the protozoan *Theileria parva* using MPSS reveals that the majority of genes are transcriptionally active in the schizont stage. Nucleic Acids Res 33:5503–5511
13. Brenner S, Johnson M, Bridgham J, Golda G, Lloyd DH, Johnson D, Luo S, McCurdy S, Foy M, Ewan M, Bridgham J, Golda G, Lloyd DH, Johnson D, Luo S, McCurdy S, Foy M, Ewan M, Roth R, George D, Eletr S, Albrecht G, Vermaas E, Williams SR, Moon K, Burcham T, Pallas M, DuBridge RB, Kirchner J, Fearon K, Mao J, Corcoran K (2000) Gene expression analysis by massively parallel signature sequencing (MPSS) on microbead arrays. Nat Biotechnol 18:630–634

14. Brenner S, Williams SR, Vermaas EH, Storck T, Moon K, McCollum C, Mao JI, Luo S, Kirchner JJ, EletrSEletr S, DuBridge RB, Burcham T, Albrecht G (2000) *In vitro* cloning of complex mixtures of DNA on microbeads: physical separation of differentially expressed cDNAs. Proc Natl Acad Sci USA 97:1665–1670
15. Chen JJ, WuR, Yang PC, Huang JY, Sher YP, Han MH, Kao WC, Lee PJ, Chiu TF, Chang F, Chu YW, Wu CW, Peck K (1998) Profiling expression patterns and isolating differentially expressed genes by cDNA microarray system with colorimetry detection. Genomics 51:313–324
16. Drmanac S, Drmanac R (1994) Processing of cDNA and genomic kilobase-size clones for massive screening mapping and sequencing by hybridization. Biotechniques 17:328–9–332–6
17. Leamon JH, Lee WL, Tartaro KR, Lanza JR, Sarkis GJ, deWinter AD, Berka J, Lohman KL (2003) A massively parallel PicoTiterPlate based platform for discrete picoliter-scale polymerase chain reactions. Electrophoresis 24:3769–3777
18. Schena M, Heller RA, Theriault TP, Konrad K, Lachenmeier E, Davis RW (1999) Microarrays: biotechnology's discovery platform for functional genomics. Trends Biotechnol 16:217–218
19. Wang D, Urisman A, LiuYT, SpringerM, Ksiazek TG, Erdman DD, Mardis ER, Hickenbotham M, Magrini V, Eldred J, Latreille JP, Wilson RK, Ganem D, DeRisi JL (2003) Viral discovery and sequence recovery using DNA microarrays. PLoS Biol 1:E2
20. Wieland I, Bolger G, Asouline G, Wigler M (1990) A method for difference cloning: gene amplification following subtractive hybridization. Proc Natl Acad Sci USA 87:2720–2724
21. Beck MT, Holle L, Chen WY (2001) Combination of PCR subtraction and cDNA microarray for differential gene expression profiling. Biotechniques 31:782–4, 786
22. Munir S, Singh S, Kaur K, Kapur V (2004) Suppression subtractive hybridization coupled with microarray analysis to examine differential expression of genes in virus infected cells. Biol Proceed Online 6:94–104
23. Pan YS, Lee YS, Lee YL, Lee WC, Hsieh SY (2006) Differentially profiling the low-expression transcriptomes of human hepatoma using a novel SSH/microarray approach. BMC Genomics 7:131
24. Yang GP, Ross DT, Kuang WW, Brown PO, Weigel RJ (1999) Combining SSH and cDNA microarrays for rapid identification of differentially expressed genes. Nucleic Acids Res 27:1517–1523
25. Bui LC, Leandri RD, Renard JP, Duranthon V (2005) SSH adequacy to preimplantation mammalian development: scarce specific transcripts cloning despite irregular normalization BMC Genomics 6:155
26. Fallsehr C, Zapletal C, Kremer M, Demir R, von Knebel DM, Klar E (2005) Identification of differentially expressed genes after partial rat liver ischemia/reperfusion by suppression subtractive hybridization. World J Gastroenterol 11:1303–1316
27. Galbraith EA, Antonopoulos DA, White BA (2004) Suppressive subtractive hybridization as a tool for identifying genetic diversity in an environmental metagenome: the rumen as a model. Environ Microbiol 6:928–937
28. Liu YB, Wei ZX, Li L, Li HS, Chen H, Li XW (2003) Construction and analysis of SSH cDNA library of human vascular endothelial cells related to gastrocarcinoma. World J Gastroenterol 9:2419–2423
29. Nishizuka S, Tsujimoto H, Stanbridge EJ (2001) Detection of differentially expressed genes in HeLa x fibroblast hybrids using subtractive suppression hybridization. Cancer Res 61:4536–4540
30. Ganova-Raeva L, Smith AW, Fields H, Khudyakov Y (2004) New Calicivirus isolated from walrus. Virus Res 102:207–213
31. Birkenmeyer LG, Desai SM, Muerhoff AS, Leary TP, Simons JN, Montes CC, Mushahwar IK (1998) Isolation of a GB virus-related genome from a chimpanzee. J Med Virol 56:44–51
32. Nishizawa T, Okamoto H, Konishi K, Yoshizawa H, Miyakawa Y, Mayumi M (1997) A novel DNA virus (TTV) associated with elevated transaminase levels in posttransfusion hepatitis of unknown etiology. Biochem Biophys Res Commun 241:92–97
33. Ji W, right MB, Cai L, Flament, Lindpaintner K (2002) Efficacy of SSH PCR in isolating differentially expressed genes. BMC Genomics 3:12

34. Szybalsky W, Kim SC, Hasan N, Podhajska AJ (1991) Class-IIS restriction enzymes – a review. Gene 100:13–26
35. Boyd AC, Charles IG, Keyte JW, Brammar WJ (1986) Isolation and computer-aided characterization of MmeI a type II restriction endonuclease from *Methylophilus methylotrophus*. Nucleic Acids Res 14:5255–5274
36. Tabor S, Richardson CC (1995) A single residue in DNA polymerase of the *E. coli* DNA polymerase I family is critical for distinguishing between deoxy and dideoxyribonucleotides. Proc Natl Acad Sci USA 92:6339–6343
37. Kotewicz ML, Sampson CM, D'Alessio JM, Gerard GF (1988) Isolation of cloned Moloney murine leukemia virus reverse transcriptase lacking ribonuclease H activity. Nucleic Acids Res 16:265–277
38. Seth D, Gorrell MD, McGuinness PH, Leo MA, Lieber CS, McCaughan GW, Haber PS (2003) SMART amplification maintains representation of relative gene expression: quantitative validation by real time PCR and application to studies of alcoholic liver disease in primate. J Biochem Biophys Methods 55:53–66
39. Swaminathan N, George D, McMaster K, Szablebski J, VanEtten JL, Mead DA (1994) Restriction generated oligonucleotides utilizing the two base recognition endonuclease CviJI*. Nucleic Acids Res 22:1470–1475
40. Swaminathan N, McMaster K, Skowron PM, Mead DA (1998) Thermal cycle labeling: Zeptomole detection sensitivity and microgram probe amplification using CviJI* restriction-generated oligonucleotides. Anal Biochem 255:133–141
41. Wright FA, Lemon WJ, Zhao WD, Sears R, Zhuo D, Wang JP, Yang HY, Baer T, Stredney D, Spitzner J, Stutz A, Krahe R, Yuan B (2001) A draft annotation and overview of the human genome. Genome Biol 2(7):1–25
42. Tucholski J, Skowron PM, Podhajska AJ (1995) MmeI a class-IIS restriction endonuclease: purification and characterization. Gene 157:87–92
43. Armengaud J, Jouanneau Y(1993) Addition of a class IIS enzyme site in the mutagenic primer to improve two-step PCR-based targeted mutagenesis. Nucleic Acids Res 21:4424–4425
44. Ganova-Raeva L, Zhang X, Cao F, Fields H, Khudyakov Y (2006) Primer extension enrichment reaction (PEER): a new subtraction method for identification of genetic differences between biological specimens. Nucleic Acids Res 34:e76
45. Biagini P, de Lamballerie X, de Micco P (2007) Effective detection of highly divergent viral genomes in infected cell lines using a new subtraction strategy (primer extension enrichment reaction – PEER). J Virol Methods 139(1):106–110
46. Sargent TD (1983) Differential gene expression in the gastrula of *Xenopus laevis*. Science 222:135–139
47. Maser RL (1995) Analysis of differential gene expression in the kidney by differential cDNA screening subtractive cloning and mRNA differential display. Semin Nephrol 15:29–42
48. Lee MT, Kaushansky K, Ralph P, Ladner MB (1990) Differential expression of M-CSF G-CSF and GM-CSF by human monocytes. J Leukoc Biol 47:275–282
49. Rothstein JL, Johnson D, Jessee J, Skowronski J, DeLoia JA, Solter D, Knowles BB (1993) Construction of primary and subtracted cDNA libraries from early embryos. Methods Enzymol 225:587–610
50. Sargent T (1987) Isolation of differentially expressed genes. Methods Enzymol 152:423–432
51. Ermolaeva OD, Sverdlov ED (1996) Subtractive hybridization a technique for extraction of DNA sequences distinguishing two closely related genomes: critical analysis. Genet Anal 13:49–58
52. Davis MM (1984) Cell-type-specific cDNA probes and the murine I region: the localization and orientation of Ad alpha. Proc Natl Acad Sci USA 81:2194–2198
53. Rubenstein JL, Brice AE, Ciaranello RD, Denney D, Porteus MH, Usdin TB (1990) Subtractive hybridization system using single-stranded phagemids with directional inserts. Nucleic Acids Res 18:4833–4842

54. Schweinfest CW, Henderson KW, Gu JR, Kottaridis SD, Besbeas S, Panotopoulou E, Papas TS (1990) Subtraction hybridization cDNA libraries from colon carcinoma and hepatic cancer. Genet Anal Tech Appl 7:64–70
55. Scott MR, Westphal K-H, Rigby PWJ (1983) Activation of mouse genes in transformed cells. Cell 34:557–567
56. Reynet C, Kahn CR (1993) Rad: a member of the Ras family overexpressed in muscle of type II diabetic humans. Science 262;1441–1444
57. Ermolaeva OD, Lukyanov SA, Sverdlov ED (1996) The mathematical model of subtractive hybridization and its practical application. Proc Int Conf Intell Syst Mol Biol 4:52–58
58. Milner JJ, Cecchini E, Dominy PJ (1995) A kinetic model for subtractive hybridization. Nucleic Acids Res 23:176–187
59. Ray A, Macwana S, Ayoubi P, Hall LT, Prade R, Mort AJ (2004) Negative subtraction hybridization: an efficient method to isolate large numbers of condition-specific cDNAs. BMC Genomics 5:22
60. Lee SW (1991) Positive selection of candidate tumor-suppressor genes by subtractive hybridization. Proc Natl Acad Sci USA 88:2825–2829
61. Konietzko U, Kuhl D (1998) A subtractive hybridisation method for the enrichment of moderately induced sequences. Nucleic Acids Res 26:1359–1361
62. Li MS, Monahan IM, Waddell SJ, Mangan JA, Martin SL, Everett MJ, Butcher PD (2001) cDNA-RNA subtractive hybridization reveals increased expression of mycocerosic acid synthase in intracellular *Mycobacterium bovis* BCG. Microbiology 147:2293–2305
63. Ishino F, Kuroiwa Y, Miyoshi N, Kobayashi S, Kohda T, Kaneko-Ishino T (2001) Subtraction-hybridization method for the identification of imprinted genes. Methods Mol Biol 181:101–112
64. Sharma S, Chang JT, Della NG, Campochiaro PA, Zack DJ (2002) Identification of novel bovine RPE and retinal genes by subtractive hybridization. Mol Vis 8:251–258
65. Zimmermann CR (1980) Molecular cloning and selection of genes regulated in Aspergillus development. Cell 21:709–715
66. Hara E, Kato T, Nakada S, Sekiya S, Oda K (1991) Subtractive cDNA cloning using oligo(dT)30-latex and PCR: isolation of cDNA clones specific to undifferentiated human embryonal carcinoma cells. Nucleic Acids Res 19:7097–7104
67. Kohchi T, Fujishige K, Ohyama K (1995) Construction of an equalized cDNA library from *Arabidopsis thaliana*. Plant J 8:771–776
68. Lopez-Fernandez LA, del Mazo J (1993) Construction of subtractive cDNA libraries from limited amounts of mRNA and multiple cycles of subtraction. Biotechnique 15:654–659
69. Laveder P, De Pitta C, Toppo S, Valle G, Lanfranchi G (2002) A two-step strategy for constructing specifically self-subtracted cDNA libraries. Nucleic Acids Res 30:e38
70. Ying SY, Lin SL (2003) Subtractive cloning of differential genes using RNA-PCR. Methods Mol Biol 221:253–259
71. Zhumabayeva B, Chang C, McKinley J, Diatchenko L, Siebert PD (2001) Generation of full-length cDNA libraries enriched for differentially expressed genes for functional genomics. Biotechniques 30:512–520
72. Zhumabayeva B, Chang C, McKinley J, Diatchenko L, Siebert PD (2003) Generation of full-length cDNA libraries enriched for differentially expressed genes. Methods Mol Biol 221:223–237
73. Welsh J, Chada K, Dalal SS, Cheng R, Ralph D, McClelland M (1992) Arbitrarily primed PCR fingerprinting of RNA. Nucleic Acids Res 20:4965–4970
74. Fuchs B, Zhang K, Bolander ME, Sarkar G (2000) Differential mRNA fingerprinting by preferential amplification of coding sequences. Gene 258:155–163
75. Kato K (1995) Description of the entire mRNA population by a 3′ end cDNA fragment generated by class IIS restriction enzymes. Nucleic Acids Res 23:3685–3690
76. Kato K (1996) RNA fingerprinting by molecular indexing. Nucleic Acids Res 24:394–395
77. Matz M, Usman N, Shagin D, Bogdanova E, Lukyanov S (1997) Ordered differential display: a simple method for systematic comparison of gene expression profiles. Nucleic Acids Res 25:2541–2542

78. Zabeau M, Vos P (1993) Selective restriction fragment amplification: a general method for DNA fingerprinting. Eur Patent Appl Publ #0534858A1
79. Pardinas JR, Combates NJ, Prouty SM, Stenn KS, Parimoo S (1998) Differential subtraction display: a unified approach for isolation of cDNAs from differentially expressed genes. Anal Biochem 257:161–168
80. SAGE (2006) Current technologies and applications. Horizon Bioscience, Norwich, England
81. Dinel S, Bolduc C, Belleau P, Boivin A, Yoshioka M, Calvo E, Piedboeuf B, Snyder EE, Labrie F, St Amand J (2005) Reproducibility bioinformatic analysis and power of the SAGE method to evaluate changes in transcriptome. Nucleic Acids Res 33:e26
82. Vos P, Hogers R, Bleeker M, Reijans M van de LT, Hornes M, Frijters A, Pot J, Peleman J, Kuiper M (1995) AFLP: a new technique for DNA fingerprinting. Nucleic Acids Res 23:4407–4414
83. Botstein D (1980) Construction of a genetic linkage map in man using restriction fragment length polymorphisms. Am J Hum Genet 32:314–331
84. Akopyants NS, Fradkov A, Diatchenko L, Hill JE, Siebert PD, Lukyanov SA, Sverdlov ED, Berg DE (1998) PCR-based subtractive hybridization and differences in gene content among strains of *Helicobacter pylori*. Proc Natl Acad Sci USA 95:13108–13113
85. Kim S, Park S, Choung S, Park HO, Choi YC (2005) DARFA: a novel technique for studying differential gene expression and bacterial comparative genomics. Biochem Biophys Res Commun 336:168–174
86. Velculescu VE, Zhang L, Vogelstein B, Kinzler KW (1995) Serial analysis of gene expression. Science 270:484–487
87. Gnatenko DV, Dunn JJ, McCorkle SR, Weissmann D, Perrotta PL, Bahou WF (2003) Transcript profiling of human platelets using microarray and serial analysis of gene expression. Blood 101:2285–2293
88. Huang ZG, Ran ZH, Lu W, Xiao SD (2006) Analysis of gene expression profile in colon cancer using the Cancer Genome Anatomy Project and RNA interference. Chin J Dig Dis 7:97–102
89. Wang X, Zhang C, Zhang L, Wang X, Xu S (2006) High-throughput assay of DNA methylation based on methylation-specific primer and SAGE. Biochem Biophys Res Commun 341:749–754
90. Shiraki T (2003) Cap analysis gene expression for high-throughput analysis of transcriptional starting point and identification of promoter usage. Proc Natl Acad Sci USA 100:15776–15781
91. Dunn JJ, McCorkle SR, Praissman LA, Hind G, Van Der LD, Bahou WF, Gnatenko DV, Krause MK (2002) Genomic signature tags (GSTs): a system for profiling genomic DNA. Genome Res 12:1756–1765
92. Chen J, Sadowski I (2005) Identification of the mismatch repair genes PMS2 and MLH1 as p53 target genes by using serial analysis of binding elements. Proc Natl Acad Sci USA 102:4813–4818
93. Lisitsyn NA (1995) Representational difference analysis: finding the differences between genomes. Trends Genet 11:303–307
94. Felske A (2002) Streamlined representational difference analysis for comprehensive studies of numerous genomes. J Microbiol Methods 50:305–311
95. Jiang H (2000) RaSH a rapid subtraction hybridization approach for identifying and cloning differentially expressed genes. Proc Natl Acad Sci USA 97:12684–12689
96. Bowler LD (2004) Representational difference analysis of cDNA. Methods Mol Med 94:49–66
97. Hubank M, Schatz DG (1999) cDNA representational difference analysis: a sensitive and flexible method for identification of differentially expressed genes. Methods Enzymol 303:325–349
98. Pastorian K, Hawel L III, Byus CV (2000) Optimization of cDNA representational difference analysis for the identification of differentially expressed mRNAs. Anal Biochem 283:89–98
99. Muerhoff AS, Leary TP, Simons JN, Pilot-Matias TJ, Dawson GJ, Erker JC, Chalmers ML, Schlauder GG, Desai SM, Mushahwar IK (1995) Genomic organization of GB viruses A and B: two new members of the Flaviviridae associated with GB agent hepatitis. J Virol 69:5621–5630

100. Simons JN, Pilot-Matias TJ, Leary TP, Dawson GJ, Desai SM, Schlauder GG, Muerhoff AS, Erker JC, Buijk SL, Chalmers ML (1995) Identification of two flavivirus-like genomes in the GB hepatitis agent. Proc Natl Acad Sci USA 92:3401–3405
101. Hubank M, Bryntesson F, Regan J, Schatz DG (2004) Cloning of apoptosis-related genes by representational difference analysis of cDNA. Methods Mol Biol 282:255–273
102. Bowler LD, Hubank M, Spratt BG (1999) Representational difference analysis of cDNA for the detection of differential gene expression in bacteria: development using a model of iron-regulated gene expression in *Neisseria meningitidis*. Microbiology 145:3529–3537
103. Hollestelle A, Schutte M (2005) Representational difference analysis as a tool in the search for new tumor suppressor genes. Methods Mol Med 103:143–159
104. Boukerche H, Su ZZ, Kang DC, Fisher PB (2004) Identification and cloning of genes displaying elevated expression as a consequence of metastatic progression in human melanoma cells by rapid subtraction hybridization. Gene 343:191–201
105. Wada J, Kumar A, Ota K, Wallner EI, Batlle DC, Kanwar YS (1997) Representational difference analysis of cDNA of genes expressed in embryonic kidney. Kidney Int 51:1629–1638
106. Hansen-Hagge TE, Trefzer U, zu Reventlow AS, Kaltoft K, Sterry W (2001) Identification of sample-specific sequences in mammalian cDNA and genomic DNA by the novel ligation-mediated subtraction (Limes). Nucleic Acids Res 29: E20
107. Zeng J, Gorski RA, Hamer D (1994) Differential cDNA cloning by enzymatic degrading subtraction (EDS). Nucleic Acids Res 22:4381–4385
108. Kohne DE (1977) Room temperature method for increasing the rate of DNA reassociation by many thousandfold: the phenol emulsion reassociation technique. Biochemistry 16:5329–5341
109. Yang M (1996) Cloning differentially expressed genes by linker capture subtraction. Anal Biochem 237:109–114
110. Lin F (2001) Myopodin a synaptopodin homologue is frequently deleted in invasive prostate cancers. Am J Pathol 159:1603–1612
111. Yu YP (2001) Identification of a novel gene with increasing rate of suppression in high grade prostate cancers. Am J Pathol 158:19–24
112. Li J, Wang F, Kashuba V, Wahlestedt C, Zabarovsky ER (2001) Cloning of deleted sequences (CODE): A genomic subtraction method for enriching and cloning deleted sequences. Biotechniques 31:788,790,792–788,790,793
113. Lin TY, Ying SY (2003) Subtractive hybridization for the identification of differentially expressed genes using uracil-dNA glycosylase and mung-bean nuclease. Methods Mol Biol 221:239–251
114. Chalaya T, Gogvadze E, Buzdin A, Kovalskaya E, Sverdlov ED (2004) Improving specificity of DNA hybridization-based methods. Nucleic Acids Res 32:e130
115. Li L, Techel D, Gretz N, Hildebrandt A (2005) A novel transcriptome subtraction method for the detection of differentially expressed genes in highly complex eukaryotes. Nucleic Acids Res 33:e136
116. Matz M, Shagin D, Bogdanova E, Britanova O, Lukyanov S, Diatchenko L, Chenchik A (1999) Amplification of cDNA ends based on template-switching effect and step-out PCR. Nucleic Acids Res 27:1558–1560
117. Hu Y, Hirshfield I (2005) Rapid approach to identify an unrecognized viral agent. J Virol Methods 127:80–86
118. Hedrick SM (1984) Isolation of cDNA clones encoding T cell-specific membrane-associated proteins. Nature 308:149–153
119. Hubank M (1994) Identifying differences in mRNA expression by representational difference analysis of cDNA. Nucleic acids research 22:5640–5648
120. Wang Z (1991) A gene expression screen. Proc Natl Acad Sci USA 88:11505–11509
121. Lukyanov KA (1995) Inverted terminal repeats permit the average length of amplified DNA fragments to be regulated during preparation of cDNA libraries by polymerase chain reaction. Anal Biochem 229:198–202

122. Chua KB, Wang LF, Lam SK, Crameri G, Yu M, Wise T, Boyle D, Hyatt AD, Eaton BT (2001) Tioman virus a novel paramyxovirus isolated from fruit bats in Malaysia. Virology 283:215–229
123. DeShazer D (2004) Genomic diversity of *Burkholderia pseudomallei* clinical isolates: subtractive hybridization reveals a *Burkholderia mallei*-specific prophage in *B. pseudomallei* 1026b. J Bacteriol 186:3938–3950
124. Diatchenko L, Lukyanov S, Lau YF, Siebert PD (1999) Suppression subtractive hybridization: a versatile method for identifying differentially expressed genes. Methods Enzymol 303:349–380
125. Diatchenko L, Slade GD, Nackley AG, Bhalang K, Sigurdsson A, Belfer I, Goldman D, Xu K, Shabalina SA, Shagin D, Max MB, Makarov SS, Maixner W (2005) Genetic basis for individual variations in pain perception and the development of a chronic pain condition. Hum Mol Genet 14:135–143
126. Gurskaya NG, Diatchenko L, Chenchik A, Siebert PD, Khaspekov GL, Lukyanov KA, Vagner LL, Ermolaeva OD, Lukyanov SA, Sverdlov ED (1996) Equalizing cDNA subtraction based on selective suppression of polymerase chain reaction: cloning of Jurkat cell transcripts induced by phytohemaglutinin and phorbol 12-myristate 13-acetate. Anal Biochem 240:90–97
127. Bonaldo MF, Lennon G, Soares MB (1996) Normalization and subtraction: two approaches to facilitate gene discovery. Genome Res 6:791–806
128. Chenchik A (1996) Full-length cDNA cloning and determination of mRNA 5′ and 3′ ends by amplification of adaptor-ligated cDNA. Biotechniques 21:526–534
129. Lukyanov K, Diatchenko L, Chenchik A, Nanisetti A, Siebert P, Usman N, Matz M, Lukyanov S (1997) Construction of cDNA libraries from small amounts of total RNA using the suppression PCR effect. Biochem Biophys Res Commun 230:285–288
130. Siebert PD (1995) An improved PCR method for walking in uncloned genomic DNA. Nucleic Acids Res 23:1087–1088
131. Rebrikov DV, Britanova OV, Gurskaya NG, Lukyanov KA, Tarabykin VS, Lukyanov SA (2000) Mirror orientation selection (MOS): a method for eliminating false positive clones from libraries generated by suppression subtractive hybridization. Nucleic Acids Res 28:E90
132. Jeung JU, Cho SK, Shin JS (2005) A partial-complementary adapter for an improved and simplified ligation-mediated suppression PCR technique. J Biochem Biophys Methods 64:110–120
133. Strauss C, Mussgnug JH, Kruse O (2001) Ligation-mediated suppression-PCR as a powerful tool to analyse nuclear gene sequences in the green alga *Chlamydomonas reinhardtii*. Photosynth Res 70:311–320
134. Yueqing C, Zhengbo H, Zhongkang W, Youping Y, Guoxiong P, Yuxian X (2006) Hybridization monitor: A method for identifying differences between complex genomes. J Microbiol Methods 64:305–315
135. Boengler K, Pipp F, Schaper W, Deindl (2003) Rapid identification of differentially expressed genes by combination of SSH and MOS. Lab Invest 83:759–761
136. Rebrikov DV, Desai SM, Siebert PD, Lukyanov SA (2004) Suppression subtractive hybridization. Methods Mol Biol 258:107–134
137. Iguchi K, Takahashi Y, Kaneto Y, Kubota M, Usui S, Hirano K (2005) Identification of differentially expressed genes in hepatic HepG2 cells treated with acetaminophen using suppression subtractive hybridization. Biol Pharm Bull 28:1148–1153
138. Larose M, Bouchard C, Chagnon YC (2001) A new gene related to human obesity identified by suppression subtractive hybridization. Int J Obes Relat Metab Disord 25:770–776
139. Beissbarth T (2006) Interpreting experimental results using gene ontologies. Methods Enzymol 411:340–352
140. Malyala A, Pattee P, Nagalla SR, Kelly MJ, Ronnekleiv OK (2004) Suppression subtractive hybridization and microarray identification of estrogen-regulated hypothalamic genes. Neurochem Res 29:1189–1200
141. Rebrikov D, Desai S, Kogan YN, Thornton AM, Diatchenko L (2002) Subtractive cloning: new genes for studying inflammatory disorders. Ann Periodontol 7:17–28

142. Shackel NA, McGuinness PH, Abbott CA, Gorrell MD, McCaughan GW (2003) Novel differential gene expression in human cirrhosis detected by suppression subtractive hybridization. Hepatology 38:577–588
143. Yokota N, Mainprize TG, Taylor MD, Kohata T, Loreto M, Ueda S Dura W, Grajkowska W, Kuo JS, Rutka JT (2004) Identification of differentially expressed and developmentally regulated genes in medulloblastoma using suppression subtraction hybridization. Oncogene 23:3444–3453
144. Wang X, Feuerstein GZ (2000) Suppression subtractive hybridization: application in the discovery of novel pharmacological targets. Pharmacogenomics 1:101–108
145. Bai GQ, Liu Y, Cheng J, Zhang SL, Yue YF, Huang YP, Zhang LY (2005) Transactivating effect of complete S protein of hepatitis B virus and cloning of genes transactivated by complete S protein using suppression subtractive hybridization technique. World J Gastroenterol 11:3893–3898
146. He N, Qin Q, Xu X (2005) Differential profile of genes expressed in hemocytes of White Spot Syndrome Virus-resistant shrimp (*Penaeus japonicus*) by combining suppression subtractive hybridization and differential hybridization. Antiviral Res 66:39–45
147. Kiss C, Nishikawa J, Dieckmann A, Takada K, Klein G, Szekely L (2003) Improved subtractive suppression hybridization combined with high density cDNA array screening identifies differentially expressed viral and cellular genes. J Virol Methods 107:195–203
148. Yin J, Chen MF, Finkel TH (2004) Differential gene expression during HIV-1 infection analyzed by suppression subtractive hybridization. AIDS 18:587–596
149. Schibler U, Rifat D, Lavery DJ (2001) The Isolation of differentially expressed mRNA sequences by selective amplification via biotin and restriction-mediated enrichment. Methods 24:3–14
150. Jeffreys AJ, May CA (2003) DNA enrichment by allele-specific hybridization (DEASH): a novel method for haplotyping and for detecting low-frequency base substitutional variants and recombinant DNA molecules. Genome Res 13:2316–2324
151. Schena M, Shalon D, Davis RW, Brown PO (1995) Quantitative monitoring of gene expression patterns with a complementary DNA microarray. Science 270:467–470
152. Schena M, Shalon D, Heller R, Chai A, Brown PO, Davis RW (1996) Parallel human genome analysis: microarray-based expression monitoring of 1000 genes. Proc Natl Acad Sci USA 93:10614–10619
153. Mutch DM, Berger A, Mansourian R, Rytz A, Roberts MA (2002) The limit fold change model: a practical approach for selecting differentially expressed genes from microarray data. BMC Bioinformatics 3:17
154. Gresham D (2006) Genome-wide detection of polymorphisms at nucleotide resolution with a single DNA microarray. Science 311:1932–1936
155. Roussel EE, Gingra MM-C (2006) High-throughput gene expression profiling—a work-in-progress with great potential for proteomics. Curr Opin Drug Discov Devel 9:332–338
156. Rubinstein R, Simon I (2005) MILANO–custom annotation of microarray results using automatic literature searches. BMC Bioinformatics 6:12
157. Semeiks JR, Rizki A, Bissell MJ, Mian IS (2006) Ensemble attribute profile clustering: discovering and characterizing groups of genes with similar patterns of biological features. BMC Bioinformatics 7:147
158. Strausberg RL, Feingold EA, Grouse LH, Derge JG, Klausner RD, Collins FS, Wagner L, Shenmen CM, Schuler GD, Altschul SF Zeeberg B, Buetow KH, Schaefer CF, Bhat NK, Hopkins RF, Jordan H, Moore T, Max SI, Wang J, Hsieh F, Diatchenko L, Marusina K, Farmer AA, Rubin GM, Hong L, Stapleton M, Soares MB, Bonaldo MF, Casavant TL, Scheetz TE, Brownstein MJ, Usdin TB, Toshiyuki S, Carninci P, Prange C, Raha SS, Loquellano NA, Peters GJ, Abramson RD, Mullahy SJ, Bosak SA, McEwan PJ, McKernan KJ, Malek JA, Gunaratne PH, Richards S, Worley KC, Hale S, Garcia AM, Gay LJ, Hulyk SW, Villalon DK, Muzny DM, Sodergren EJ, Lu X, Gibbs RA, Fahey J, Helton E, Ketteman M, Madan A, Rodrigues S, Sanchez A, Whiting M, Madan A, Young AC, Shevchenko Y,

Bouffard GG, Blakesley RW, Touchman JW, Green ED, Dickson MC, Rodriguez AC, Grimwood J, Schmutz J, Myers RM, Butterfield YS, Krzywinski MI, Skalska U, Smailus DE, Schnerch A, Schein JE, Jones SJ, Marra MA (2002) Generation and initial analysis of more than 15000 full-length human and mouse cDNA sequences. Proc Natl Acad Sci USA 99:16899–16903
159. van den Berg N, Crampton BG, Hein I, Birch PR, Berger DK (2004) High-throughput screening of suppression subtractive hybridization cDNA libraries using DNA microarray analysis. Biotechniques 37:818–824
160. Cao W, Epstein C, Liu H, DeLoughery C, Ge N, Lin J, Diao R, Cao H, Long F, Zhang X, Zhang X, Chen Y, Wright PS, Busch S, Wenck M, Wong K, Saltzman AG, Tang Z, Liu L, Zilberstein A (2004) Comparing gene discovery from Affymetrix GeneChip microarrays and Clontech PCR-select cDNA subtraction: a case study. BMC Genomics 5:26
161. Rajeevan MS, Ranamukhaarachchi DG, Vernon SD, Unger ER (2001) Use of real-time quantitative PCR to validate the results of cDNA array and differential display PCR technologies. Methods 25:443–451
162. Gastel JA, Sutter TR (1996) A control system for cDNA enrichment reactions. Biotechniques 20:870–875
163. Gonzalez JM, Portillo MC, Saiz-Jimenez C (2005) Multiple displacement amplification as a pre-polymerase chain reaction (pre-PCR) to process difficult to amplify samples and low copy number sequences from natural environments. Environ Microbiol 7:1024–1028
164. Lukyanov KA, Matz MV, Bogdanova EA, Gurskaya NG, Lukyanov SA (1996) Molecule by molecule PCR amplification of complex DNA mixtures for direct sequencing: an approach to in vitro cloning. Nucleic Acids Res 24:2194–2195
165. Makrigiorgos GM, Chakrabarti S, Zhang Y, Kaur M, Price BD (2002) A PCR-based amplification method retaining the quantitative difference between two complex genomes. Nat Biotechnol 20:936–939
166. Reyes GR, Kim JP (1991) Sequence-independent single-primer amplification (SISPA) of complex DNA populations. Mol Cell Probes 5:473–481
167. Soares MB, Bonaldo MF, Jelene P, Su L, Lawton L, Efstratiadis A 1994) Construction and characterization of a normalized cDNA library. Proc Natl Acad Sci USA 91:9228–9232
168. Ying SY (2004) Complementary DNA libraries: an overview. Mol Biotechnol 27:245–252

CHAPTER 7

SUBTRACTIVE HYBRIDIZATION WITH COVALENTLY MODIFIED OLIGONUCLEOTIDES

SHI-LUNG LIN, DONALD CHANG, JOSEPH D. MILLER, SHAO-YAO YING*

Department of Cell and Neurobiology, Keck School of Medicine, University of Southern California, 1333 San Pablo Street, BMT-403, Los Angeles, CA 90033.
*Corresponding author
Phone: 002-1-323-442-1856; Fax: 002-1-323-442-3466; E-mail: sying@usc.edu or lins@usc.edu*

Abstract:	The ability to compare two different nucleic acid libraries has permitted inquiries into the role of differentially expressed genes or deleted/inserted genomic sequences involved in the mechanisms of neoplastic transformation, developmental regulation, physiological processing, pathological disorder, and therapeutic efficacy. Subtractive hybridization between two complementary DNA (cDNA) libraries is a powerful tool for identifying differentially expressed genes. In principle, an excess amount of modified subtracter cDNAs derived from cells of a control group are used to bind with tester messenger RNAs (mRNAs) or cDNAs isolated from the cells of interest. Because the subtracter cDNAs are modified to interfere with the amplification processes of reverse transcription (RT) and/or polymerase chain reaction (PCR), all subtracter-bound tester sequences are degraded and only the differentially expressed genes in the tester can be preserved for RT–PCR amplification. To improve the efficiency of subtractive hybridization, we have developed a chemical modification procedure to generate covalently binding cDNAs as the subtracter to capture the homologous tester sequences. We have also proved that the covalently bound duplex hybrids cannot be separated in PCR and thus are removed from the amplified differential gene sequences. Using the novel principle of covalently hybridized subtraction (CHS), we provide an easy, fast, and effective subtractive hybridization method for understanding the alterations of gene expression and/or chromosomal rearrangement in disordered cells in comparison with normal ones, which may reveal targets for gene therapy, eugenic improvement, pharmaceutical drug design, and investigation of etiological mechanisms.
Keywords:	Subtracter, covalently modified subtracter, deamination of purines, carboxylation of pyrimidine bases, differential DNA, uracil–DNA glycosylase (UDG), aziridinyl-benzoquinone (AZQ), covalent modification, modified pyrimidine, modified purine, covalently hybridized subtraction (CHS), acetic anhydride, alkaline acetic chloride, alkaline potassium permanganate, sodium cyanide/sulfuric acid mixture.

Abbreviations: AZQ, aziridinylbenzoquinone; cDNA, complementary DNA; CHS, covalently hybridized subtraction; dNTP, deoxyribonucleotidetriphosphate; dUTP, deoxyuridine triphosphate; mRNA, messenger RNA; PCR, polymerase chain reaction; RB, retinoblastoma; RT, reverse transcription; UDG, uracil–DNA glycosylase.

TABLE OF CONTENTS

1. Introduction 168
2. Subtractive Hybridization Methods 169
3. Covalent Modification 171
4. Subtractive Hybridization with Covalently Modified
    Subtracters 176
5. Applications 179
6. Protocols 180
    6.1 Preparation of Subtracter and Tester DNA Libraries 180
    6.2 Covalent Modification of Subtracter DNAs 180
    6.3 Subtractive Hybridization and CHS–PCR Amplification 181
    6.4 Covalent Binding Efficiency and Subtractive
         Stringency of CHS 183
    6.5 Identification of Genomic Deletion Using CHS 184
References 185

1. INTRODUCTION

Cells respond to environmental changes by altering their gene expression patterns to produce the proteins required for cell adaptation to the new environments. Alterations of gene expression patterns in a variety of biological processes as well as in response to external stimuli determine the cell fate and development of all life forms. Therefore, differentially expressed genes often encode the cryptic signals essential for cell adaptation and survival. To find these differentially expressed genes, several methods have been designed to detect and isolate different DNA sequences that are present in one expressed gene library but absent in the other. One of the most commonly used methods to accomplish this purpose is subtractive hybridization, involving the elimination of homologous (common) sequences from the mixture of two mutually compared DNA libraries. This kind of selective isolation can be done either between two complementary DNA (cDNA) libraries [1], or between two genomic DNA libraries [2]. In brief, this method relies upon the generation of cDNA or genomic DNA libraries from both control cells (subtracter DNAs) and cells after experiment, treatment, disorder, or change (tester DNAs). The two DNA libraries are then denatured and hybridized to each other, resulting in subtracter–tester hybrid formation if a sequence is common to both DNA populations. By removing the subtracter-bound common sequences, the remaining DNAs are the desired differential sequences, which are only present in the tester and expressive of the treatment, disorder, or change of interest.

Subtractive hybridization

Subtractive hybridization has been successfully used in the discovery of many functional genes and crucial genomic loci, such as T_4 and T_8 lymphocyte-surface glycoproteins [3,4], gamma–interferon-induced cytokines in monocytes [5], choroidermia loci [6], Duchenne muscular dystrophy-related loci [7], and human Y-chromosome-specific DNA [2].

As used here, tester DNAs refer to the DNA library isolated from a treated, mutated, infected, differentiated, or abnormal cell source, while subtracter DNAs refer to the DNA library isolated from a cell source with different status, such as nontreated, undifferentiated, or relatively normal cells (or tissues containing homogeneous cells). The tester library contains desired DNA sequences that are abundant in the tester but very limited in the subtracter. The differential DNA sequences represent the differences between tester and subtracter gene expression patterns (if two cDNA libraries are used as samples for comparison), or those of two compared genomic complexities (if two genomic DNA libraries are used). In this chapter, the isolation of the differential DNA sequences is achieved by using covalently modified subtracter DNAs to remove the tester homologues through covalent hybridization, which refers to a strong heat-stable binding interaction between the modified subtracter and nonmodified tester DNAs. The covalently modified subtracter refers to a DNA library that is chemically modified and thus capable of forming covalent bonds with homologous tester DNAs. The covalent modification comprises two chemical reactions, deamination of purines and carboxylation of pyrimidine bases. An amino-blocking agent is used to block or remove the amino-group of purine bases, such as acetic anhydride and alkaline acetic chloride. Then, a carboxylating agent is used to generate a carboxyl-group on the base structure of the subtracter's pyrimidines, such as a sodium cyanide/sulfuric acid mixture or hot alkaline potassium permanganate. The term "homologues" means DNA sequences that are common to both tester and subtracter DNA libraries.

In some cases, the isolated desired DNAs are so abundant in a cellular source that they can be directly detected and isolated without any enrichment. However, in most cases, the desired differential DNAs are too limited in amount to be detected and a polymerase chain reaction (PCR) is used to enrich the desired DNAs after subtractive hybridization [8,9]. When starting materials are limited, PCR is also used to enrich the subtracter and tester DNA libraries, via amplicon DNAs [10]. In short, such amplification is achieved by ligating a sequence-specific adapter to the ends of an endonuclease-restricted DNA library (amplicon), resulting in the generation of a primer-annealing region in the DNAs for subsequent PCR amplification.

2. SUBTRACTIVE HYBRIDIZATION METHODS

Several methods have been designed to detect and isolate differential DNA sequences which are present in one complementary [11] or genomic DNA [12] library but absent in the other. Representational difference analysis (RDA) was one of the first subtractive hybridization methods, and is particularly effective in elimination of homologous DNA sequences from two mutually compared DNA libraries. This method relies upon the generation of representative DNA libraries from both control cells (subtracter) and cells of experimental treatments, disorders, or morphological/functional changes (tester). The two DNA libraries are then denatured and mixed, resulting in the formation of subtracter–tester hybrid duplexes when the sequences are common to both tester and subtracter DNA libraries. By removing the common sequences and surplus subtracter, the DNAs remaining in the mixture are the desired DNA sequences only present in the tester library, which is related to the treatment, disorder, or change of interest.

The use of biotinylated subtracter DNAs is an improvement to increase the specificity of subtractive hybridization via streptavidin-based chromatography and to reduce the amount of subtracter needed for repeated hybridization. For example, Straus et al. [13] hybridized biotinylated-deletion-mutant genomic DNAs with restricted-wild-type genomic DNAs, and then subtracted the unwanted common hybrid duplexes with avidin-coated beads. The remaining sequences were ligated to a specific adapter and amplified by PCR, resulting in discovery of genomic deletions present in the mutant but absent in the wild type. Concurrently, Duguid et al. [14] performed a similar experiment but using a biotinylated double-stranded cDNA library isolated from a normal hamster brain to hybridize with a nonmodified cDNA library from a scrapie-infected hamster brain, generating biotinylated hybrid complexes that were removed by biotin-binding avidin resins. The cDNAs remaining in the suspension were amplified and confirmed to be scrapie-infected specific gene sequences. These experiments often require several cycles of subtractive hybridization because of the low efficiency of formation of the biotin–avidin complex and the contamination of subtracter fragments. These drawbacks cause an unfortunate increase in lab work and potential loss of desired DNA sequences during the necessary repeated subtraction steps.

Lin et al. [15] and Bjourson et al. [16] devised a further improvement in subtractive hybridization that employed a biotinylated primer and a uracil-containing deoxynucleotide mixture (e.g. mixture of dATP, dCTP, dGTP, and dUTP) to generate biotinylated uracil-containing subtracter DNAs (U-DNAs) for enzymatic subtraction. In these cases, control and experimental DNA libraries were isolated from cells under different conditions, restricted by an endonuclease, and ligated to different primer-specific adapters. Then, a special PCR, using the uracil-containing deoxynucleotide mixture, was performed to produce the biotinylated subtracter U-DNAs, which were then hybridized with nonmodified tester DNAs, resulting in the formation of biotinylated and uracil-containing heterohybrid duplexes that were common to both libraries. Because the biotinylated heterohybrids were removed by streptavidin–phenol–chloroform extraction and surplus subtracter U-DNAs were further digested by uracil–DNA glycosylase (UDG), the remaining tester DNAs were the desired differential DNA sequences. However, this method still required tedious work in biotinylation and at least two rounds of extraction and chromatography.

Subtractive hybridization

Finally, subtraction with covalent affinity was invented to simplify the process of subtractive hybridization, in which an aziridinylbenzoquinone (AZQ) interstrand cross-linking agent was used to covalently subtract common sequences in both tester and subtracter libraries [17,18]. Single-stranded tester was firstly hybridized with single-stranded subtracter to form hybrid duplexes, and then the AZQ was added to generate covalent bonds between the hybridized duplexes, caused by the cross-linking interaction of guanine and cytosine. Because the AZQ cross-links all double-stranded nucleic acid sequences, this kind of external covalent binding greatly facilitates homolog subtraction after hybridization. However, during subtractive hybridization only the single-stranded tester and subtracter can be used as starting materials due to the interstrand cross-linking action of the AZQ-like agents, which prevents analysis of genomic DNA samples, limits the experiment to the starting materials, and prevents adapter-specific amplification of the final results. These disadvantages impose more restrictions in sample selection, less stability of sample storage and less sensitivity in the final detection step in comparison with traditional subtraction hybridization. Further, detection of the final desired DNA sequences is accomplished by a nonspecific random-primer extension reaction, which lowers the specificity of the final results.

To reduce the drawbacks of AZQ-like cross-linking agents, we have developed a chemical modification procedure to generate covalent binding DNAs as the subtracter to capture complementary tester mRNAs or DNAs, as shown in Figures 1 and 2 [19]. Modified subtracter DNAs were generated by carboxylating the base structures of certain subtracter DNA nucleotides to introduce strong covalent affinity between the modified subtracter and the homologous tester DNAs. After that, the desired differential (heterologous) DNA sequences remained in the hydrogen-binding form, whereas the hybridized common (homologous) DNA sequences were covalently bound. Since the covalent binding cannot be broken in a PCR, there is no amplification of the homologous sequences but great amplification of the desired differential sequences. The desired DNA sequences found after such a covalently hybridized subtraction (CHS) and subsequent selective amplification are the DNA sequences that only exist in the tester but not in the subtracter DNA library. This technique is designed for the subtractive hybridization of differential sequences between two DNA libraries from distinct cell sources and will allow more efficient isolations in experiments on cancer formation, development of gene therapy, and understanding of pathological status and developmental regulation.

3. COVALENT MODIFICATION

As used here, covalent modification refers to a chemical reaction in which direct covalent bonding with nonmodified tester sequences and modified subtracter sequences is generated by amino-blocking and carboxylating reagents. The amino-blocking reagent is a chemical, which can block or remove the amino-group of a

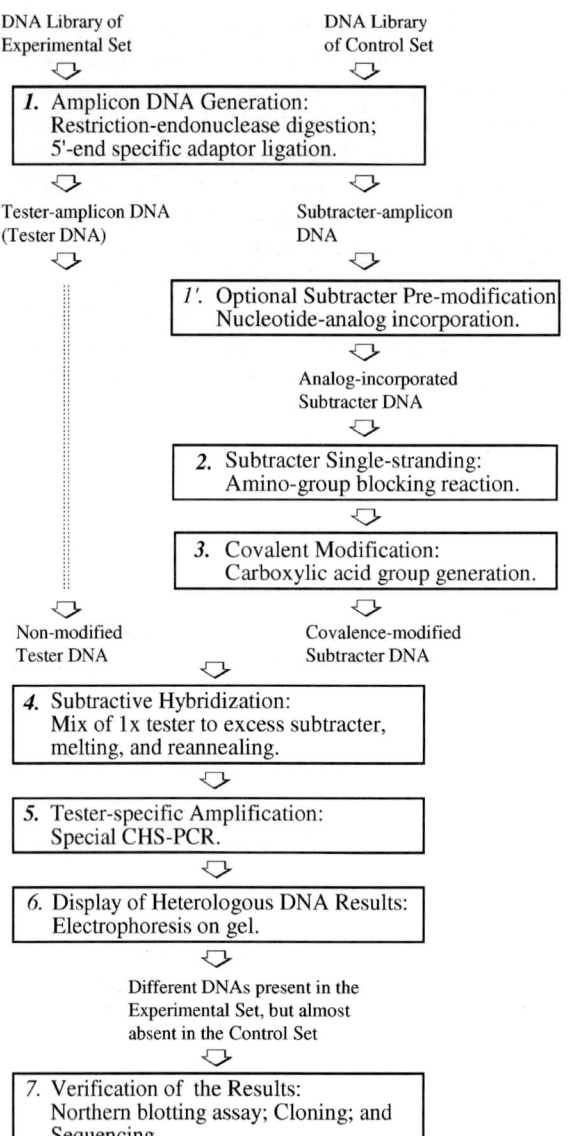

Figure 1. A flowchart protocol for the covalently hybridized subtraction (CHS) assay, illustrating the covalently bonded hybrid formation between tester and subtracter DNAs, and differential amplification steps after subtractive hybridization. The process is shown up to the final products of the first round CHS. To iterate another round of subtraction, the first difference products are used as tester following the same scheme to generate the second difference products and so on.

Subtractive hybridization

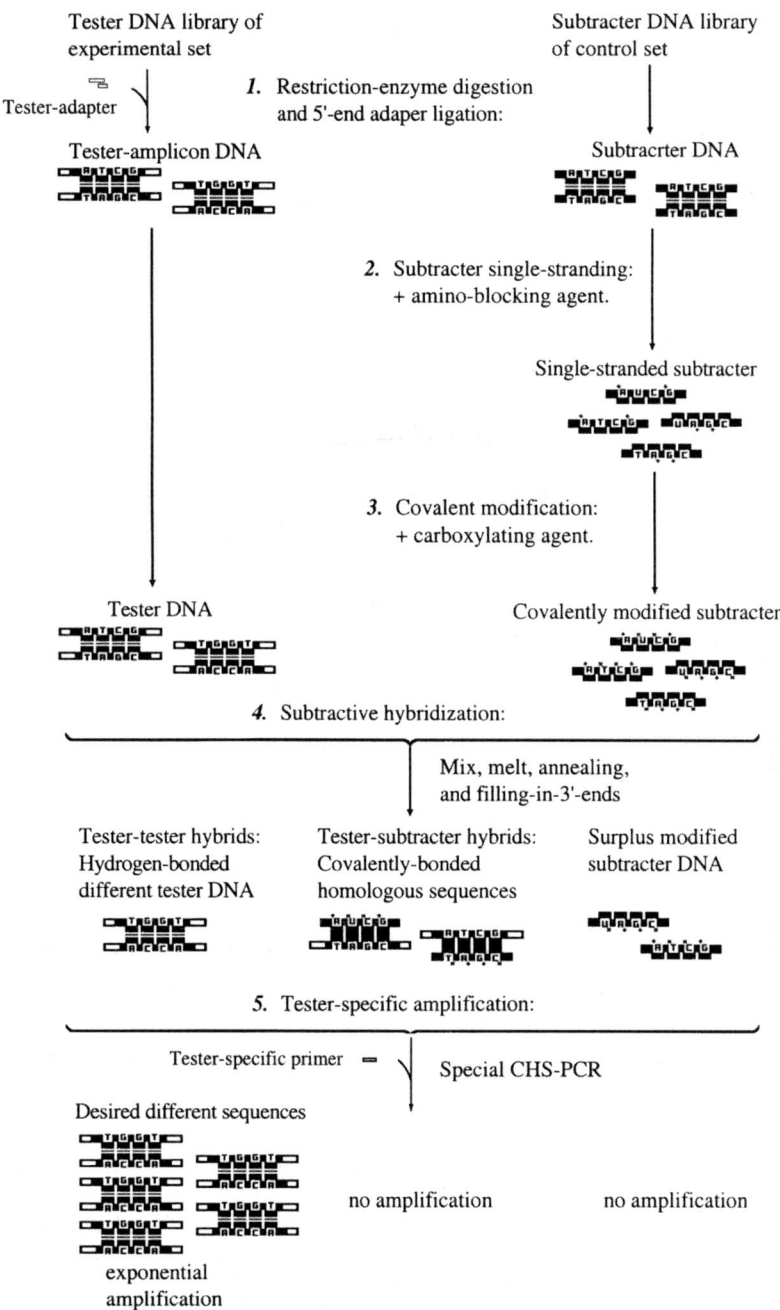

Figure 2. A schematic illustration of CHS subtractive hybridization in Figure 1.

nucleotide base, such as acetic anhydride or alkaline acetic chloride, while the carboxylating reagent is a chemical such as sodium cyanide/sulfuric acid mixture or hot alkaline potassium permanganate that can generate a carboxyl-group on the base structure of a modified subtracter sequence to allow covalent bonding with a nonmodified tester sequence.

The advantages of covalently modified subtracter sequences are as follows: First, during hybridization, the affinity between homologues can be greatly enhanced by covalent modification, such as the carboxyl-group on the C-5/C-6 of modified pyrimidines, resulting in peptide-like binding with the activating amino-group on the C-6/C-2 of nonmodified purines, respectively (Figure 3). Such covalent bonding between homologues fully inhibits any further reaction of the homologues and therefore reduces contamination with common homologues and surplus subtracter sequences. Second, the covalently modified subtracter sequences are single-stranded and inert to each other, resulting in a high binding efficiency in heterohybrid formation between the modified subtracter and nonmodified tester DNA strands rather than two modified subtracter strands. Third, because the covalent binding is an interstrand interaction occurring either between adenine and modified uracil or between guanine and modified cytosine, covalently pairing significantly increases the specificity of CHS, which occurs only between tester and subtracter sequences with highly matched base pairs.

In experiments (Figures 1 and 2), a subtracter DNA library is first prepared from the control samples, in the following steps: (a) restricting the initial DNA library with a restriction enzyme to generate 5′-cohesive termini on both ends; (b) ligating a specific adapter to the ends of the restricted DNAs to form a short template for binding with a specific PCR primer; and (c) incubating the adapter-ligated DNAs in PCR to permit the primer-dependent enrichment of the

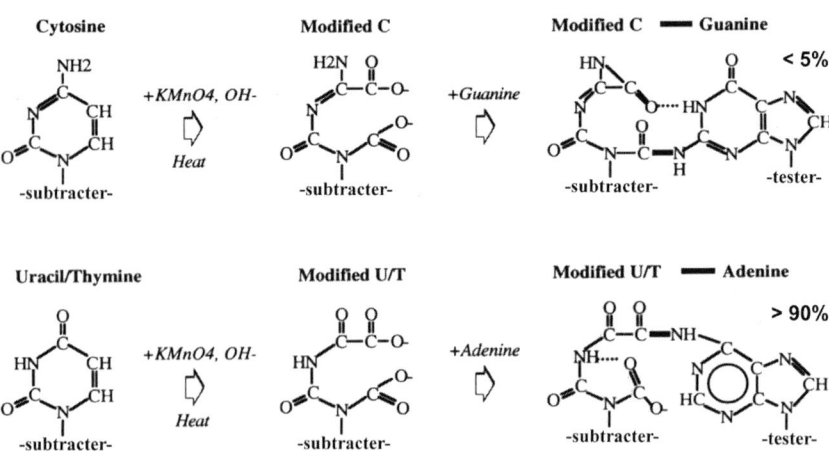

Figure 3. A detailed illustration of interstrand covalent bond formation in step 3 of Figure 2.

Subtractive hybridization

subtracter amplicon library. A subtracter amplicon library can be made from either a cDNA library or a genomic DNA library. The specific adapters and primers for PCR amplification are shown in Table 1.

Because covalent modification can be greatly facilitated by using some nucleotide analogs in the subtracter, we preferably incorporate deoxyuridine triphosphates (dUTP) into the subtracter amplicon sequences during PCR. For example, when 2'-deoxy-dUTPs instead of deoxythymidine triphosphates is used to generate the subtracter amplicon, the carboxylation reaction will occur only on the C-4 of uracil rather than the C-2, which is sometimes carboxylated if deoxythymidine triphosphates are used. Some alternative analog formulae are shown below, in which A, B, D, E, and F are selected from either a N or a CH group, G is a 2'-deoxy-D-ribose triphosphates, and X is a methyl group while Y is a H group and vice versa.

Table 1. The adapters and primers used in the CHS assay

Name	Application	Sequence
T-dpn2–24 mer	5'-ligation adapter; PCR specific primer for tester genomic DNA	5'-GCCACCAGAAGAGCGTG TACGCCA-3'
T-dpn2–12 mer	5'-ligation linker for tester genomic DNA	5'-GATCTGGCGTAC-3' (5'dephosphorylated)
S-dpn2–24 mer	5'-ligation adapter; PCR specific primer for subtracter genomic DNA	5'-CGGTAGTGACTCGGT TAAGATCGA-3'
S-dpn2–12 mer	5'-ligation linker for subtracter genomic DNA	5'-GATCTCGATCTT-3' (5'-dephosphorylated)
T-hpa2–24 mer	5'-ligation adapter; PCR specific primer for tester cDNA	5'-GCCACCAGAAGAGCGTG TACGTCC-3'
T-hpa2–11mer	5'-ligation linker for tester cDNA	5'-CGGGACGTACA-3' (5'-dephosphorylated)
S-hpa2–24 mer	5'-ligation adapter; PCR specific primer for subtracter cDNA	5'-CGGTAGTGACTCGGT TAAGATCGC-3'
S-hpa2–11 mer	5'-ligation linker for subtracter cDNA	5'-CGGCGATCTTA-3' (5'-dephosphorylated)

To prevent the reassociation of undesired subtracter–subtracter duplexes during hybridization, the amino-groups of subtracter DNAs must be blocked or removed by chemical blocking agents before covalent modification. The blocking reaction is preferably carried out by acetylating the amino-groups of the subtracter purines to form inactive acetamido-groups [20], which are incapable of binding to the carboxyl-groups of another modified sequence, which results in single-stranding the subtracter sequences. Acetic anhydride and alkaline acetic chloride are major ingredients in the amino-blocking reagent for CHS. Because the single-stranded subtracter DNAs will no longer protect their pyrimidine bases from oxidative modification, a carboxylating agent can easily oxidize the alkene, carbonyl or sometimes methyl groups [20] of the pyrimidine bases into activating carboxyl-groups, which are able to form covalent peptide-like bonds with the activating amino-groups of nonmodified tester sequences. Hot alkaline potassium permanganate is a major ingredient in carboxylating reagents due to the reaction of nucleophilic addition. Although adenine (A), guanine (G), cytosine (C), thymine (T), and uracil (U) bases were first used in the generation of covalently modified subtracter sequences, any nucleotide or its analog capable of being incorporated and modified into nucleotide sequences may be used as well. For example, such possible substitutes could be 2′-deoxy-uracil derivatives, para-toluene derivatives etc. that have the same capability of being covalently modified.

To increase subtraction efficiency, the subtracter is carboxylated on C_4 of uracil/thymine or C_5/C_6 of pyrimidines to generate sufficient affinity for peptide-bond formation with the C_6/C_2 amino-groups of the tester purines, respectively. Most frequently, the carboxylated group is generated on the C_5 of uracil and covalently bound to the C_6-amino-group of adenine. These covalent bonds cannot be broken during PCR amplification; therefore, unbound tester can be amplified with thermostable DNA polymerases like Taq DNA polymerases.

After covalent modification, the denatured and modified subtracter DNAs only covalently hybridize with the homologous tester sequences in a mild alkaline condition, resulting in an increase of binding efficiency amend more complete subtraction. The preferred medium is a heat-stable EPPS/EDTA buffer (pH 8.5) in which the blocked amino-groups of the subtracter are released. The homologous sequences are reassociated at a temperature sufficient to inhibit nonspecific hybridization, preferably between about 60–80°C, most preferably about 68–74°C. The ratio of the modified subtracter to the nonmodified tester is preferably between about 5:1 and 10:1.

4. SUBTRACTIVE HYBRIDIZATION WITH COVALENTLY MODIFIED SUBTRACTERS

The present protocol describes an improved subtractive hybridization method, called the CHS assay, for finding sequences which differ between two cDNA, or genomic DNA libraries. This method is primarily designed for quickly isolating

Subtractive hybridization

differentially expressed genes (either up- or down-regulated), easily detecting large genomic deletions/insertions, and precisely searching chromosome-specific loci. The principle of CHS is dependent on the subtraction efficiency of covalent binding between common sequences (homologues) during PCR or cloning, resulting in no amplification of the homologues. The principle is based on: single-stranding of subtracter DNAs, covalent modification of the subtracter base structures, hybridization of the modified subtracter DNAs with other nonmodified tester DNAs to subtract covalently bound common sequences in both DNA libraries, and then amplification of remaining heterogeneous tester DNAs to quantify the differentially expressed genes in the tester. In conjunction with adapter-ligation and adapter-specific PCR amplification, very small subtracter and tester libraries can be used as starting materials for comparison.

Covalently modified subtractive hybridization provides an easy, fast, and effective isolation of desired differentially expressed sequences from either cDNA or genomic DNA libraries, following the steps shown in Figure 2: (a) providing a library of tester DNAs, which is ligated to a tester-specific adapter for selective amplification; (b) mixing the denatured tester DNAs with a library of denatured subtracter DNAs, which have been modified by chemical agents so as to covalently bond with the tester homologues to form a denatured product; (c) permitting both tester and subtracter DNAs in the denatured mixture to form double-stranded hybrid duplexes composed of hydrogen-bonded homoduplexes and covalently-bonded heteroduplexes; and (d) amplifying the hydrogen-bonded homoduplexes with tester-specific primers, thereby providing a differential DNA library enriched in DNAs unique to the tester condition. Steps b–e can be repeated on the enriched DNA library for more differential enrichment.

The utilization of covalently modified subtracter DNAs avoids several limitations of subtractive hybridization. First, during subtractive hybridization, the affinity of the subtracter to its homologous tester can be greatly enhanced by covalent modification, resulting in peptide-like binding to the tester amino-groups (Figure 3). Such covalent peptide-like binding between subtracter and tester homologues fully inhibit their amplification in PCR and therefore minimizes the needed subtractive hybridization cycle. Second, the covalently modified subtracter DNAs are single-stranded, resulting in strong binding to tester but not subtracter DNAs. Third, because the covalent binding occurs only in interstrand base pairing either between adenine–thymine (–uracil) or between guanine–cytosine, this feature significantly increases the specificity of hybridized subtraction and the sensitivity of differential sequence detection.

For generation of tester amplicon DNAs, the DNA library of experimental cells is digested by a restriction endonuclease on both ends, preferably a four-cutter restriction enzyme, and ligated to a specific 5′-adapter. This ligated DNA library called a tester-amplicon is then used to generate tester DNA by a template-dependent primer-extension reaction in the presence of a tester-specific primer, preferably using the adapters and primers listed in Table 1. On the other hand, subtracter amplicon DNAs are amplified by a similar procedure but with a subtracter-specific adapter and primer, which share no affinity to the tester-specific ones. However, when the starting materials are abundant, the subtracter amplicon DNAs can be made by digestion with the four-cutter restriction enzyme but without the adapter ligation.

As shown in step 2 of Figure 2, blocking the activating amino-groups of the subtracter must be completed before covalent modification in order to prevent the formation of covalent bonds between subtracter and subtracter sequences. This blocking reaction is carried out by acetylating the amino-groups of subtracter purines to form inactive acetamido-groups [20], which are incapable of binding to the modified subtracter, resulting in single-stranded subtracter sequences. Acetic anhydride and alkaline acetic chloride are two preferred amino-blocking agents. Since the single-stranded subtracter cannot protect its base structures from oxidative agents, a carboxylating agent (as shown in step 3 of Figure 2) can easily oxidize the alkene, carbonyl, or sometimes methyl groups [20] of the subtracter bases to form carboxyl-groups, which are capable of forming covalent peptide-like bonds with the nonmodified amino-groups of the tester (Figure 3). Alkaline potassium permanganate and sodium cyanide/sulfuric acid mixtures are two preferred carboxylating agents based on the principles of oxidation and nucleophilic addition, respectively.

Following step 4 of Figure 2, tester DNAs are denatured and then hybridized with an excess amount of covalently modified subtracter at 68–72°C. The ratio of subtracter to tester DNAs is preferably in the range of about 5:1–10:1.

If the ratio is too high, enrichment of rare DNA sequences that only exist in the tester will not be obtained. If the ratio is too low, common nonspecific sequences will not be completely subtracted, and may cause false-positive contamination. The optimal ratio will vary depending on the stringency of subtractive hybridization between compared DNA libraries.

During the subtractive hybridization step, two kinds of hybrid duplexes are formed as follows: First, the tester–tester homohybrid duplexes, which consist of desired heterologous (differential) sequences only present in tester but almost absent in subtracter; And, the tester–subtracter heterohybrid duplexes which consist of common homologues present in both tester and subtracter. Because the linkage between tester and subtracter homologues is formed via covalent bonds, the resulting tester–subtracter heterohybrid duplexes cannot be amplified by PCR or vector cloning, in which each round of amplification requires the separation of the hybridized duplexes. Contrarily, the binding of tester–tester homohybrid duplexes is hydrogen-bonding (H-bond), which can be amplified by PCR or vector cloning. Therefore, the amounts of the desired differentially expressed sequences are greatly increased, whereas the contribution of common sequences will be negligible.

The subtracted tester DNAs can be subjected to another round of subtractive hybridization and amplification. The identified differential tester sequences are useful for DNA library selection assay and cloning analysis, representing the desired differential DNA sequences which are stimulated or up-regulated in the treated, mutated, infected, differentiated, or abnormal cells. By the same token, the roles of tester and subtracter DNAs can be performed in a reverse order to isolate the suppressed or down-regulated sequences. These identified sequences can also be used to probe the full-length mRNAs or cDNAs from the tester library (if cDNA tester is used as a sample), or to locate the deleted/inserted loci in a specific chromosome by *in situ* hybridization (if genomic DNA is used). The information so obtained will provide further understanding of a variety of diseases, physiological phenomena, and genetic functions.

5. APPLICATIONS

The CHS will be very useful in the identification of genes specifically involved in development, cell differentiation, aging, and a variety of pathological disorders, such as cancer, genetic defects, autoimmune diseases, and any other disorders related to genetic malfunction. The identification of these differentially expressed genes will lead to the determination of their open-reading frames and translated polypeptide products, which may contribute to specific drug-design or therapy for regulation of these genes. Such therapeutic approaches include transcription inhibitors, monoclonal antibodies against the expressed protein, anti-sense RNA, and chemicals that can interact with the gene or its protein product to cure or alleviate related disorders. For example, the methods of the present invention can be used to screen candidate genes for gene therapy to correct inherent defects. When a defect is caused by stimulation of a specific unknown gene, the identification of this gene will help the design of antisense oligonucleotides against the gene or production of monoclonal antibodies against the corresponding protein product.

Alternatively, the CHS subtractive hybridization can also be used to screen some types of chromosomal abnormalities, such as deletion and insertion. Because genomic DNA fragments of less than 1 kb are prepared by restriction enzyme digestion before subtractive hybridization [11], the target deletion or insertion must be larger than this size for efficient amplification. The identification of these chromosomal deletions or insertions may contribute to the diagnosis or prognosis of certain virus infections, inherent problems, or developmental defects. For example, retinoblastoma (RB) gene deletion occurs in hereditary RB. If the deletion can be identified early, this information might allow therapeutic intervention to prevent the onset of RB.

Although the CHS assay is primarily designed for medical and biological research, the method will also be useful in pharmaceutical, agricultural, and environmental research, which involving biological systems. For example, when gene expression is compared between drug-treated and nontreated cells, the results may indicate the mechanism by which the drug acts. For another example, when the genomic DNA from disease-resistant plant cells is compared with that from disease-susceptible plant cells, the results will indicate the candidate loci for the resistance gene(s). Thus, CHS can provide a variety of information critical to understanding changes in gene expression across different genomes.

With the high efficiency of covalent modification and CHS, the labor- and time-consuming factors in subtractive hybridization assay can be reduced to the minimum. Also, the preparation of covalently modified subtracter is cheaper and more efficient than that of other modification methods. Most importantly, covalent modification can be carried out continuously with only a few changes of buffers. Taken together, these special features make CHS a fast, simple, effective, and inexpensive protocol for quickly isolating differentially expressed gene sequences of interest.

6. PROTOCOLS

6.1 Preparation of Subtracter and Tester DNA Libraries

For example, LNCaP cells, a prostate cancer cell line, were grown in Dulbecco's Modified Eagle's medium (DMEM) supplemented with 2% fetal calf serum. For a 3-day activin treatment, six dishes of experimental cells were treated with 1.5 ml of 200 μgl^{-1} activin per day, while two dishes of control cells were not treated. On the fifth day after the first treatment, a 55% reduction in growth was observed in the experimental (tester) cells compared to the control (subtracter) cells by both microscopy and cell counting. All cells were trypsinized and total RNAs were isolated with a TRIzol reagent (GIBCO/BRL), respectively. After 1 μg of RNAs were mixed with the oligo-dT primer and heated to 65°C (10 min), a reverse transcription (RT) reaction was performed using a cDNA cycle kit (Invitrogen, CA), and all RT products (2 μg) were double-stranded with a DNA polymerase–ligase–RNase cocktail mixture [21]. About 1 μg of subtracter cDNAs was then digested by a four-cutting enzyme, such as Hpa2 (20 U for 5 h, 37°C), and ligated with a subtracter-specific primer 5'-pCGGTAGTGAC TCGGT TAAGA TCGC-3' in the 5'-end, while tester cDNAs were ligated with a tester-specific primer 5'-pGCCACCAGAA GAGCGTGTAC GTCC-3' in the same manner. This produced the subtracter and tester cDNA libraries, respectively.

6.2 Covalent Modification of Subtracter DNAs

Subtracter cDNAs were diluted and amplified by the PCR with the subtracter-specific primer. During PCR, the recessed 3'-ends of the subtracter were filled by a Taq-like thermostable DNA polymerase (7 min, 72°C) with dATP (2 mM), dCTP (2 mM), dGTP (2 mM), dTTP (0.5 mM), and dUTP (3.5 mM). A 30-cycle amplification was performed (1 min, 95°C; 1 min, 72°C; 3 min, 68°C), and the amplified products, namely U-DNA sequences, were recovered by a Micropure–EZ column (Microcon) and resuspended in 10 μl of ddH$_2$O. About 50 μl of pure acetic anhydride was added (3 min, 94°C) into the U-DNAs of the subtracter to block the activating amino-groups by acetylation, by which the subtracter sequences also become single-stranded, and then the reaction was neutralized by 500 μl of Tris buffer (10 mM, pH 7.4). After the acetylated U-DNAs were recovered by a Micropure–EZ column and resuspended in a total of

Subtractive hybridization 181

10 µl of Tris buffer (10 mM, pH 7.4), 20 µl of an alkaline potassium permanganate reagent (1 mM $KMnO_4$, 1 mM NaCl, pH 10.0) was added (3 min, 80°C) to generate carboxyl-groups on C-5/C-6 of uracil/cytosine in the subtracter which can covalently bind to the amino-groups on C-6/C-2 of adenine/guanine of the tester, respectively. The carboxylated subtracter was finally recovered by a Micropure–EZ column and resuspended in a total of 10 µl of 10 mM N-[2-hydroxyethyl]piperazine-N'-[3-propanesulfonic acid] and [ethylenedinitrilo] tetraacetic acid (EPPS/EDTA) mixture buffer (pH 8.5). The modified subtracter must be used immediately for subtractive hybridization with the tester.

6.3 Subtractive Hybridization and CHS–PCR Amplification

For hybridization, the tester DNAs (500 ng) from the experimental cells were mixed with the covalently modified subtracter (3 µg) in 10 µl of the EPPS/EDTA buffer and denatured at 94°C for 5 min. The mixture was then vortexed, added to 1 µl of 5 M NaCl to adjust salt concentration, and incubated at 70°C (16 h). The hybridized DNAs were finally diluted with 20 µl $MgCl_2$ solution (2.5 mM) and amplified by PCR with the B-specific primer. A 20-cycle amplification was performed (1 min, 95°C; 3 min, 73°C) after nick translation with *Escherichia coli* DNA polymerase 1 plus T_4 DNA polymerase 3:1 mixture (5 min, 37°C without dNTPs; 35 min, 37°C with dNTPs), and the resulting products were phenol-extracted, isopropanol-precipitated, and resuspended in 15 µl of 10 mM Tris buffer for display on a 3% agarose electrophoresis gel (Figure 4, upper panel). The DNA bands shown on the gel were excised and recovered by a gel-extraction kit (Qiagen) to give the final difference products, and then further purified by a 4% nondenatured polyacryamide gel. The processes of subtractive hybridization with modified subtracter and selective amplification was repeated until clear bands are observed on the gel. Both cDNAs and genomic DNAs were processed in the same way as mentioned above.

As shown in the Figure 4 (upper panel), the subtracter DNA sequences were amplified with subtracter-specific primer (lane 2), the tester DNA sequences were amplified with tester-specific primer (lane 3), the subtracter DNAs were amplified with tester-specific primer (lane 4), the tester DNAs were amplified with subtracter-specific primer (lane 5), the subtracter DNAs were self-subtracted by modified subtracter U-DNA and amplified with subtracter-specific primer (lane 6), the tester DNAs were self-subtracted by modified tester U-DNA and amplified with tester-specific primer (lane 7), the subtracter DNAs were subtracted by modified tester U-DNA sequences and amplified with subtracter-specific primer (lane 8), and the tester DNAs were subtracted by modified subtracter U-DNA sequences and amplified with tester-specific primer (lane 9). The self-subtraction of subtracter to subtracter (lane 6) and tester to tester (lane 7) shows complete elimination of all sequences, while the mutual subtraction between tester and subtracter (lanes 8 and 9) presents different final results on the electrophoresis gel, indicating different gene expression in

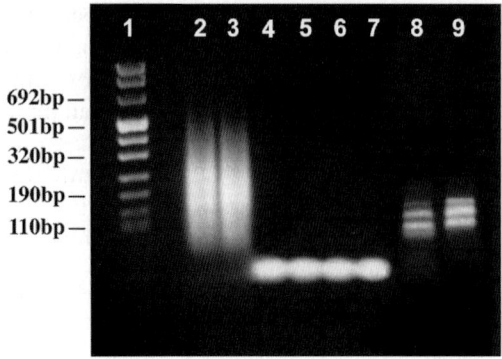

CHS probes	Gene (size)	Homology	Change % (σ)	Function
Down-regulated				
probe 1 (LC2)	Myosin-like (1.1 kb)	99%	−47.6 (1.46)*	cytoskeleton
probe 2 (LC3)	CD168 (2.8 kb)	95%	−55.9 (1.54)*	cytoskeleton
probe 3 (LC8)	novel (2.0 kb)		−64.9 (3.09)**	?
probe 4 (LC9)	Helicase motif-like (1.3 kb)	95%	−60.5 (4.66)*	replication
probe 5 (LC12)	Pax2 (3.7 kb)	97%	−77.0 (2.37)**	proliferation
probe 6 (LC13)	eIf-4A1 (1.7 kb)	100%	−53.5 (0.00)*	translation
Up-regulated				
probe 7 (LT1)	novel (0.8 kb)		+728 (1.53)**	?
probe 8 (LT6)	rBub1-like (1.8 kb)	100%	+265 (4.38)**	spindle lesion apoptosis
probe 9 (LT11)	p53 (1.7 kb)	100%	+213 (5.35)**	G1 arrest

* $n = 3$, $p < 0.01$
** $n = 4$, $p < 0.01$

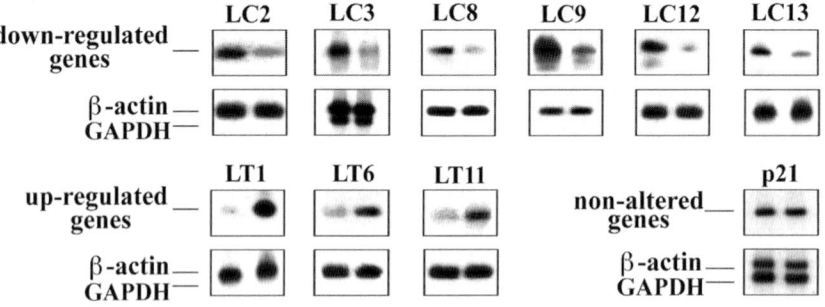

Figure 4. Identification of differentially expressed genes in human prostate cancer cells. LNCaP, after activin treatment.

Subtractive hybridization

the tester and the subtracter. The misuse of PCR primer (lanes 4 and 5) causes no amplification due to the specific affinity of the primer for its own adapter. Thus, based on the results of Figure 4, the CHS assay is sensitive and specific enough to subtract all homologous DNAs and distinguish the differential gene transcripts between two strands of DNA libraries after PCR amplification. Compared to RDA, the CHS assay has the advantages of low background and high efficiency, and usually the same results were obtained after one round of subtractive hybridization.

Sequence results for the final differentially expressed genes are shown in Figure 4, middle panel. The p53 gene (LT11) was previously known to be up-regulated in the activin-treated LNCaP cells. The known down-regulated genes (LC2, 3, 9, 12, 13) in the upper lane are related to cellular physiological functions, while the known up-regulated genes (LT6, 11) are involved in either cell-cycle regulation or apoptosis or both. All genes listed are transcriptionally altered by at least twofold. The size of each identified gene transcript is deduced from individual Northern blots, and the homology shown here indicates the sequence homology between the identified fragment and its deduced gene, rather than the entire identified sequence. Figure 4, bottom panel, shows an autoradiogram of positive Northern blots hybridized to the final differentially expressed genes displayed in CHS. The upper row (LC2–LC13) indicates six down-regulated genes mainly present in untreated LNCaP cells but not in the activin-treated cells, while the lower row (LT1, LT6, and LT11) shows three up-regulated genes significantly increased after activin treatment. The Northern blot of p21 is a negative control for activin-induced transcriptional alteration in LNCaP cells.

6.4 Covalent Binding Efficiency and Subtractive Stringency of CHS

To confirm the binding efficiency and subtraction efficacy of covalently modified subtracter, we used an apoptosin fragment as a target tester homologue, sharing about 300 base nucleotides with 70% homology to a subtracter sequence. Equal amounts of the tester fragment and subtracter were mixed, denatured, and subjected to DNA nuclease digestion, with or without hybridization. Hybridization was performed at 94°C for 3 min and then 70°C for 16 h in EEx3 buffer (30 mM EPPS, pH 8.5 at 20°C; 3 mM EDTA). Nuclease digestion was performed with a mixture of DNase I and nuclease S1 (50 U each, Roche) at 25°C for 10 min in 1x NS1 buffer (0.2 M NaCl, 50 mM sodium acetate, pH 4.5; 1 mM $ZnSO_4$, 0.5% glycerol). The results were electrophoresed on a 2% agarose gel as shown in Figure 5, left panel, showing lane 1, double-stranded apoptosin DNA fragments (200 ng); lane 2, single-stranded antisense subtracter (100 ng); lane 3, hybridization of the targeted DNA fragment and subtracter after nuclease digestion; lane 4, same as lane 3 but without digestion; and lane 5, hybridization of the targeted DNA to the subtracter (100 ng each) after nuclease digestion.

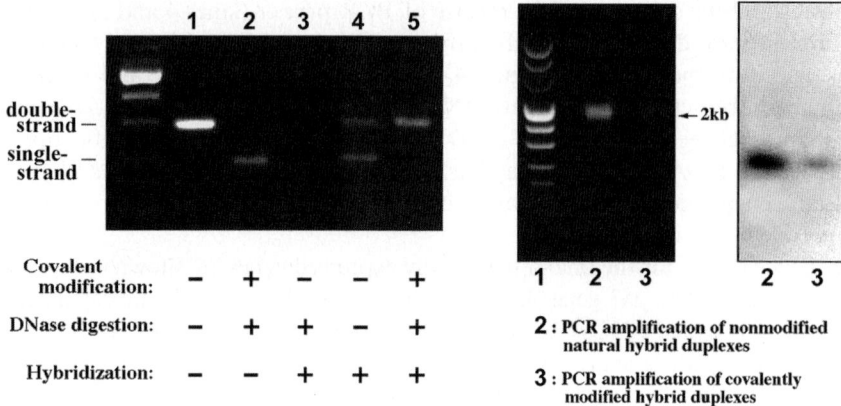

Figure 5. Detection of genomic deletion in retinoblastoma cells, Y-79.

The affinity of the subtracter for its homologous tester sequence is greatly enhanced by the covalent modification. As shown in Figure 5, left panel, the subtracter provides 100% binding efficiency (lane 5) compared to 53% in traditional probes (lane 4). Also, the interaction between subtracter and subtracter is prevented by acetylation of the amino-group of its purines, resulting in high binding efficiency between the subtracter and its targeted tester. Moreover, because of the covalent modification, the modified structures of the subtracter are highly resistant to nuclease digestion (lane 2), even after binding with the targeted sequences (lane 5). Such selective covalent bonding fully inhibits the functional activity of the targeted gene. Since covalently bound hybrid duplexes cannot be separated in cells, any enzymatic activity requiring single-stranded nucleotide templates will be effectively shut down. It has been shown that even a PCR cannot be performed through the covalently bound hybrid duplexes [22].

6.5 Identification of Genomic Deletion Using CHS

We also used the CHS assay to screen the genomic deletion in Y-79 cells and successfully identified a genomic fragment existing in the chromosome of normal retina cells but not in that of Y-79 cells (Figure 5). Y-79, a RB cell, has been known to contain an RB gene-deletion in its genome [23]. As a model of genomic subtraction by CHS, the genomic DNAs of normal retina and Y-79 cells were isolated by the IsoQuick nucleic acid extraction kit (Microprobe), respectively, restricted with Hpa2, and ligated to T-hpa-adapter and S-hpa-adapter, respectively, to give the tester (normal retina cells) and subtracter (Y-79). The sizes of restricted genomic DNAs were about 1–3 kb, which can be efficiently amplified by PCR. The uridine-analog was incorporated and covalently modified in subtracter genomic DNAs as described in Section 6.2. The subtractive hybridization and selective amplification were performed as described in Section 6.3. The resulting differential genomic DNA sequence(s) of the tester were fractionized on a 1% agarose gel and confirmed by Southern blot analysis as shown in Figure 5, right panel. A signal was detected on the Southern blots of normal DNAs but not Y-79 DNAs, indicating an at least 2 kb deletion in RB exon 2 of the Y-79 genome.

REFERENCES

1. Davis et al. (1987) Expression of a single transfected cDNA converts fibroblasts to myoblasts. Cell 51:987–1000
2. Lamar et al. (1984) Y-encoded, species-specific DNA in mice: evidence that the Y chromosome exists in two polymorphic forms in inbred strains. Cell 37:171–177
3. Maddon et al. (1985) The isolation and nucleotide sequence of a cDNA encoding the T cell surface protein T4: A new member of the immunoglobulin gene family. Cell 42:93–104
4. Littman et al. (1985) The isolation and sequence of the gene encoding T8: a molecule defining functional classes of T lymphocytes. Cell 40:237–246
5. Chang et al. (1989) Cloning and expression of a gamma-interferon-inducible gene in monocytes: a new member of a cytokine gene family. Int Immunol 1:388–397
6. Nussbaum et al. (1987) Isolation of anonymous DNA sequences from within a submicroscopic X chromosomal deletion in a patient with choroideremia, deafness, and mental retardation. Proc Natl Acad Sci USA 84:6521–6525
7. Kunkel et al. (1985) Specific cloning of DNA fragments absent from the DNA of a male patient with an X chromosome deletion. Proc Natl Acad Sci USA 82:4778–4782
8. Wang et al. (1991) A gene expression screen. Proc Natl Acad Sci USA 88:11505–11509
9. Coochini et al. (1993) Identification of genes up-regulated in differentiating nicotania glauca pith tissue, using an improved method for construction a subtractive cDNA library. Nucleic Acids Res 21:5742–5747
10. Wicland et al. (1990) A method for differential cloning; Gene amplification following subtractive hybridization. Proc Natl Acad Sci USA 87:2720–2724
11. Lisitsyn N, Wigler M (1993) Cloning the differences between two complex genomes. Science 259:946–951
12. Hubank M, Schatz DG (1994) Identifying differences in mRNA expression by representational difference analysis of cDNA. Nucleic Acids Res 22:5640–5648
13. Straus et al. (1990) Genomic subtraction for cloning DNA corresponding to deletion mutations. Proc Natl Acad Sci USA 87:1889–1893
14. Duguid et al. (1990) Library subtraction of in vitro cDNA libraries to identify differentially expressed genes in scapic infection. Nucleic Acids Res 18:2789–2792
15. Lin SL, Ying SY (1999) Differentially expressed genes in activin-induced apoptotic LNCaP cells. Biochem Biophys Res Commun 257:187–192
16. Bjourson et al. (1992) Combined subtraction hybridization and polymerase chain reaction amplification procedure for isolation of strain-specific rhizobium DNA sequences. Appl Environ Microbiol 58:2296–2301
17. Hartley JA, Berardini M, Ponti M, Gibson NW, Thompson AS, Thurston DE, Hoey BM, Butler J (1991) DNA cross-linking and sequence selectivity of aziridinylbenzoquinones. Biochemistry 30:11719–11724
18. Hampson IN, Pope L, Cowling GJ, Dexter TM (1992) Chemical crosslinking subtraction; A new method for the generation of subtractive hybridization probes. Nucleic Acids Res 20:2899
19. Ying SY, Lin SL (1999) High performance subtractive hybridization of cDNAs by covalent bonding between specific complementary nucleotides. BioTechniques 26, 966–979
20. Solomons et al. (1996) Organic chemistry, 6th edn. Wiley, New York, pp 693, 803–804

21. Ueli et al. (1983) A simple and very efficient method for generating cDNA libraries. Gene 25:263–269
22. Ying SY, Chuong CM, Lin SL (1999) Suppression of activin-induced apoptosisby a novel antisense strategy. Biochem Biophys Res Commun 265:669–673
23. Lee, EY, Bookstein R, Young LJ, Lin CJ, Rosenfeld MG, Lee WH (1988) Molecular mechanism of retinoblastoma gene inactivation in retinoblastoma cell line Y79. Proc Natl Acad Sci USA 85:6017–6021

CHAPTER 8

COINCIDENCE CLONING: ROBUST TECHNIQUE FOR ISOLATION OF COMMON SEQUENCES

ANTON A. BUZDIN

Shemyakin-Ovchinnikov Institute of Bioorganic Chemistry, Russian Academy of Sciences, 16/10 Miklukho-Maklaya, 117997 Moscow, Russia
Phone: +(7495) 3306329; Fax: +(7495) 3306538; E-mail: anton@humgen.siobc.ras.ru

Abstract:	Coincidence cloning (CC) is aimed at finding DNA fragments, which are common to the samples under study. The nature of these input DNAs may be genomic or cDNA, cloned or uncloned. The approach is based on cloning identical (or almost identical) nucleotide sequences belonging to different fragmented genomic DNA or cDNA pools, while discarding sequences that are not common to both. Early versions of the CC technique were not very efficient. Their most serious disadvantage was rather low selectivity, so that the resulting libraries of the fragments contained large amounts of sequences unique to one of the two sets of DNA fragments under comparison. To avoid this, Azhikina and colleagues were the first to exploit the technique of selective polymerase chain reaction (PCR) suppression, which strongly increased the efficiency of CC. Another important problem is the "nonspecific" imperfect hybridization between nonorthologous repetitive elements or short sequence-similar sequences, which produces chimeric clones representing an impressive fraction of the libraries (up to 60%, when complex genomic mixtures). An important improvement in this technique comprises treatment of the hybrids with the nucleases, specifically recognizing single-nucleotide mismatches or more extended loop regions. This results in digestion of improperly matched hybrids (primarily chimeras), whereas perfect, nonchimeric heteroduplexes are greatly enriched in the final mixture (up to 96% or more).
Keywords:	Common sequences, cloning selection, physical separation, PCR-only-based approaches, repetitive element, repetitive sequence, chimeric clone, chimera, chimeric duplex, nonrepetitive DNA, true genomic sequence representation, competitor DNA, enrichment factor, evolutionary conserved sequence, nonmethylated genomic sites coincidence cloning (NGSCC), RIDGES, methylation site, unmethylated CpG, genomic repeat expression monitor (GREM), promoter-active repeats.
Abbreviations:	BAC, bacterial artificial chromosome; CC, coincidence cloning; ELT, expressed LTR tag; dNTP, deoxyribonucleotidetriphosphate; GREM, genomic repeat expression monitor; LTR, long terminal repeat; NGSCC, nonmethylated genomic sites

coincidence cloning; PCR, polymerase chain reaction; PS, PCR suppression effect; SAGE, serial analysis of gene expression.

TABLE OF CONTENTS

1. Introduction 188
2. Cloning Selection of Heteroduplexes 191
3. Physical Separation of Hybrids 193
4. PCR-only-based Approaches 193
5. Future Prospects 202
6. Protocols 202
   6.1 Cloning Similarities in Genomic DNAs 202
   6.2 Cloning and Presice Mapping of Transcribed
       Repetitive Elements 204
   6.3 Finding Methylated or Unmethylated CpGs
       in Large Genomic Contigs 207
References 209

1. INTRODUCTION

Unlike subtractive hybridization, which is aimed at the recovery of differential sequences, the approach termed "coincidence cloning" (CC) was developed to find DNA fragments, which are common to the samples under study. According to Rebeca Devon and Anthony Brookes, who were among the inventors of this method, "the term *coincidence cloning* encompasses a wide range of methodologies, the aim of which is to isolate DNA sequences which occur in both of two input DNA sources. The nature of these input DNAs may be genomic or cDNA, cloned or uncloned. If the input DNAs are genomic then the product will be enriched for useful markers co-occurring between the two. If the input DNAs comprise one genomic resource and one cDNA resource the product will contain genes mapping to that particular genomic region" (Devon and Brookes 1996).

The approach is based on cloning identical (or almost identical) nucleotide sequences belonging to different fragmented genomic DNA or cDNA pools, while discarding sequences that are not common to both (Devon and Brookes 1996). By comparing genomic DNA fragments with fragments of any similarly fragmented locus (cloned in the form of bacterial artificial chromosomes – BACs, cosmids etc.), one can select and identify the genomic fragments belonging to this locus.

To this end, both fragmented DNAs under comparison are specifically tagged (e.g. by ligating different terminal adapter oligonucleotides), mixed, denatured, and hybridized, followed by the isolation of duplexes having both specific tags (i.e. "heterohybrid" products derived from both samples, which are common to both tagged DNA mixtures). The former step is the key stage of the whole procedure, as an efficient isolation of proper hybrids provides construction of CC libraries, truly enriched in common sequences (Figure 1).

Three major approaches for distinguishing heterohybrids from homohybrids can be mentioned. First (*cloning selection*), hybridizing input DNAs (source A and B) may be flanked by different restriction enzyme recognition sites (e.g. by introducing such sites in ligated terminal adapters). Restriction endonuclease-treated hybridization products are ligated into the vector predigested to produce sticky ends complementary to those appeared in heterohybrids, and random transformants are screened. Second (*physical separation*), one of the source DNAs can be immobilized on a solid support (e.g. source A DNA bound to magnetic beads) and, following hybridization, the support-attached fraction (which contains unhybridized source A molecules, source A homohybrids, and heterohybrids source A/B) is separated from the liquid-phase fraction (all other types of DNAs). The solid-phase fraction is further polymerase chain reaction

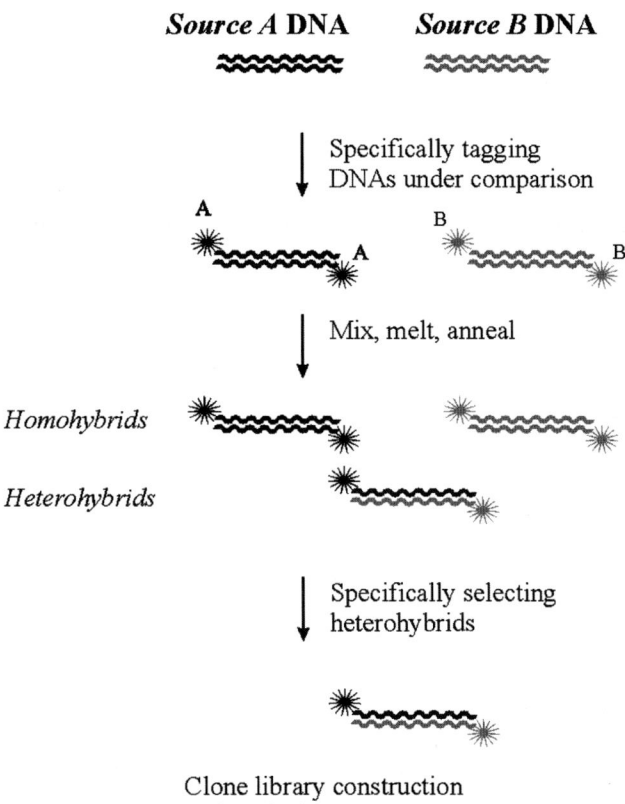

Figure 1. Generalized scheme for coincidence cloning (CC) approaches. Proper isolation of source 1–2 heteroduplexes is the key procedure in all CC-based techniques.

(PCR) amplified with primers specific to source B linkers, resulting in the more or less specific amplification of source A/B heterohybrids. Third (*PCR-only-based approaches*), source A and B DNAs are ligated to different oligonucleotide linkers, and further hybridization products are amplified with pairs of source A- and B-specific primers (described below in this chapter).

Early versions of the CC technique were not very efficient and, therefore, have not been widely used. Their most serious disadvantage was rather low selectivity, so that the resulting libraries of the fragments contained large amounts of sequences unique to one of the two sets of DNA fragments under comparison. To avoid this, Azhikina and colleagues exploited the technique of selective PCR suppression (PCR suppression effect is described in detail in Chapter 2; Diatchenko et al. 1996), which strongly increased the efficiency of CC (Azhikina et al. 2004; Azhikina and Sverdlov 2005; Azhikina et al. 2006).

Another important problem is the "nonspecific" imperfect hybridization between nonorthologous repetitive elements or short sequence-similar sequences, which produces chimeric clones representing an impressive fraction of the libraries (up to 60%, when complex genomic mixtures like mammalian DNAs are hybridized). Such chimeric clones make the library analysis problematic, as (1) it is frequently difficult to distinguish chimeras from "true" hybrids, especially for unsequenced genomes, and (2) more sequencing work is needed to get substancial data sets. The recent paper by Chalaya et al. (2004) describes an important improvement in this technique, which comprises treatment of the hybrids with the nucleases, specifically recognizing single-nucleotide mismatches, or more extended loop regions. This results in digestion of improperly matched hybrids (primarily chimeras), whereas perfect, nonchimeric heteroduplexes are greatly enriched in the final mixture (up to 96% or more).

Also, when considering CC, one has to clearly realize that kinetical requirements are extremely important for the success of the whole procedure (see Chapters 1 and 10). When very complex DNA mixtures are hybridized, it is very difficult to obtain satisfactory reassociation values for a reasonable time. In particular, when genome size increases beyond 5×10^8 bp (complexity comparable with that of arabidopsis or drosophila genomes), the kinetics of hybridization start to become an increasingly important factor limiting reassociation (Milner et al. 1995). To enhance the kinetics of hybridization, increased hybridization times, higher DNA concentrations, longer DNA fragments, and the use of techniques that enhance the rate of reassociation (reviewed in Chapter 10, Section 2), are recommended.

The last (but not the least) problem is a very low reassociation rate of nonrepetitive DNAs. Indeed, DNA reassociation rate for each particular fragment is proportional to the square of its concentration; therefore, repetitive elements presented in a genome by ~10 (some pseudogenes), ~1000 (several mammalian endogenous retroviral families), or ~1,000,000 (human *Alu* retrotransposons) copies will hybridize, respectively, 10^2-, 10^6-, and 10^{12}-fold faster than fragments representing unique genomic sequences. As the latter's reassociation rate is incomparably lower, ~99% reassociation may take months or years, an enormous background of repetitive sequences appears when reasonable (approximately days) hybridization time is used. In this case, a great majority of double-stranded molecules in solution are hybridized repeats, whereas unique sequences mostly remain in a single-stranded form. Interestingly, the famous primate-specific *Alu* retrotransposons were discovered first using CC of sequences co-occurring in a set of comparing genomes (Nelson et al. 1989; Aslanidis and de Jong 1991). Therefore, the true genomic sequence representations will be enormously biased in the resulting libraries.

A rather efficient attempt to improve the situation is the addition of competitor DNA fractions enriched in genomic repeats (Sambrook and Russell 2001) into hybridization mixture. Such competitors are mostly fractions of quickly reassociating double-stranded DNA, purified from single-stranded DNA using hydroxyapatite column chromatography. Such fractions, for example, commercially available "C_0t A" and "C_0t B" DNAs from Gibco BRL (USA) are greatly enriched in genomic repeats and may be used to decrease the background of repetitive sequences in cloned libraries. To this end, the initial genomic DNAs to be hybridized must be fragmented (either by sonication or digestion with restriction endonucleases), tagged (through the ligation of adapter sequences, by incorporation of biotin or other signal molecules), denatured and allowed to hybridize in the presence of competitor DNA, taken in a 100–1000-fold weight excess.

The major part of genomic repeats presenting in the sample DNA will hybridize to competitor DNA. At the next stage, it is crucial to isolate the "proper" hybrids (those formed by original genomic DNA fragments) while discarding genomic–competitor DNA duplexes and single-stranded DNAs. This can be done, for example, by using selective PCR amplification of the proper hybrids (see protocol in Chapter 10, Section 4.2), or using biotin–streptavidin systems. To our experience, the use of C_0t A DNA taken in 100-fold weight excess results in a decrease of genomic repeat-containing clones from ~93% to 76–78% of the libraries, when nonsimplified frequent-cutter endonuclease-digested human genomic DNA is hybridized (Chalaya et al. 2004), which is significantly closer to the natural genomic occurrence of repetitive elements, occupying approximately two thirds of human DNA (Lander et al. 2001; Venter et al. 2001).

To measure CC efficacy, the characteristic value termed *enrichment factor* was proposed. The enrichment factor is given by the ratio of the relative mass representation of a target sequence in the product DNA mixture, over its relative mass representation in the most complex of the two DNA sources (Devon and Brookes 1996).

2. CLONING SELECTION OF HETERODUPLEXES

This approach utilizes preliminary labeling of hybridizing DNA sources with ligated oligonucleotide adapters, harboring different restriction sites. For example (Figure 2), source A DNA is tagged by linkers having *Not*I restriction site,

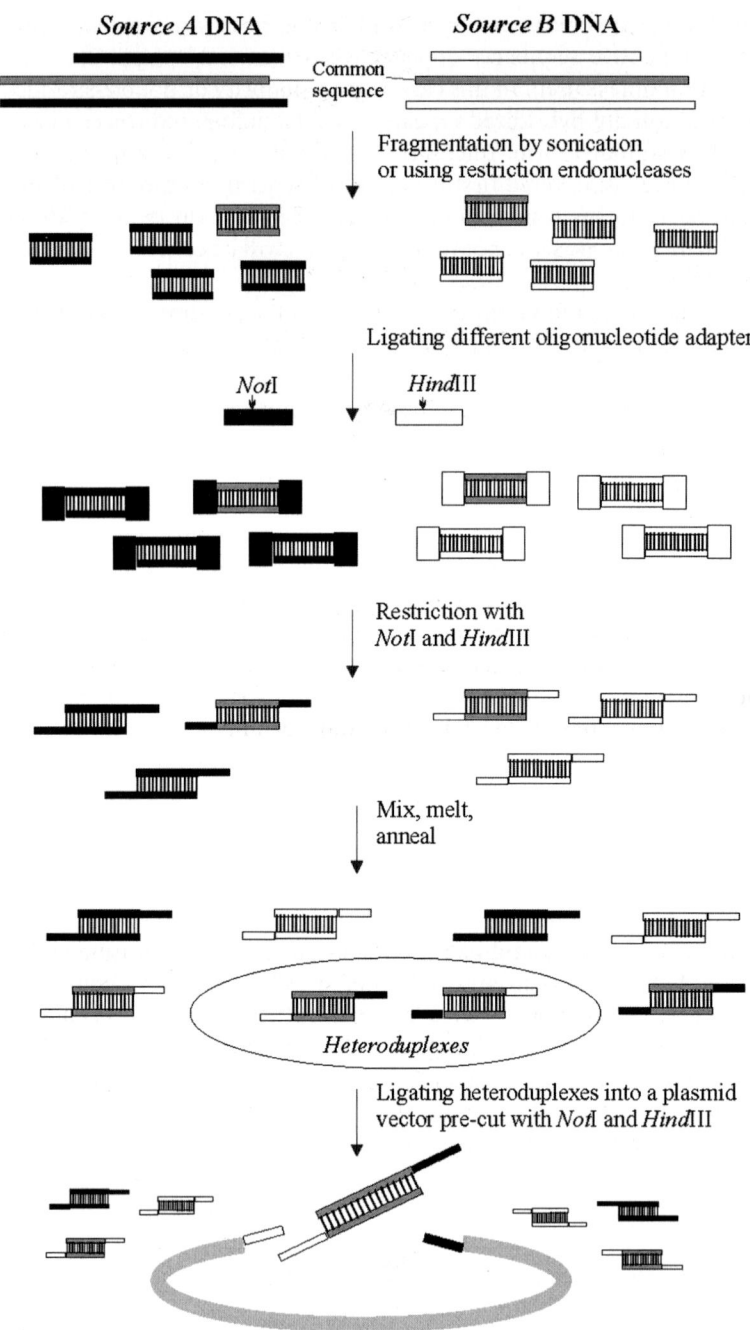

Figure 2. Variant of coincidence cloning (CC) utilizing selective cloning of the heteroduplexes using their unique combination of terminal restriction sites, differing them from all types of homoduplexes.

whereas linkers for source B have recognition sequence for *Hind*III enzyme. Both DNA samples are mixed, denatured, and allowed to hybridize. Following filling-in the DNA termini, the mixture is treated with both restriction enzymes and ligated into the plasmid vector having sticky ends compatible to those produced by *Not*I and *Hind*III restriction endonucleases, respectively. Theoretically, only heterohybrids derived from both sources will be ligated and further cloned into *Escherichia coli*. Using this approach, an enrichment factor of 10–20-fold could be achieved (Nelson et al. 1989). Cloning selection is limited by the recovery of significant levels of background products, mostly due to the cloning of source A- or B-only hybrids into an incompletely digested vector and the cloning of chimeric duplexes discussed above (Devon and Brookes 1996). For these reasons, other CC modifications have been developed.

3. PHYSICAL SEPARATION OF HYBRIDS

To improve the separation of hybrids, one of the source DNAs can be immobilized on a solid carrier (e.g. source A DNA bound to nylon filter or to streptavidin-coated magnetic beads via biotinylated primer). Prior hybridization, one or both DNA sources can be preblocked for repeats by adding an excess of $C_o t$ fractions. Following hybridization (Figure 3), the support-attached fraction (which contains unhybridized source A molecules, source A homohybrids, and heterohybrids source A/B) is separated from the soluble fraction (all other types of DNAs). The optimization of hybridization and washing conditions are critical in this type of experiment. The solid-phase fraction is further PCR amplified with primers specific to source B linkers, resulting in the more or less specific amplification of source A/B heterohybrids. The enrichment factor value (~1000-fold) increased dramatically comparing to the previous group of methods (Lovett et al. 1991; Parimoo et al. 1991), but, however, was still insufficient to reliably recover unique sequences or rare transcripts (Devon and Brookes 1996). Regardless of further development of some improved related versions of CC (Brookes et al. 1994), heterohybrid isolation remained puzzling in many applications, and a problem of unwanted chimeric hybrid formation has not been solved in these techniques. The CC in that form, therefore, did not become popular among the research community. The use of the PS effect, which will be described in the next section, revolutionerized CC and made it a method of choice: heterohybrid selection became an effective, easy, and quick procedure.

4. PCR-ONLY-BASED APPROACHES

These approaches, relying on the PS effect, have been developed recently by Tatyana Azhikina and colleagues (Azhikina et al. 2004; Azhikina and Sverdlov 2005; Azhikina et al. 2006). In general, in all its applications, PS allows the amplification of wanted sequences and simultaneously suppresses the amplification of unwanted ones. Pan handle-like stem-loop DNA constructs for PS are

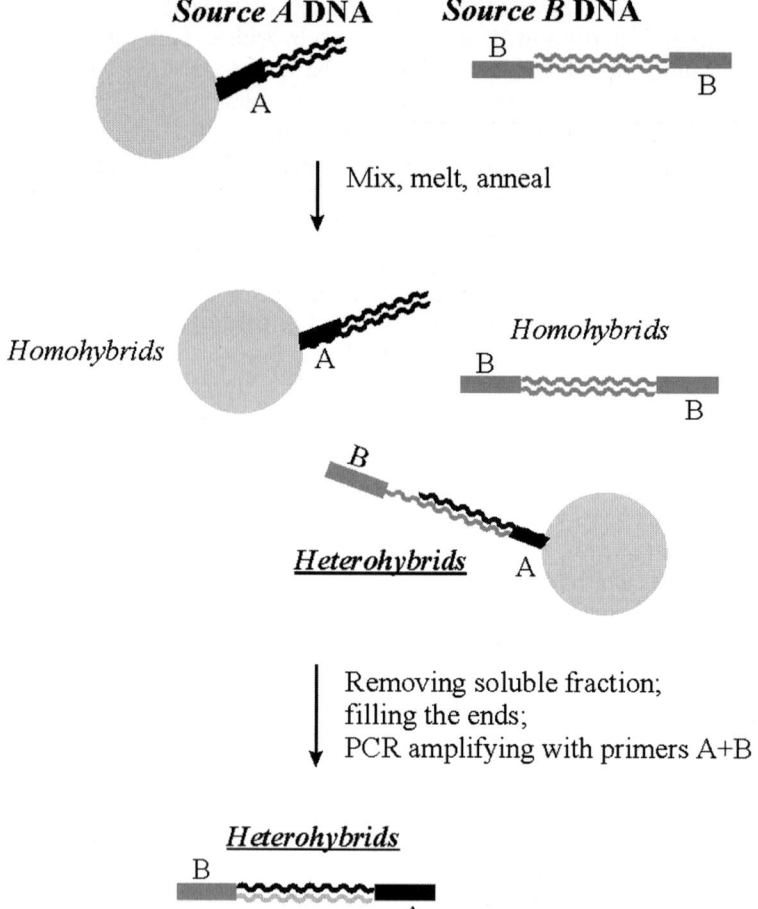

Figure 3. Selection of heteroduplexes based on both (1) attaching source A DNA to a solid carrier (magnetic beads) and (2) PCR amplification using combination of source A- and B-specific primers.

created by ligation of long guanine–cytosine (GC)-rich adapters to DNA or cDNA restriction fragments (Figure 4) (Luk'ianov et al. 1994). As a result, each single-stranded DNA fragment is flanked by terminal inverted repeats (i.e. by self-complementary ends). During PCR, on denaturing and annealing, the self-complementary ends of each single strand form duplex stems, converting each fragment into a large pan handle-like stem-loop structure. The formation of stable duplex structures at the fragment ends makes the PCR with the adapter-primer (A-primer) alone relatively inefficient, because the intramolecular annealing of the complementary termini is kinetically favored and more stable than the intermolecular annealing of shorter A-primers. This effect is therefore

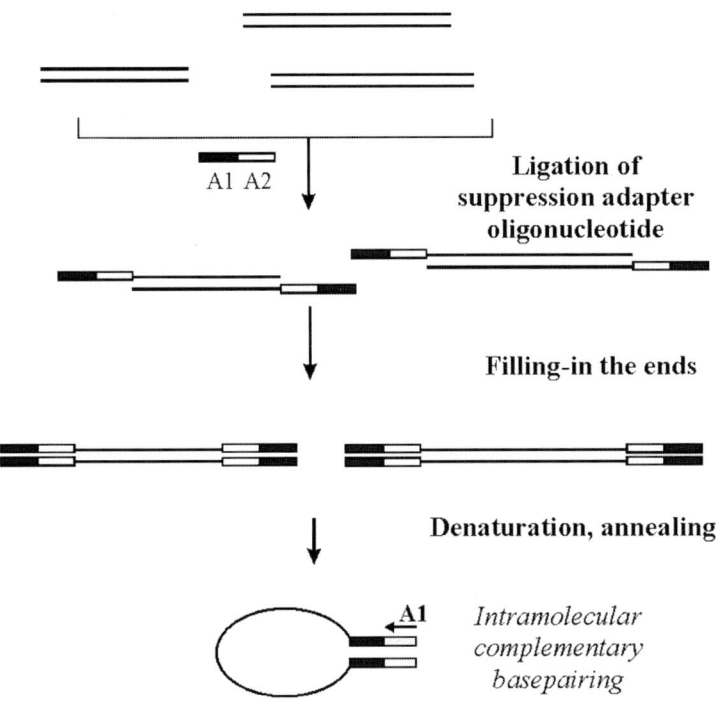

Figure 4. Principle of the PCR suppression (PS) effect. DNA molecules flanked by GC-rich ligated adapter oligonucleotides termed suppression adapters (~40 bp long artificial inverted terminal repeats) form intramolecular terminal duplexes, thus preventing terminal adapter-specific primer annealing and, consequently, inhibiting the PCR.

called PCR suppression (Luk'ianov et al. 1994). However, PCR is efficient in the presence of both A-primers and target-primers (T-primers, targeted at the specific sequences in the single-stranded loops). The T-primer anneals to its target and is used by DNA polymerase to initiate DNA synthesis. The newly synthesized product has two termini, which are not complementary and, thus, cannot fold into a stem-loop structure. This fragment is, therefore, not subject to the PS effect, and is efficiently amplified. Consequently, only the fragments containing the target are exponentially amplified by PCR, while the background fragments without the target remain inert.

Figure 5 represents a simple model of the use of CC for isolation of evolutionary conserved sequences shared by comparing genomes, recently reported by Chalaya et al. (Chalaya et al. 2004). Genomic DNAs of human and of New World monkey marmoset *Callithrix pigmaea* were digested with frequent-cutter

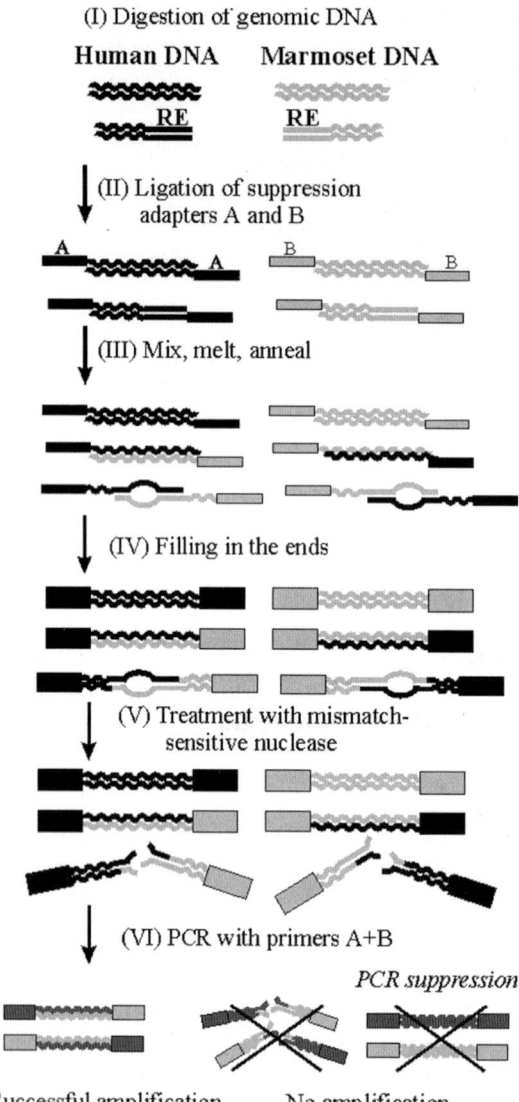

Figure 5. Schematic representation of the mispaired DNA rejection technique rationale. Not only exactly matched identical sequences, but also a number of background chimeric duplexes, which are usually products of hybridization between repetitive elements (REs), are generally PCR amplified and appear in DNA libraries. The addition of mismatch-sensitive nucleases makes it possible to selectively cleave background duplexes containing imperfectly matched regions and, therefore, to enrich the resulting library in target sequences.

restriction endonuclease, and two different sets of suppression adapters were ligated to them. Samples were then mixed, denatured, and allowed to reanneal, followed by filling the ends with DNA polymerase (Figure 5) and treatment with mismatch-specific nucleases to decrease production of chimeric hybrids. These enzymes recognize improperly matched double-stranded DNAs and cut such "wrong" hybrids, thus clearly enhancing hybridization efficacy (see also Chapters 9 and 10). At the next stage, hybridization products are subjected to PCR with primers specific to the suppression adapters used, so that only human–*C. pigmaea* hybrid molecules are amplified. As a result, we managed to create a genomic library highly enriched in evolutionary conserved sequences shared by human and *C. pigmaea* genomes.

Another successful application of the CC is the new technique called "nonmethylated genomic sites coincidence cloning (NGSCC)", which results in a set of sequences that are derived from the genomic locus of interest and contain an unmethylated CpG site. In this case, annealing mixture complexities were not as high as for whole-genome digest hybridizations, and treatment with mismatch-sensitive nucleases was not needed. The technique is based on the initial fragmentation with a methyl-sensitive restriction enzyme (Figure 6). To simplify the DNA sets to be compared, they can be additionally digested with a frequent-cutter that is not sensitive to methylation of its target site, e.g. *Alu*I. As a result, the lengths of the fragments can be restricted to a size that is optimal for subsequent PCR amplifications, usually up to 1.5 kb. Different suppression adapters are then ligated to sticky ends produced by methyl-sensitive restriction enzyme and to blunt ends created by *Alu*I. Further, PCR amplification with primers specific to both adapters used results in the amplicon of genomic fragments having unmethylated CpG site at one terminus and *Alu*I restriction site in another one.

This amplicon is further hybridized to a new-suppression-adapter-ligated fragmented DNA from the genomic locus of the interest (the authors analyzed methylation profiles of a ~1 Mb long human genomic locus *D19S208-COX7A1* from chromosome 19). In the following nested PCR, only those unmethylated CpG-containing fragments that match to D19S208-COX7A1 genomic locus, were amplified. Sequencing of the resulting libraries derived from initial genomic DNAs from healthy and cancerous tissues enabled authors to create the first large-scale comprehensive tissue- and cancer-specific methylation map for that locus (Azhikina and Sverdlov 2005). More recently, the same group of authors combined NGSCC with serial analysis of gene expression (SAGE) thus creating new technique termed "RIDGES", which is significantly more informative than NGSCC, as its outcome in 10–20-fold more information about methylation sites per one sequenced clone (Azhikina et al. 2006).

As mentioned above, the use of CC is not restricted to genomic DNA analysis. In particular, recently published technique termed genomic repeat expression monitor (GREM) utilizes CC of preamplified 3′-terminal genomic flanking regions of the repetitive elements with the set of cDNA 5′-terminal parts. This results in construction of a hybrid genomic DNA/cDNA library,

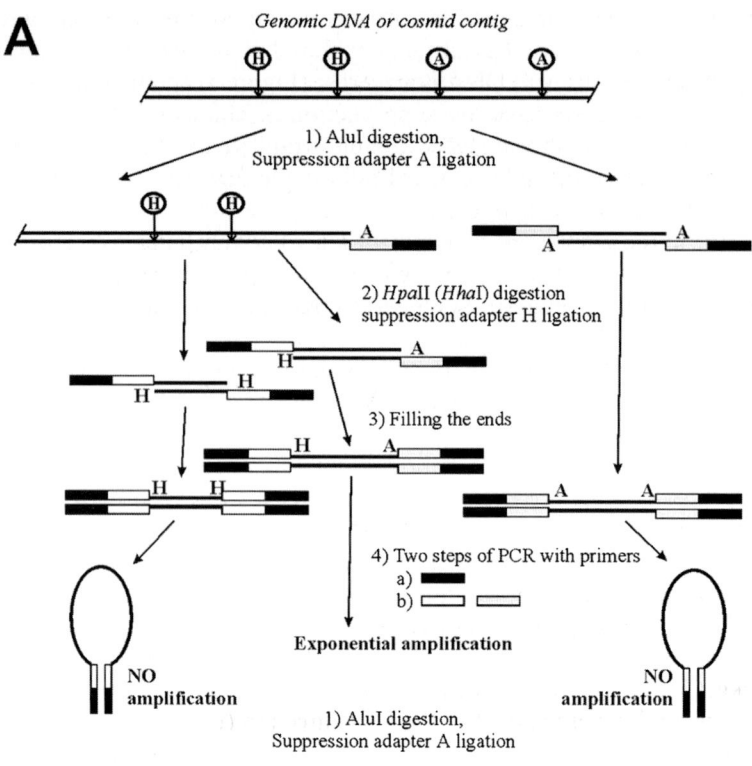
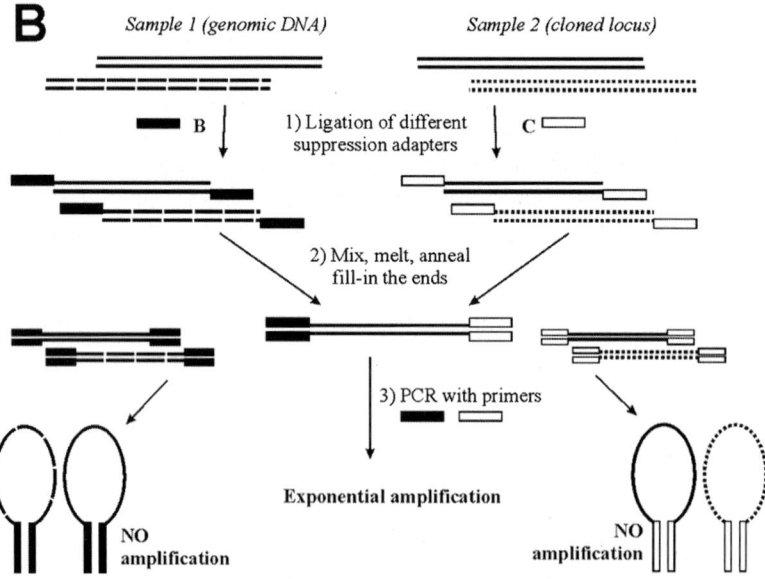

Figure 6.

(*Continued*)

enriched in promoter-active repeats, thus making it possible to create a comprehensive genome-wide map of such repetitive elements (Buzdin et al. 2006b; Buzdin et al. 2006a). We applied GREM for the analysis of long terminal repeats (LTRs) of mostly human-specific family of endogenous retroviruses called *HS LTRs* (Buzdin et al. 2002). The GREM technique outlined in Figure 7 consists of three major stages: (1) synthesis of full-length cDNA libraries whose clones include specific oligonucleotide adapters exactly tagging the cDNA 5′-ends, (2) selective PCR amplification of genomic repeat-flanking regions, and (3) hybridization of the genomic repeat-flanking regions to the cDNA with a subsequent PCR amplification of the genome–cDNA heteroduplexes.

The first stage of GREM is aimed at the amplification of full-length cDNAs tagged at the 5′-ends with a specific adapter oligonucleotide (CS in our case). The tagging is achieved due to the "cap-switch" effect in the process of cDNA synthesis. Having reached the 5′-end of the mRNA template, oligo(dT)-primed reverse transcriptase adds a few additional deoxycytidine nucleotides to the 3′-end of the cDNA. An oligonucleotide with an oligo-ribo(G) sequence at its 3′-end hybridizes to the deoxycytidine stretch to form a primer, which allows reverse transcriptase to switch templates and to continue replicating to the end of the oligonucleotide. This technique allows one to precisely tag the cDNA 5′-ends that correspond to transcription start sites (Figure 7). Prior to the hybridization at stage (3), the cDNA was digested with *Alu*I restriction endonuclease to get shorter fragments and to avoid further background amplification of hybrids with read-through transcripts driven in the sense orientation with respect to the LTR direction (Figure 7, stage 1, step "*Alu*I digestion"). *Alu*I was chosen because the HS LTR consensus sequence lacks restriction sites of this frequent-cutter endonuclease. The treatment of cDNA with *Alu*I (Figure 7) suppresses the yield of sense read-through LTR containing products at the following stage (see below).

At the second stage, we selectively PCR amplified genomic regions flanking the 3′-termini of HS LTRs. The cDNA hybridization with the amplicon obtained was used to select the cDNA molecules that contain HS LTRs at their 5′-termini. The amplification of genomic flanking regions is a critical step ensuring the specificity of the whole procedure. Nested PCRs result in selective amplification of all target repeat-flanking sequences, whereas cDNA

Figure 6. cont'd. Summary of the NGSCC procedure. (A) Selective amplification of *Alu* I-*Hpa* II/*Hha* I fragments. Cosmid contig/genomic DNA is depicted as horizontal lines. *Alu* I and *Hpa* II/*Hha* I recognition sites are marked by "A" and "H", respectively. A and H suppression adapters are shown as boxes in identical (external) parts are unshaded, and different (internal) parts are shaded black (for A) or gray (for H). Complementary sequences are indicated by hatching. (B) General scheme for the Selective Suppression of PCR (SSP)-assisted coincidence cloning (CC) procedure. Fragments unique for each of the samples under comparison are shown as dotted and dashed lines. Identical fragments are depicted as continuous lines. B and C suppression adapters are represented by blank and filled boxes, respectively.

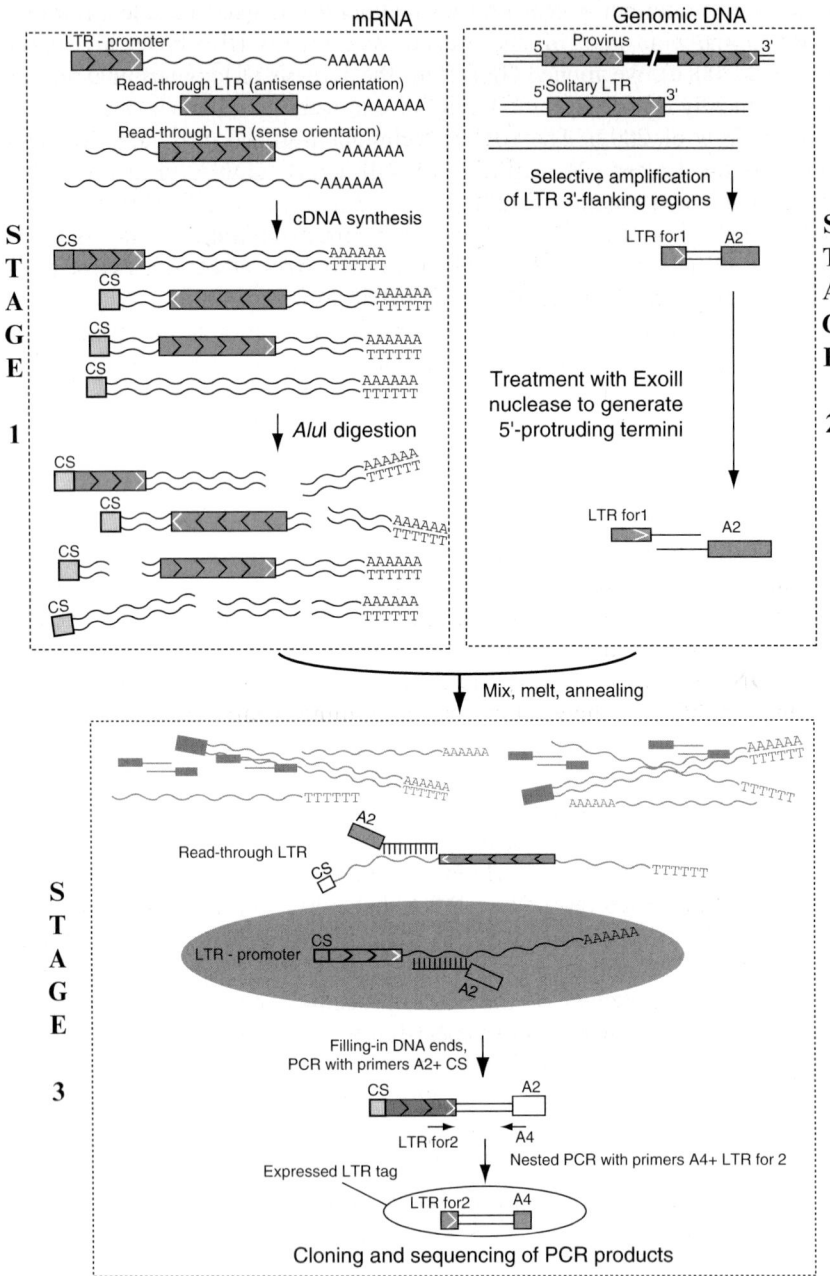

Figure 7. Schematic representation of the GREM technique (for details, see Section 6.2). The procedure includes three major stages: (1) genome-wide amplification of the genomic DNA flanking the 3'-ends of target repetitive elements (here, HS LTRs). Treatment of the resulting amplicon with *Exo*III generates 5'-protruding ends to be used at the third stage. (2) A double-stranded oligo d(T)-primed cDNA library is synthesized for tissues where expression of repetitive elements is to be studied.

(*Continued*)

amplification would not provide similar selectivity, as the exact locations of transcription start sites within repeat sequences may vary for different individual repetitive elements and, therefore, the design of suitable primers for PCR would be problematic.

To amplify genomic LTR flanking regions, we digested human genomic DNA with *Alu*I, ligated the fragments obtained to a 45 nt long GC-rich synthetic linker oligonucleotide (A1A2), and performed a series of nested PCR amplifications using HS LTR specific and adapter specific primers. As mentioned above, the HS LTR consensus sequence lacks *Alu*I restriction sites, whereas this endonuclease normally produces DNA fragments too short to be subject to PCR fragment size selection (Rebrikov et al. 2000). As mentioned above, the use of GC-rich suppression adapters minimizes background PCR amplification and results in almost 100% selective amplification of the expected fraction of the genome. The amplified LTR flanking sequences were treated with *Exo*III exonuclease to generate 5′-protruding termini required at stage (3) of GREM and to avoid any background cross-hybridization between LTR-containing sequences. We have recently demonstrated (Buzdin et al. 2002; Buzdin et al. 2003a) that *Exo*III may be used to remove adapter sequences from hybridizing mixtures. Under the conditions used, *Exo*III removes nucleotides slowly enough (~5 nucleotides per minute) to more or less precisely excise ~30 HS LTR 3′-terminal nucleotides from the amplicons. At the last step, the digested cDNA was hybridized to the LTR 3′-flanking genomic fragments. To selectively amplify the heteroduplexes containing genomic LTR flanking regions and cDNA 5′-terminal fragments generated due to LTR promoter activity, we used PCR with the CS primer against 5′-cDNA tags and A2 primer specific to the adapters ligated to the genomic DNA. This PCR step was followed by an additional nested PCR with primers A4 and LTRfor3 to increase the specificity of amplification (Figure 7).

As a result, only heteroduplexes, but not duplexes of cDNA not relevant to LTR expression or containing read-through LTRs, were amplified. A potential background of transcripts containing LTRs read-through in the sense direction was supposed to be negligible. A careful inspection of human transcribed sequence databases revealed in total 38 transcripts containing read-through HS LTRs, among them only four LTRs in the sense orientation. An *in silico* simulation of *Alu*I digestion suggested a complete removal of all such transcripts from GREM libraries.

◄───

Figure 7. cont'd. At this stage cDNAs are tagged by a linker oligonucleotide (CS) at the RNA transcription start sites using the "cap-switch" effect. cDNAs are then digested with *Alu* I restriction endonuclease that has no recognition sites within HS LTRs. This step precludes amplification of LTR sequences read-through in the sense orientation. (3) Finally, the genomic DNA amplicon (stage 1) is hybridized to the 5′-tagged cDNAs (stage 2). The protruding DNA ends are filled in with DNA polymerase, and the hybrids obtained (expressed LTR tags – ELTs) are nested PCR amplified with primers specific to the flanking genomic DNA adapter and cDNA 5′-terminal tag sequence, respectively.

The finally obtained amplified heteroduplexes were further cloned and sequenced. Every particular heteroduplex contained a 3′-HS LTR terminal portion, a fragment of the 3′-flanking genomic DNA, and an adapter sequence. Importantly, GREM makes it possible to characterize promoter activity of the repetitive elements in both qualitative and quantitative ways, as the number of particular heteroduplexes linearly correlates with the transcriptional activity of the corresponding promoters.

5. FUTURE PROSPECTS

The use of PCR suppression effect makes CC a method of choice for a number of applications involving finding a common fraction of nucleic acids in the samples under study. Incorporation of a stage of treatment with mismatch-sensitive nucleases significantly increases the fidelity of CC, which is especially important when complex sources like mammalian genomic DNAs are analyzed. The comparison of only two sources per one experiment may seem important limitation, but pooling several DNA samples in one source may, at least partly, solve this problem. CC, therefore, has bright perspectives to be widely used for (1) finding common transcripts in the analyzing tissue probes (e.g. for the recovery of genes coexpressed in cancer samples), (2) for experimental mapping of known or unknown transcripts on both characterized or uncharacterized genomic contigs, (3) for genome wide recovery of individual transcriptionally active repetitive elements, their mapping and quantification of their promoter activity, (4) for finding methylated or unmethylated CpGs in megabase-scale genomic contigs and, finally, (5) for the facile identification of evolutionary conserved sequences, shared by the comparing genomes.

6. PROTOCOLS

6.1 Cloning Similarities in Genomic DNAs

6.1.1 Starting material

DNA samples. In our experiments, we extracted DNA from four mixed human blood samples, or from blood samples of chimpanzee *Pan paniscus* and marmoset *C. pigmaea* using a genomic DNA purification kit (Promega lot #A7710) according to the manufacturers' recommendations.

Oligonucleotides. We used the standard suppression adapters A1A2 (5′-*GTAATACGACTCACTATAGGGCAGCGTGGTCGCGGCCGAGGT*-3′) and B1B2 (5′-*CGACGTGGACTATCCATGAACGCATCGAGCGGCCGCCCGGGCAGGT*-3′). For nested PCR amplifications, the following primers specific for the suppression adapter set were used: A1, 5′-*GTAATACGACTCACTATAGGGC*-3′, and B1, 5′-*CGACGTGGACTATCCATGAACGCA*-3′. A2, 5′-*AGCGTGGTCGCGCCGAGGT*-3′, and B2, 5′-*TCGAGCGGCCGCCCGGGCAGGT*-3′. Oligonucleotides were synthesized using an ASM-102U DNA synthesizer (Biosan, Novosibirsk, Russia).

6.1.2 DNA preparation for hybridization

Digestion of genomic DNA. About 1µg of genomic DNA was digested with 10 units of frequent-cutter blunt end-producing restriction endonuclease *Alu*I (Fermentas) at 37°C, for 2 h. DNA was phenol–chloroform extracted, ethanol precipitated, and dissolved in 25 µl of sterile water.

Ligation of the suppression adapters. The suppression adapter ligation was done as described previously in this book (Lavrentieva et al. 1999). We used T4 DNA ligase (Promega) and suppression adapters A1A2 and B1B2 (see above), annealed to 10 nt long oligonucleotide complementary to the adapter 3′-terminal part, A3 and B3, respectively). Ligated DNA was purified using Quiagen PCR product purification kit, ethanol precipitated and dissolved in 5 µl of hybridization buffer (0.5 M NaCl, 50 mM Hepes, pH 8.3, 0.2 mM EDTA).

6.1.3 DNA hybridization

We mixed 800 ng of each of both DNA samples assigned for hybridization in a volume of 8 µl of 1x hybridization buffer, denatured at 95°C for 10 min, and hybridized at 65°C or 85°C for 50 h. The final 8 µl mixture was diluted with 72 µl of dilution buffer (50 mM NaCl, 5 mM Hepes, pH 8.3, 0.2 mM EDTA). In some experiments, $C_0 tA$ fraction competitor DNA (Gibco BRL, USA) was added in 100x weight excess to the hybridization mixture.

Filling in the termini of hybridized DNA. We used AmpliTaq DNA polymerase (1 unit per 1 µg of hybridized DNA) to fill in the ends of DNA duplexes at 72°C for 20 min.

6.1.4 Hybridized DNA treatment with mismatch sensitive nucleases

About 100 ng aliquots of hybridized DNA were digested with 1 µl Surveyor nuclease (Transgenomic, USA) in 20 µl of 1x buffer supplied by the manufacturer, overnight incubation at 42°C, or treated with 0.1 unit of mung bean nuclease (Promega) at 37°C for 15 min. DNA samples were phenol–chloroform extracted and ethanol precipitated.

6.1.5 PCR amplification of hybridization products and library construction

Nested PCR amplification. DNA samples were dissolved in 100 µl of water and 1µl was PCR amplified with 0.2 µM primers specific for the used suppression adapter set: A1 and B1. The PCR conditions were as follows: 95°C for 15″, 65°C for 10″, 72°C for 90″, 15 cycles. To increase the amplification specificity, we used an additional round of nested PCR for 500-fold dissolved products of the latter PCR with 0.2 µM primers A2 and B2, under the same cycling conditions. The number of nested PCR cycles varied substantially depending on the particular hybridization.

Clone library construction. The PCR products obtained were cloned in *E. coli* strain DH5α using a TA-cloning system (Promega). We sequenced positive clones by the dye termination method using an applied biosystems 373 automatic DNA sequencer.

DNA sequence analysis. We used BLAT search (http://genome.ucsc.edu/cgi-bin/hgBLAT) to map clone inserts within human and chimpanzee genomes. Homology searches against GenBank were done using the BLAST web server at National Center for Biotechnology Information (NCBI; http://www.ncbi.nlm.nih.gov/BLAST) (Altschul et al. 1990). For multiple alignments the ClustalW program (Thompson et al. 1994) was used.

6.1.6 PCR amplification of evolutionary conserved sequences

As much as 40 ng of Old World monkey *C. pigmaea* blood DNA sample were PCR amplified using multiple sets of 0.2 µM unique genomic primers flanking the presumable conserved genomic loci. The resulting PCR products were analyzed on 1.2% agarose gels and sequenced.

6.2 Cloning and Presice Mapping of Transcribed Repetitive Elements

We applied this approach termed GREM to identify at a genome, wide scale of promoter active human-specific endogenous retroviruses and their LTRs (HS LTRs, see Section 4 (Figure 7). Below both adapter and LTR-specific primer structures are listed, to be adopted by the user depending on the specific task (which type of genomic repeats will be studied and which suppression adapter set will be used).

6.2.1 Starting material

Oligonucleotides. For linker ligation we used the standard suppression adapter A1A2 (5'-*GTAATACGACTCACTATAGGGCAGTCGACGCGTGCCCGGT-CCGAC*-3') annealed to short oligonucleotide complementary to its 3'-end A3, 5'-*GTCGGACCGGGC*-3'. For nested PCR amplifications, the following primers specific for the suppression adapter set were used: A1, 5'-*GTAATACGAC-TCACTATAGGGC*-3', A2, 5'-*AGCGTGGTCGCGGCCGAGGT*-3', and A4, 5'-*TCGACGCGTGCCCGGTCGACCT*-3'. LTR-specific primers were as follows: LTRfor1 (5'-*GTCTTGTGACCCTGACACATCC*-3'), LTRfor2, 5'-*CCTCCATATGCTGAACGCTG*-3', and LTRfor3, 5'-*GGGGCAACCCACCC-CTAC*-3'. Oligonucleotides were synthesized using an ASM-102U DNA synthesizer (Biosan, Novosibirsk, Russia). For cap-swith based cDNA synthesis, the following oligonucleotides purchased from Clontech (USA) were used: CDS (5'-*AAGCAGTGGTATCAACGCAGAGTAC(T)$_{30}$*-3'), riboCS, 5'-*TAACAACGCA-GAGTACGC$_R$G$_R$G$_R$G*-3' with three 3'-terminal ribonucleotides, and CS, 5'-*TAACAACGCAGAGTACGCGG*-3'.

DNA samples. We extracted DNA from three mixed human placentas using a genomic DNA purification kit (Promega lot #A7710) according to the manufacturers' recommendations.

Tissue sampling. Testicular parenchyma and seminoma were sampled from a surgical specimen under non-neoplastic conditions. Representative samples were divided into two parts, one of which was immediately frozen in liquid nitrogen, and the other was formalin-fixed and paraffin-embedded for histological analysis.

6.2.2 RNA isolation and cDNA synthesis

Total RNA was isolated from frozen samples pulverized in liquid nitrogen using an RNeasy mini RNA purification kit (Qiagen). All RNA samples were further treated with DNase I to remove residual DNA. Full-length cDNA samples were obtained according to a cap switch effect-based SMART cDNA synthesis protocol (Clontech, BD Biosciences) using an oligo(dT)-containing primer (CDS), PowerScript reverse transcriptase (Clontech, BD Biosciences), and a riboCS oligonucleotide. When PowerScript reverse transcriptase reaches the 5'-end of the mRNA, the enzyme's terminal transferase activity adds a few additional deoxycytidine nucleotides to the 3'-end of the cDNA. The riboCS oligonucleotide, which contains three guanineribonucleotide residues at its 3'-end, basepairs with the deoxycytidine stretch, creating an extended template. Reverse transcriptase then switches templates and continues the replication to the end of the oligonucleotide. The resulting full-length single-stranded cDNA contains 5'-terminal sequences complementary to the riboCS oligonucleotide. An Advantage 2 Polymerase mix (Clontech), CS, and CDS oligonucleotides were used to synthesize the second cDNA strands and to PCR-amplify double-stranded cDNA. Prior to further hybridization in the GREM procedure, 1 µg of cDNA was digested with 10 units of *Alu*I restriction endonuclease (Fermentas) for 3 h at 37°C. This enzyme was used because the HS LTR consensus sequence lacks *Alu*I recognition sites.

6.2.3 Selective amplification of genomic regions flanking HS LTRs

Selective amplification of LTR 3'-flanking regions was based on the PCR suppression effect described in detail elsewhere (Siebert et al. 1995; Lavrentieva et al. 1999; Buzdin et al. 2002). Human genomic DNA (1 µg) was digested with 10 units of *Alu*I (Fermentas) restriction endonuclease, ethanol precipitated and dissolved in 20 µl of sterile water. Then, 100 pmol of annealed suppression adapters A1A2/A3 were ligated overnight to 300 ng of the digested DNA using three units of T4 DNA ligase (Promega) at 16°C. The ligated DNA was purified using Quiaquick purification columns (Quiagen) and eluted with 50 µl of water. About 1 µl of the eluted DNA was PCR amplified with the HS LTR-specific primer LTRfor1 and adapter-specific primer A1 using the following cycling program: (1) 72°C, 1', (2) 95°C, 1', and (3) 95°C, 15"; 65°C, 15"; 72°C, 1' for 20 cycles. The PCR products were 500-fold diluted and used as templates for nested

PCR with the downstream HS LTR-specific primer LTRfor2 and adapter-specific primer A2 under the same cycling conditions, for 22 cycles. The amplified LTR flanking sequences were treated with *Exo*III exonuclease (Promega) to generate 5′-protruding termini exactly as described in (Buzdin et al. 2002; Buzdin et al. 2003a).

6.2.4 GREM procedure

The technique includes hybridization of PCR amplified genomic sequences flanking repetitive elements (HS LTRs in our case) with cDNA, followed by selective amplification and cloning of hybrid DNA duplexes (see Figure 7). About 100 ng of *Exo*III-treated LTR flanking sequences, obtained as described above, were mixed with 300 ng of cDNA in 4 μl of hybridization buffer (0.5 M NaCl, 50 mM HEPES, pH 8.3, 0.2 mM EDTA), overlaid with mineral oil, denatured at 95°C for 5 min and hybridized at 68°C for 14 h. The final mixture was diluted with 36 μl of dilution buffer (50 mM NaCl, 5 mM HEPES, pH 8.3, 0.2 mM EDTA), and 1 μl of the diluted hybridization mixture [BG1] was PCR-amplified with 0.2 μM adapter-specific primer A2 and 0.2 μM cDNA 5′-end-specific primer CS under the following conditions: (1) 72°C for 5 min to fill in the ends of DNA duplexes, (2) 95°C for 15″, 65°C for 15″, 72°C for 1′30″, 8 cycles. The PCR products were 500-fold diluted and reamplified by nested PCR for 20 cycles (95°C, 15″, 65°C, 15″, 72°C, 1′30″) with 0.2 μM nested adapter-specific primer A4, and 0.2 μM HS LTR 3′-end-specific primer LTRfor3. The final PCR products were cloned in *E. coli* using a pGEM-T vector system (Promega) and sequenced by the dye termination method using an Applied Biosystems 373 automatic DNA sequencer.

6.2.5 DNA sequence analysis and repetitive element transcriptional status control

DNA sequence analysis. The human specific HERV-K LTR group (HS) consensus sequence was taken from our previous work (Buzdin et al. 2003b). LTR flanking regions were investigated with the RepeatMasker program (http://ftp.genome.washington.edu/cgi-bin/RepeatMasker; Smit AFA and Green P, unpublished data). Homology searches against GenBank were done using the BLAST web server at NCBI (http://www.ncbi.nlm.nih.gov/BLAST) (Altschul et al. 1990). To determine genomic locations of LTR flanking regions, the UCSC genome browser and BLAT searches (http://genome.ucsc.edu/cgi-bin/hgBLAT) were used.

Transcriptional status control. For RT–PCR control of LTR transcriptional status, we used pairs of primers, one of which was specific to the 3′-terminal part of a particular HS LTR (primer sequences not shown), and the other specific to a unique sequence within the corresponding genomic LTR 3′-flanking region. Prior to the RT–PCR analysis, the priming efficiency of the primers was preexamined by genomic PCRs at temperatures varying depending on the primer combination used. These PCRs were done for 19, 22, 25, and 28 cycles, with 40 ng of the human genomic DNA template

Coincidence cloning: robust technique for isolation 207

isolated from testicular parenchyma. The RT–PCR was done with cDNA samples of the same tissue, an equivalent of 20 ng total RNA being used as template in each PCR reaction performed in a final volume of 40 µl. About 5 µl aliquots of the reaction mixture after 21, 24, 27, 30, 33, 36, and 39 cycles of the amplification were analyzed by electrophoresis in 1.5% agarose gels. In all cases, the transcriptional status was determined from the number of PCR cycles needed to detect a PCR product of the expected length and the PCR product concentration measured using a Photomat system and the Gel Pro Analyzer software.

6.3 Finding Methylated or Unmethylated CpGs in Large Genomic Contigs

6.3.1 Materials and trivial protocols

Growth and transformation of *E. coli* cells, preparation of plasmid DNA, polyacrylamide gel electrophoresis (PAGE) and agarose gel electrophoresis, as well as other standard manipulations are performed as described by Sambrook et al. (1989). A model cosmid library representing the *D19S208-COX7A1* locus on human chromosome 19 was provided by Dr. Lisa Stubbs (Lawrence Livermore National Laboratory). The cells were grown overnight at 37°C in 5 ml of LB medium supplemented with kanamycin (20 mg/ml). Cosmid DNA was isolated using a Wizard Plus Miniprep DNA Purification System (Promega) according to the manufacturers' recommendations. DNA and samples from normal testis and seminoma tissues were kindly provided by Dr. Lyudmila Leppik. Clone inserts were sequenced with an Amersham Biosciences/Molecular Dynamics MegaBACE4000 Capillary Sequencer.

6.3.2 Selective amplification of AluI-HpaII fragments

Amplicon preparation from AluI-HpaII fragments obtained from cosmids. Cosmid DNA samples were digested with *Alu*I (MBI Fermentas, Vilnius, Lithuania), and suppression adapter A1A2 from the Section 6.2.1 was ligated to the resulting fragments as described in Chapter 2. The ligated fragments were then further digested with *Hpa*II (MBI Fermentas) and then ligated to suppression adapter Hpa (obtained by annealing an equimolar mixture of oligonucleotides 1 and 2: *5′-GTAATACGACTCACTATAGGGCAGGGCGTG GTGCGGAGGGCGGC-3′*(1) and 5′-*CGGCCGCCCTCC-3′* (2)). Suppression fragments were used as templates in a two-step selective amplication. The first PCR was performed using 10 ng of ligated DNA as template in a 25 µl volume containing 10 pmol of external primer I (Advantage 2 PCR kit,Clontech; 5′-*GTAATACGACTCACTATAGGGC*-3′). After a 5 min incubation at 72°C the mixture was PCR amplified for 25 cycles (94°C for 30″, 66°C for 30″, and 72°C for 90″). The second-step PCR was performed for 10 cycles (94°C for 30″, 68°C for 30″, 72°C for 90″) using 1 µl of the one tenths diluted amplicon as template and internal primers I-Alu (5'-AGCGTG-GTCGCGGCCGAGAG-3') and I-Hpa (5'-AGGGCGTGGTGCGGAGGGCGGC-3'). In order to remove flanking sequences corresponding to the primers used in the second PCR step, the amplicons were digested with AluI and HpaII. To fill in the ends after the HpaII reaction, the samples were treated with the Klenow fragment of E. coli DNA polymerase I (Promega) and purified by phenol–chloroform extraction followed by ethanol precipitation. The suppression adapter B1B2 (obtained by annealing an equimolar mixture of oligonucleotides 3 and 4: 5'-TGTAGCGTGAAGACGACAGAATCGAGCG-GCCGCCCGGGCAGGT-3' (3) and 5'-ACCTGCCC-3' (4)) was ligated to the set of fragments so obtained.

Amplicon preparation from genomic AluI-HpaII fragments. The amplicons were prepared as described for cosmid DNA with two exceptions: (1) the second PCR step was performed with the 5'-phosphorylated primer I-Hpa, and (2) the PCR product was digested with AluI only. After digestion, the set of fragments was purified and ligated to suppression adapter C (obtained by annealing an equimolar mixture of oligonucleotides 5 and 6: 5'-TGTAGCGTGAAGAC-GACAGAAAGTCGACGCGTGCCCGGGCTGGT-3' (5) and 5'-ACCAGC-CC-3' (6)).

6.3.3 The NGSCC procedure

Two mixes, relating HpaII/normal and HpaII/tumor genomic DNAs, respectively, were prepared, each containing 200 ng of the cosmid amplicon and 6 µl of the appropriate genomic DNA amplicon in 3 µl of HB hybridization buffer (50 mM HEPES pH 8.3; 0.5 M NaCl, 0.02 mM EDTA pH 8.0, 10% w/v PEG 8000). The mixtures were overlaid with ~30 µl of mineral oil and incubated for 3 min at 99°C (denaturing) and then for 18 h at 68°C (reannealing). After this, 200 µl of prewarmed HB (68°C) was added to each tube and 1 µl thereof was used in the following PCR as template. Reactions were performed in a 25 µl volume containing 10 pmol of external primer II (5'-TGTAGCGTGAAGAC-GACAGAA-3'). After preincubation for 5 min at 72°C, reaction mixtures were subjected to PCR for 25 cycles of 94°C C for 30" s, 66°C C for 30' and 72°C for 90". The second PCR was performed for 10 cycles of 94°C for 30", 68°C for 30" and 72°C for 90", using 1 µl of the one tenths diluted first PCR product as the template in a 25 µl volume containing 10 pmol of the internal primers II-B and I-Hpa (5'-TCGAGCGGCCGCCCGGGCAGGT-3' and 5'-GTAATAC-GACTCACTATAGGGC-3', respectively). The two PCR product mixtures were ligated into a pGEM-T vector (Promega) and cloned in E. coli. The clones were arrayed in microtiter plates, and the inserts were sequenced. The sequences obtained were mapped by comparison with those deposited in GenBank using the BLAST (Altschul et al. 1990) web server at NCBI (http://www.ncbi.nlm. nih.-gov/BLAST). The data were further analyzed using the Draft Human Genome Browser (http://genome.ucsc.edu/ goldenPath/hgTracks.html).

Bisulfite sequencing. To approve the true DNA methylation status, bisulfite sequencing was performed for some differential CpGs found using NGSCC. Primers specific for bisulfite modified DNA were designed using MethPrimer software (http://itsa.ucsf.edu/_urolab/methprimer/). Bisulfite modification was performed according to (Olek et al. 1996). Agarose beads were used directly in two successive PCRs. The first and second reactions were performed for 40 and 25 cycles, respectively (95°C for 20″, Annealing temperature varied upon individual primer combinations – for 20″, 72°C for 40″), in a 25 µl volume containing 10 pmol of each primer and 1 U of Taq DNA Polymerase (GibcoBRL). The fragments obtained were sequenced using the standard PCR products sequencing protocol.

REFERENCES

Altschul SF, Gish W, Miller W, Myers EW, Lipman DJ (1990) Basic local alignment search tool. J Mol Biol 215:403–410

Aslanidis C, de Jong PJ (1991) Coincidence cloning of Alu PCR products. Proc Natl Acad Sci USA 88:6765–6769

Azhikina T, Gainetdinov I, Skvortsova Y, Batrak A, Dmitrieva N, Sverdlov E (2004) Non-methylated genomic sites coincidence cloning (NGSCC): an approach to large scale analysis of hypomethylated CpG patterns at predetermined genomic loci. Mol Genet Genomics 271:22–32

Azhikina T, Gainetdinov I, Skvortsova Y, Sverdlov E (2006) Methylation-free site patterns along a 1-Mb locus on Chr19 in cancerous and normal cells are similar. A new fast approach for analyzing unmethylated CCGG sites distribution. Mol Genet Genomics 275:615–622

Azhikina TL, Sverdlov ED (2005) Study of tissue-specific CpG methylation of DNA in extended genomic loci. Biochemistry (Mosc) 70:596–603

Brookes AJ, Slorach EM, Morrison KE, Qureshi SJ, Blake D, Davies K, Porteous DJ (1994) Cloning the shared components of complex DNA resources. Hum Mol Genet 3:2011–2017

Buzdin A, Khodosevich K, Mamedov I, Vinogradova T, Lebedev Y, Hunsmann G, Sverdlov E (2002) A technique for genome-wide identification of differences in the interspersed repeats integrations between closely related genomes and its application to detection of human-specific integrations of HERV-K LTRs. Genomics 79:413–422

Buzdin A, Kovalskaya-Alexandrova E, Gogvadze E, Sverdlov E (2006a) At least 50% of human-specific HERV-K (HML-2) long terminal repeats serve in vivo as active promoters for host nonrepetitive DNA transcription. J Virol 80:10752–10762

Buzdin A, Kovalskaya-Alexandrova E, Gogvadze E, Sverdlov E (2006b) GREM, a technique for genome-wide isolation and quantitative analysis of promoter active repeats. Nucleic Acids Res 34:e67

Buzdin A, Ustyugova S, Gogvadze E, Lebedev Y, Hunsmann G, Sverdlov E (2003a) Genome-wide targeted search for human specific and polymorphic L1 integrations. Hum Genet 112:527–533

Buzdin A, Ustyugova S, Khodosevich K, Mamedov I, Lebedev Y, Hunsmann G, Sverdlov E (2003b) Human-specific subfamilies of HERV-K (HML-2) long terminal repeats: three master genes were active simultaneously during branching of hominoid lineages. Genomics 81:149–156

Chalaya T, Gogvadze E, Buzdin A, Kovalskaya E, Sverdlov ED (2004) Improving specificity of DNA hybridization-based methods. Nucleic Acids Res 32:e130

Devon RS, Brookes AJ (1996) Coincidence cloning. Taking the coincidences out of genome analysis. Mol Biotechnol 5:243–252

Diatchenko L, Lau YF, Campbell AP, Chenchik A, Moqadam F, Huang B, Lukyanov S, Lukyanov K, Gurskaya N, Sverdlov ED, Siebert PD (1996) Suppression subtractive hybridization: a method for generating differentially regulated or tissue-specific cDNA probes and libraries. Proc Natl Acad Sci USA 93:6025–6030

Lander ES, Linton LM, Birren B, Nusbaum C, Zody MC, Baldwin J, Devon K, et al. (2001) Initial sequencing and analysis of the human genome. Nature 409:860–921

Lavrentieva I, Broude NE, Lebedev Y, Gottesman II, Lukyanov SA, Smith CL, Sverdlov ED (1999) High polymorphism level of genomic sequences flanking insertion sites of human endogenous retroviral long terminal repeats. FEBS Lett 443:341–347

Lovett M, Kere J, Hinton LM (1991) Direct selection: a method for the isolation of cDNAs encoded by large genomic regions. Proc Natl Acad Sci USA 88:9628–9632

Luk'ianov SA, Gurskaia NG, Luk'ianov KA, Tarabykin VS, Sverdlov ED (1994) [Highly-effective subtractive hybridization of cDNA]. Bioorg Khim 20:701–704

Milner JJ, Cecchini E, Dominy PJ (1995) A kinetic model for subtractive hybridization. Nucleic Acids Res 23:176–187

Nelson DL, Ledbetter SA, Corbo L, Victoria MF, Ramirez-Solis R, Webster TD, Ledbetter DH, Caskey CT (1989) Alu polymerase chain reaction: a method for rapid isolation of human-specific sequences from complex DNA sources. Proc Natl Acad Sci USA 86:6686–6690

Olek A, Oswald J, Walter J (1996) A modified and improved method for bisulphite based cytosine methylation analysis. Nucleic Acids Res 24:5064–5066

Parimoo S, Patanjali SR, Shukla H, Chaplin DD, Weissman SM (1991) cDNA selection: efficient PCR approach for the selection of cDNAs encoded in large chromosomal DNA fragments. Proc Natl Acad Sci USA 88:9623–9627

Rebrikov DV, Britanova OV, Gurskaya NG, Lukyanov KA, Tarabykin VS, Lukyanov SA (2000) Mirror orientation selection (MOS): a method for eliminating false positive clones from libraries generated by suppression subtractive hybridization. Nucleic Acids Res 28:e90

Sambrook J, Russell DW (2001) Molecular cloning: a practical manual. CSHL Press, Cold Spring Harbour, New York

Siebert PD, Chenchik A, Kellogg DE, Lukyanov KA, Lukyanov SA (1995) An improved PCR method for walking in uncloned genomic DNA. Nucleic Acids Res 23:1087–1088

Thompson JD, Higgins DG, Gibson TJ (1994) CLUSTAL W: improving the sensitivity of progressive multiple sequence alignment through sequence weighting, position-specific gap penalties and weight matrix choice. Nucleic Acids Res 22:4673–4680

Venter JC, Adams MD, Myers EW, Li PW, Mural RJ, Sutton GG, Smith HO, et al. (2001) The sequence of the human genome. Science 291:1304–1351

CHAPTER 9

DNA HYBRIDIZATION IN SOLUTION FOR MUTATION DETECTION

ANTON A. BUZDIN

Shemyakin-Ovchinnikov Institute of Bioorganic Chemistry, Russian Academy of Sciences, 16/10 Miklukho-Maklaya, 117997 Moscow, Russia
Phone: + (7495) 3306329; Fax: + (7495) 3306538; E-mail: anton@humgen.siobc.ras.ru

Abstract:	This group of methods is aimed at the identification of single nucleotide-scale differences between the comparing DNA samples. Mutation and polymorphism detection is of increasing importance in the field of molecular genetics because the study of mutations reveals the normal functions of genes, proteins, noncoding RNAs, the causes of many malignancies, and the variability of responses among individuals. A plethora of single nucleotide polymorphisms (SNPs) are not deleterious by themselves, but are linked to phenotypes associated with diseases and drug responses, thus providing a great opportunity for their use in large-scale association and population studies. Millions of SNPs have been identified in recent years. However, this figure seems negligible compared to the real number of SNPs and other mutations presented in the genomes. Many mutation discovery methods quickly and effectively indicate the presence of a mutation in a sample region, but fail to resolve its characterization and localization; another family of methods permits precise mutation mapping, but in a greatly more laborious and expensive way. The group of novel approaches for mutation detection, which combines high performance, cost-efficiency, reliability, and detailed mutation characterization, will be reviewed in this chapter.
Keywords:	Mutation detection, chemical cleavage, chemical modification of mispaired nucleotides, hydroxylamine, osmium tetroxide, potassium permanganate, nuclease-based mutation scanning, resolvase-like endonucleases, artificial nucleases, single-stranded DNA specific nucleases, T4 endonuclease VII, T7 endonuclease I, CEL I, Surveyor, endonuclease V, enzymatic mismatch cleavage (EMC), RNase cleavage of mismatched nucleotides, single-base extension (SBE), duplex-specific nuclease preference (DSNP), allele-specific PCR, allele-specific competitive blocker–polymerase chain reaction (ACB–PCR), LigAmp, MutS, glycosylase mediated polymorphism detection, physical isolation of imperfectly matched DNA, single-strand conformational polymorphism (SSCP), enzyme-amplified electronic transduction, QCM.

Abbreviations: ACB–PCR, allele-specific competitive blocker–polymerase chain reaction; dNTP, deoxyribonucleotidetriphosphate; DSN, duplex-specific nuclease; DSNP, duplex-specific nuclease preference; EMC, enzymatic mismatch cleavage; FRET, fluorescent resonance energy transfer; MALDI, matrix-assisted laser desorption ionization; PCR, polymerase chain reaction; QCM, quartz crystal microbalance; RT-PCR, reverse transcription polymerase chain reaction; SNP, single nucleotide polymorphisms; SPR, surface plasmon resonance; SSCP, single-strand conformational polymorphism.

TABLE OF CONTENTS

1. Introduction 212
2. Chemical Approaches 215
3. Enzymatic Approaches 217
   3.1 Nuclease-Based Mutation Scanning 218
   3.2 Allele-Specific PCR-Based Approaches 229
   3.3 Other Enzymatic Approaches for Mutation Scanning 230
4. Physical Approaches 233
5. Bioinformatical Approaches 235
References 236

1. INTRODUCTION

Unlike subtractive hybridization, which generally deals with finding relatively long differential DNA fragments, this group of methods is aimed at the identification of very small, single nucleotide-scale differences between the comparing DNA samples. Mutation and polymorphism detection is of increasing importance in the field of molecular genetics. Mutations are believed to contribute strongly to the genetic variability in living beings, in particular their disease or drug side-effect predispositions. The study of mutations reveals the normal functions of genes, proteins, noncoding RNAs, the causes of many malignancies, and the variability of responses among individuals. Recent mutations that have not yet become polymorphisms are often deleterious and pertinent to the disease history of afflicted individuals. Small insertions or deletions of nucleotides are common polymorphic variations in the human genome and mutation-induced sequence variations are playing an important role in the development of cancer, among others. A plethora of single nucleotide polymorphisms (SNPs) are not deleterious by themselves, but are linked to phenotypes associated with diseases and drug responses, thus providing a great opportunity for their use in large-scale association and population studies. Moreover, SNPs are increasingly recognized as important diagnostic markers for the detection of drug-resistant strains of hazardous microorganisms like bacterium *Bacillus anthracis* and for differentiation of virulent strains from their nonvirulent counterparts. From this, it is clear that SNP and mutation discovery is of great interest in today's life sciences.

DNA hybridization in solution for mutation detection 213

Millions of SNPs have been identified in recent years. However, this figure seems negligible compared to the real number of SNPs and other mutations presented in the genomes. Therefore, detection of mutations, what requires the ability to detect differences in DNA structure with single nucleotide specificity, in a cheap, 100% effective manner is one of the major objectives in modern molecular genetics. However, this ideal is a little way off, and many methods are used, each with their own particular advantages and disadvantages. The ideal method would detect mutations in large fragments of DNA and position them to single base-pair (bp) accuracy and would be sensitive, precise, and robust. Currently, the need in mutation detection is reflected by the plethora of chemical, enzymatic, bioinformatical, and physically based techniques. Many mutation discovery methods quickly and effectively indicate the presence of a mutation in a sample region, but fail to resolve its characterization and localization; another family of methods permits precise mutation mapping, but in a greatly more laborious and expensive way. The group of novel approaches for mutation detection, which combines together high performance, cost efficiency, reliability, and detailed mutation characterization, will be reviewed in this chapter.

At present, mutation discovery is often performed utilizing Sanger Sequencing, which is often thought of as the "gold standard" for mutation detection. This perception is distorted due to the fact that this is the *only* method of mutation identification but this does not mean it is the best for mutation detection. The fact that many scanning methods detect 5–10% of mutant molecules in a wild-type environment immediately indicates these methods are advantageous over sequencing, at least for some purposes. Using bioinformatical approaches, discovery of a great number of mutations (mostly SNPs) was recently performed.

However, these methodologies require prior knowledge of target sequences, normally obtained through DNA sequencing, and mutation recovery in such case is usually performed by multiple sequence alignment of publicly available sequence data. Recent studies indicate that only a small percentage of mutations can be discovered using this approach and, in particular, that SNPs with low frequency are often missed. It is a serious problem for the software to detect mutations when four-color fluorescence is used when mutations are 50% (heterozygous) or less in a sample (mixed genomic samples, mitochondrial mutations, and tumor samples). These rare base substitutions within populations of DNA molecules are valuable tools for studying the DNA-damaging effects of chemicals and for pool screening for disease-associated polymorphisms.

It is clear now that high throughput methods for detecting these variations are needed for in-population screening for complex genetic diseases in which extended genomic loci, large genes, and/or several genes may be affected. Obviously, automation of the mutation detection analysis is desirable. In this respect, the idea that direct quantitation of SNPs from DNA sequencing raw data will save time and money for large amount sample analysis may seem advantageous. However, this high throughput sequencing is limited to a small number of samples, and each mutation detected in such a way, which is not considered sequencing error needs experimental confirmation and, thus, cloning and resequencing in several replicates. Therefore, such high-throughput SNP typing techniques require expensive and dedicated instruments, which render them out of reach for many laboratories. Moreover, most part of mutations is missed (few alleles analyzed). Probably, this problem will be solved by further advances in pyrosequencing-related approaches (Guo et al. 2003; Qiu et al. 2003), but the current state of the art requires recruitment of other techniques for effective large-scale mutation scanning. To meet the need of these studies, several groups of approaches have been developed. All of them are based on the rationale that mutation containing DNA molecule will form mismatches at the mutation site when hybridized to the reference wild-type DNA (Figure 1).

Thus, when mutant and wild-type DNAs are hybridized together, two complementary mismatches are formed. For example, with a T→C substitution, T–G and C–A mismatches are formed in the two heteroduplexes. Therefore, the detection and correct position of such mismatches is the key for mutation recovery. Such mispaired nucleotides can be identified directly or indirectly using very different chemical, enzymatic, or physical approaches, which offer excellent detection efficiencies coupled with high throughput and low unit cost. It should be noted that definition of the mutational change obviously requires a sequencing step, at least to confirm the results. But in this case sequencing is targeted, not a "fishing expedition", when the region, where mutation occurred, is unknown, and plenty of sequencing work is absolutely required. As a result, these methods are able to cut the costs of detecting a mutation one order of magnitude or more. Richard Cotton, one of the leading scientists in this field, says: "There are a handful of scanning methods, each having their own problems. . . . All of these methods are being improved constantly by their creators and users, thus the life

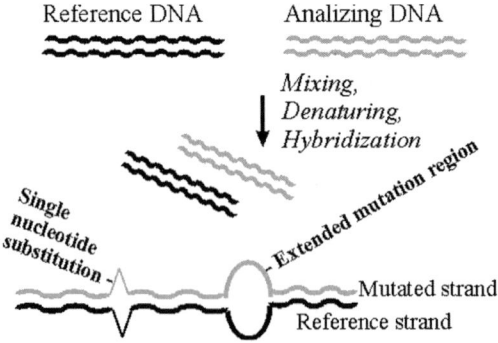

Figure 1. When hybridized to reference wild-type DNA, mutated DNAs form heteroduplexes with one or more mispaired nucleotides. These mismatch sites are targets for the majority of hybridization-based mutation detection techniques.

of those needing to detect mutations is slowly improving without the magic universal method being visible in the near future" (Cotton 1999).

The majority of such mutation detection methods are polymerase chain reaction (PCR)-based techniques dependent on the formation of heteroduplexes between wild-type and mutant strands of DNA. Briefly, chemical approaches utilize chemical cleavage or modification of the mispaired nucleotides, enzymatic ones employ enzymatic recognition of the mismatch (with further binding, cleavage, modification, or ligation of the DNA at the mispaired nucleotide(s)), whereas physical methods look for a physical difference between the mutant strand and wild-type strands of DNA, being based either on physical isolation of imperfectly matched DNA hybrids (like electrophoretic separation), or on finding differences in mismatched versus perfect DNA hybrid physical peculiarities. All these approaches utilize nucleic acids hybridization in solution and, therefore, will be described below in more detail, in comparison with each other and with direct sequencing-based approaches.

2. CHEMICAL APPROACHES

Chemical approaches for mutation detection (reviewed by Cotton (1999) and Taylor (1999)) are based on the chemical modification of mispaired nucleotides with the subsequent cleavage exactly at the position of a mismatch. Chemical cleavage, which is one of few methods capable of detecting nearly all single base mismatches, was developed in 1988 by Cotton et al. (1988) and has been widely used in research and diagnosis of many inherited diseases.

This approach, outlined in Figure 2, utilizes PCR amplification of a genomic locus of interest (alternatively, cDNA region) using as the templates reference wild-type DNA and a sample DNA that may have mutations in this region (T→C substitution in the figure). Following mixing, denaturing, and hybridizing PCR fragments, these are treated with chemical reagents, which modify mispaired nucleotides. The chemicals modify preferentially at mismatched T bases and mismatched C bases. For C bases modification, hydroxylamine (NH_2OH) is used, whereas mispaired T was originally modifying with osmium tetroxide (OsO_4), which is now replaced by potassium permanganate ($KMnO_4$) solution with a coadditive triethylammonium chloride due to high toxicity of the former (Roberts et al. 1997). Thus, it can be seen that if all four strands are labeled this mutation (as represented by the two heteroduplexes) has two chances of being detected. It should be noted that this second chance can become vital as in approximately one third of all T-G mismatches the T is unreactive, presumably due to sequence context effects (Cotton 1999). Modified DNAs are further simultaneously cleaved by piperidine, purified and analyzed either on sequencing gels or using capillary electrophoresis.

The main advantages of this approach include nearly 100% efficiency in detecting mutations in the DNA scanned and the possibility to precisely locate mutation (Figure 2) when the reference and sample DNAs are differentially

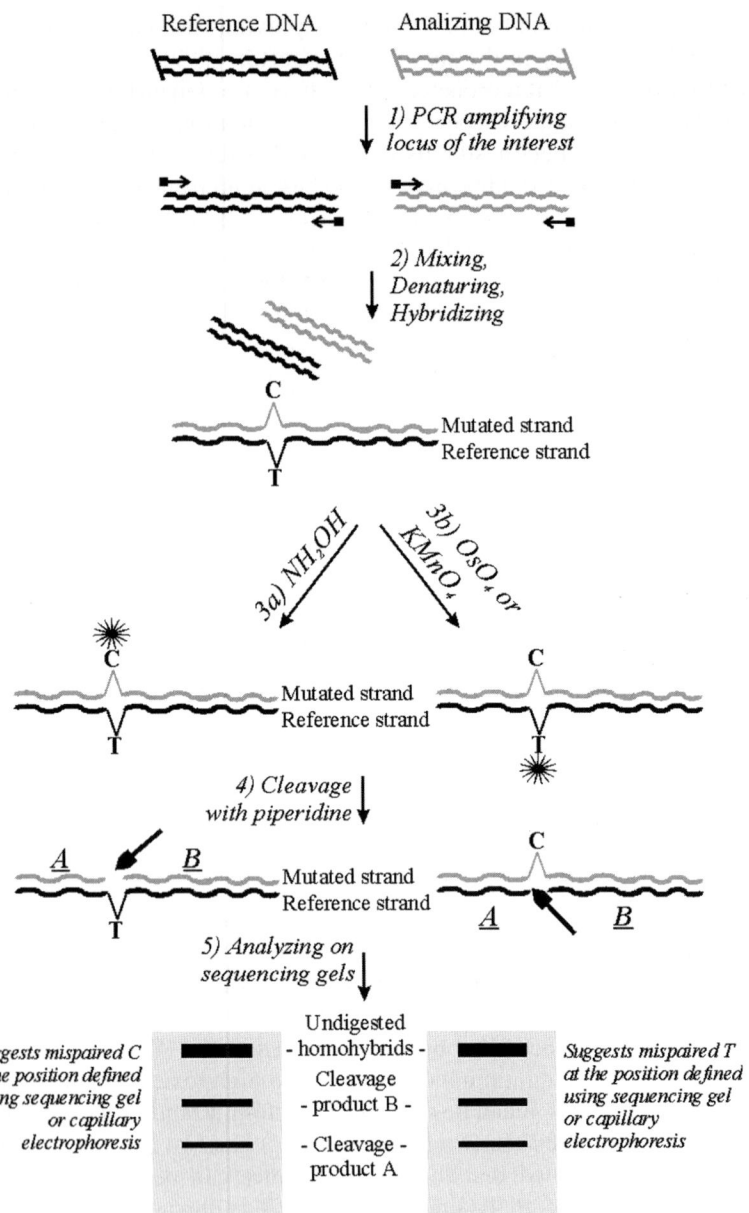

Figure 2. Outline of the chemical cleavage approach for mutation detection. In this example, mutated sample differs from the reference wild-type DNA in a single T→C substitution. Unlike enzymatic approaches, the chemical cleavage enables double detection of such types of mutations, thus significantly increasing sensitivity of the assay.

labeled (e.g. by fluorescent reagents). The most serious shortcomings of the method in its initial form include the multiple manipulations and the fact that toxic chemicals are required. This technique was greatly improved when chemical cleavage on solid support became practice: first, using biotinylated PCR primers, which made it possible to attach the DNA to streptavidin-coated beads (Rowley et al. 1995), and, second, utilizing nonspecific DNA binding to commercially available silica solid support (Bui et al. 2003). Operating with solid phase-bound DNA provides a great advantage over having to perform multiple ethanol washes and makes the protocol much simpler. Among the two variants mentioned above, the latter seems advantageous, as it somewhat economizes cost of an experiment and increases the efficacy of the whole procedure, as all DNA strands can be labeled. Another important improvement of this method is multiplexing (Rowley et al. 1995), when several analyzing DNAs may be run in the same gel lane or capillary (e.g. by using different fluorophores labeling different analyzing DNAs).

This method is currently in use in many laboratories, and many successful applications of chemical cleavage for mutation detection can be mentioned. For example, in 2004 this method was used for the recovery of mutations in the tumor suppressor p53 gene *TP53* in a group of 89 breast cancer patients, and three previously unknown mutations in protein coding sequence have been established (Lambrinakos et al. 2004). Another example was the first prenatal diagnosis performed on chorionic villi biopsy of a pregnant woman affected by a severe form of autosomal dominant transmitted retinitis pigmentosa, due to the Arg135Trp substitution in rhodopsin (Tessitore et al. 2002).

However, the chemical cleavage method is far less popular than its most serious competitor, enzymatic recognition-based approach, probably, due to its labor intensity and use of hazardous chemicals. As to sensitivity and accuracy of mutation detection, in the experiments by Deeble et al. (1997), chemical cleavage was found substantially more reliable and advantageous than producing higher background enzymatic cleavage approach.

3. ENZYMATIC APPROACHES

This group of methods is significantly more diverse than chemical cleavage-based techniques. Enzymatic approaches can be classified into three major groups: (1) those using nuclease cleavage of mispaired nucleotides in heteroduplexes (reviewed by Yeung et al. 2005)), (2) allele-specific PCR-based, and (3) methods utilizing mismatched DNA specific biding by some proteins. The first group of methods is the most widely used due to its relative simplicity, good reproducibility of the results, and the ability to find and precisely locate unknown mutations, thus permitting performing large-scale (even genome wide for some modifications) mutation scanning.

3.1 Nuclease-Based Mutation Scanning

A great variety of nucleases with very different specificities has been identified in the living organisms. Some of them have already found their application(s) in molecular biology and biomedicine, some did not find yet (e.g. the recently isolated human nuclease with an exotic function of tetraplex DNA cleavage (Sun et al. 2001)). Nuclease-based methods for mutation recovery utilize a simple rationale that some nucleases are able to bind and to preferentially digest double-stranded DNA at the mismatched nucleotide positions. Depending on the nature of the nuclease used, these techniques can be subdivided into (1) those using resolvase-like endonucleases, (2) those employing restriction endonucleases, single-stranded DNA specific nucleases and RNases, and (3) those based on the action of artificial nucleases.

3.1.1 Resolvase-like endonucleases

Bacteriophage resolvases T7 endonuclease I and T4 endonuclease VII are able to produce breaks in the double-stranded DNA duplexes at the mispaired nucleotide positions. It was in 1995, when three independently working research teams, Mashal et al. (1995), Birkenkamp and Kemper (1995), and Youil et al. (1995) simultaneously published the same idea of treatment of duplexes reference DNA-analyzing DNA with bacteriophage resolvases for the mutation recovery. The resolvases are an important group of enzymes that are responsible for catalyzing the resolution of branched DNA intermediates that form during genetic recombination. Their mode of action is directed by bends, kinks, or DNA deviations. These enzymes have their effect close to the actual site of DNA distortion (Birkenkamp and Kemper 1995). T4 endonuclease VII, the product of gene 49 of the bacteriophage T4, was the first enzyme shown to resolve Holliday structures. It has also been shown to recognize cruciforms and loops. It may also be involved in very short patch repair. Its cleavage characteristics involve it cleaving 3′ and within 6 nt from the point of DNA perturbation-causing double-stranded breakage (Youil et al. 1995).

As early as in 1990, in model experiments with synthetic oligonucleotides, T4 endonuclease VII has been shown to cleave single base-pair mismatches by Kosak and Kemper (1990). More recently, Youil and colleagues demonstrated that single mismatched nucleotide cleavage by T4 endonuclease VII depends greatly on the mispaired nucleotides: G-A and G-G mismatches are processed less effective than other types of mispaired nucleotides (Youil et al. 1995). However, as shown by Mashal et al. (1995), the use of a combination of T4 endonuclease VII with T7 endonuclease I provides very efficient digestion of all possible single nucleotide mismatches and of short (few nucleotide long) loops. More recently, cleavage with T4 endonuclease VII was used for the identification of variable number of short tandem repeats polymorphisms (Surdi et al. 1999). It should be also kept in mind that some background cleavage may occur, probably, due to unexpected secondary structures in DNA heteroduplexes (for instance, mispaired duplex cleavage specificity using this approach was ~80%, as measured by Inganas et al. (2000)). To solve this problem, Golz and colleagues have developed improved reaction conditions which can increase the selectivity of the enzyme for mismatches up to 500-fold, as demonstrated with a mutation in a 247 nt long fragment from exon 7 of human gene for p53 protein. The new conditions involve replacement of Tris/HCl buffer by phosphate buffer and change from pH 8.0 to pH 6.5. To achieve the best results, the authors recommend trying various concentrations of phosphate ions to meet individual requirements of the substrate (Golz et al. 1998b).

The basic protocols for mutation scanning using resolvases are simple and quite similar to those utilizing chemical cleavage (Figure 3; reviewed by Babon et al. (2003)). Genomic region of interest (usually 0.1–2 kb long, but in some cases up to 4 kb long, according to Del Tito et al. (1998)) is amplified with unique primers from reference and sample DNA templates, PCR products are mixed, melted, and hybridized. When the hybridizing molecules are too long, correct heteroduplex formation may be problematic and DNA fragmentation is recommended (Smith et al. 2000). Duplexes are further treated with endonuclease(s) and analyzed on sequencing gels, capillary electrophoresis, or using chromatography. The size of the digestion products indicates the location of the mutation, which is then confirmed and characterized by sequencing. As for chemical cleavage, fluorescent labeling (Del Tito et al. 1998) and multiplexing (Schmalzing et al. 2000; Shi et al. 2006) were recruited to improve the robustness of this technique.

Also, at least two new enzymes of unknown natural function mimicking resolvase activities became known: CEL I nuclease from celery (Oleykowski et al. 1998) and closely related Surveyor nuclease (Qiu et al. 2004; Mitani et al. 2006). CEL I produces single-strand breaks, whereas Surveyor nuclease cleaves with high specificity at the 3'-side of any mismatch site in both DNA strands, including all base substitutions and insertion/deletions up to at least 12 nucleotides (however, being taken in 100-fold excess, at a 1:1 to 1:4 w/w enzyme/DNA ratio, CEL I produces double-strand breaks as well (Sokurenko et al. 2001)). The use of CEL I and Surveyor nucleases is advantageous, as they can detect 100% of the sequence variants present, including deletions, insertions, and missense alterations, without having to finely tune buffer conditions (Oleykowski et al. 1998; Comai et al. 2004).

As compared with chemical cleavage, resolvases-based approach became far more popular, probably, due to its simplicity, improved safety for the researchers, and commercial availability of some mismatch cleavage enzymes and even of special kits for mutation scanning. For example, T7 endonuclease I can be purchased from New England Biolabs (USA), and another mismatch sensitive enzyme, Surveyor nuclease, is produced by Transgenomic (USA).

The efficacy of this approach has been demonstrated in a number of successful applications. For example, a number of diagnostic mutations in the cationic trypsinogen gene was found in a group of 29 hereditary pancreatitis patients

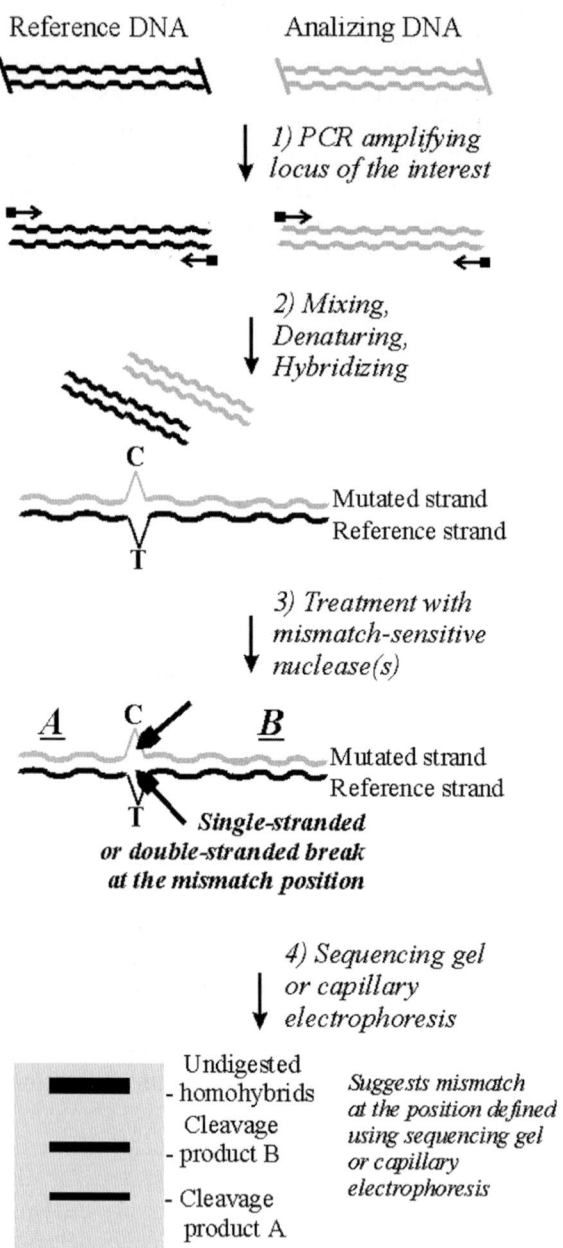

Figure 3. Generalized scheme for the enzymatic hybridization-based mutation detection approaches.

(Ford and Whitcomb 1999), the analysis of 178 colorectal cancer samples identified deleterious mutations in p53 coding sequence in 51 of them (Inganas et al. 2000); method was fruitful for the large-scale detection of mutations in two mitochondrial tRNA genes even when they were present at levels as low as 3% in DNA samples derived from patients with respiratory chain defects (Bannwarth et al. 2005); in very good agreement with these results, method was capable to detect mutations in *p53* gene when 20:1 ratio of normal versus mutated DNA was analyzed (Del Tito et al. 1998).

When the large and complex gene for fibrillin 1 (*FBN1*) was scanned in a cohort of six patients diagnosed with connective tissue disorders (four of them being diagnosed with classic Marfan syndrome), two causative mutations that result in premature translation termination, that were missed by other methods, have been identified (Youil et al. 2000). Similarly, 10 multidrug-resistant and 10 drug-susceptible clinical isolates of *Mycobacterium tuberculosis* were quickly investigated for point mutations in drug-resistant genes, *katG*, *rpoB*, *embB*, *gyrA*, *pncA*, and *rpsL* genes, which are known to be responsible for antibiotic resistance (Shi et al. 2004). Also, the use of mismatch sensitive nucleases (Surveyor nuclease in this example) provides an alternative to a laborious standard selection of desired clones from site-directed mutagenesis and PCR-based cloning methods without the necessity of sequencing DNAs purified from multiple clones. This approach was used to identify error-free clones of three genes from celery cDNA (Qiu et al. 2005).

Another application of the same idea is the removal of mismatched bases from synthetic genes by enzymatic mismatch cleavage (EMC) (Fuhrmann et al. 2005). The success of long polynucleotide *de novo* synthesis is largely dependent on the quality and purity of the oligonucleotides used. Generally, the primary product of any synthesis reaction is directly cloned, and clones with correct products have to be identified. Using mismatch sensitive nucleases like T7 endonuclease I, T4 endonuclease VII, and *Escherichia coli* endonuclease V, a novel strategy has been established for removing undesired sequence variants from primary gene synthesis products. As a model, a synthetic polynucleotide encoding the bacterial chloramphenicol-acetyltransferase (*cat*) was synthesized using different methods for one-step polynucleotide synthesis based on ligation of oligonucleotides. The influence of EMC as an error correction step on the frequency of correct products was analyzed by functional cloning of the synthetic *cat* and comparing the error rate with that of untreated products. Significant reduction of all mutation types was observed. The treatment with nucleases was successful especially in the removal of deletions and insertions from the primary ligation products.

The classical mismatch sensitive nuclease cleavage protocol has greatly evolved now, thus giving rise to a variety of new experimental techniques. First, a genome-wide approach published by Sokurenko et al. (2001), that is suitable for comparison of small genomes like bacterial DNAs (Figure 4). The method has six stages: (1) mixing two compared genomes, (2) complete restriction

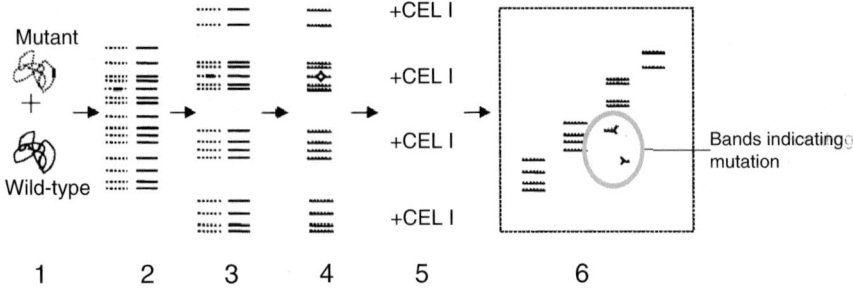

Figure 4. Scheme for the large-scale genomic SNP screening in bacterial genomes or genomic contigs of comparable lengths. Stage 1: total genomic DNA from two bacterial strains is purified and combined (a possible polymorphic site is indicated as a solid bar). Stage 2: complete endonuclease restriction of the combined genomic DNA. Stage 3: size-fractionation of the restricted DNA fragments. Stage 4: DNA heteroduplex formation by heat denaturation and reannealing of the fractionated DNA fragments (the mutation-induced mispaired region is shown as an open square). Stage 5: CEL 1 treatment of the reannealed DNA fragments. Stage 6: agarose gel analysis of CEL 1-treated DNA fractions (the mismatch-cleaved fragments are indicated as shortened fragments with halved squares on the ends).

endonuclease digestion of the combined genomic DNA, (3) size-fractionation of the restricted DNA fragments, (4) DNA heteroduplex formation by heat denaturation and reannealing of the fractionated DNA, (5) CEL I treatment of the reannealed DNA fragments and, finally, (6) agarose gel analysis of CEL I-treated DNA fractions. If a mutation occurs, two new lower molecular weight bands appear on gel that can be isolated and sequenced to locate the mutation. Using this approach, the authors managed to detect various simple mutations directly in the genomic DNA of isogenic pairs of recombinant *Pseudomonas aeruginosa*, *E. coli*, and *Salmonella* isolates. Also, by using a cosmid DNA library and genomic fractions as hybridization probes, they compared total genomic DNA of two clinical *P. aeruginosa* clones isolated from the same patient, but exhibiting divergent phenotypes. This multistep method, although efficient in hands of Sokurenko and colleagues did not become popular, probably, due to its labor intensity.

Another method, recently published by Huang et al. (2002), is the elegant modification of the standard enzymatic mismatch digestion protocol, aimed to increase the cleavage specificity and thus to decrease background signaling, which sometimes makes mutation detection problematic. The authors have developed a mutation scanning method that combines thermostable bacterial Endonuclease V (Endo V) and DNA ligase (Figure 5). Variant and wild-type PCR amplicons are generated using fluorescently labeled primers, and heteroduplexed. *Thermotoga maritima* EndoV recognizes and primarily cleaves heteroduplex DNA one base 3' to the mismatch, as well as nicking matched DNA at low levels. *Thermus* sp. DNA ligase reseals the background nicks to create a highly sensitive and specific assay.

DNA hybridization in solution for mutation detection 223

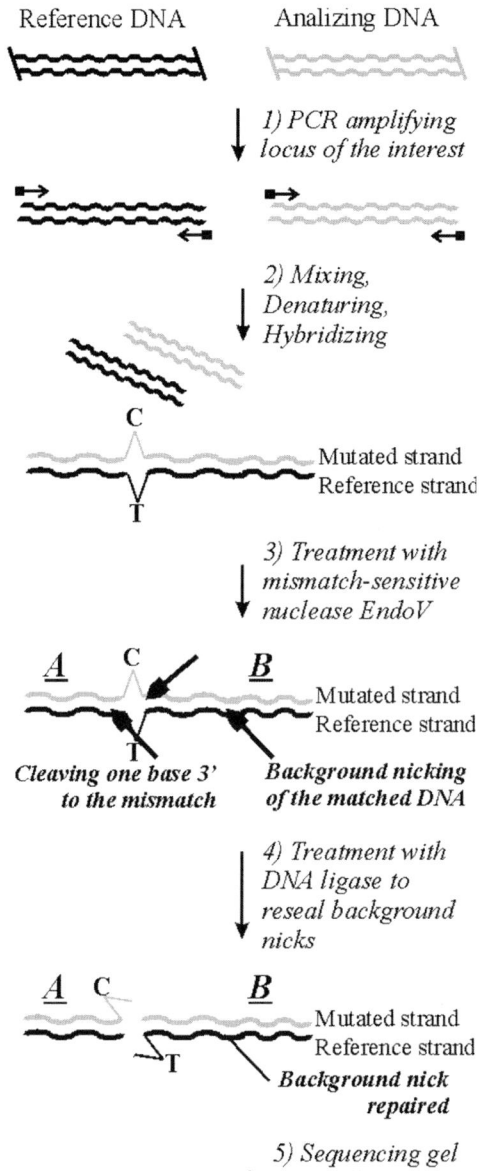

Figure 5. Enzymatic mutation detection technique of an increased accuracy, mediated by DNA ligase repair of background nicks. At stage 4, DNA ligase reseals background nicks produced by mismatch sensitive nuclease(s) employed at the previous step. Note that true mismatches remain unrepaired.

The fragment mobility on a DNA sequencing gel reveals the approximate position of the mutation. This method identified 31/35 and 8/8 unique point mutations and insertions/deletions, respectively, in the *p53*, *VHL*, *K-ras*, *APC*, *BRCA1*, and *BRCA2* genes. The technique has the sensitivity to detect unknown mutations diluted 1:20 with wild-type DNA, This method is well suited for scanning low-frequency mutations in pooled samples and for analyzing tumor DNA containing a minority of the unknown mutation.

3.1.2 Restriction endonucleases, single-stranded DNA-specific nucleases and RNases.

Not only mismatch specific nucleases may be employed for enzymatic mutation detection. RNase cleavage of mismatched nucleotides in single-stranded RNA probes hybridized to reference DNA sequences became a practice over 20 years ago (reviewed by Goldrick (2001)) The original methods relied on RNase A for mismatch cleavage; however, this enzyme fails to cleave many mismatches and more recently it was replaced by other enzymes like RNase 1 and RNase T1 to cleave mismatches in duplex RNA targets. The detection is improved, when these enzymes are used in conjunction with nucleic acid intercalating dyes. This method is being used to detect mutations and SNPs in a wide variety of genes involved in human genetic disease and cancer, as well as in disease-related viral and bacterial genes (Goldrick 2001). The most serious drawback of this method is that large amount of RNA probe is needed; moreover, nonspecific RNA degradation frequently makes such an analysis problematic.

Recently, two research groups proposed using single-strand-specific nucleases for DNA heterohybrids processing. Chalaya et al. (2004) used mung bean nuclease for cleavage of the loop regions produced during heterohybrid formation. In the conditions used by the authors, the enzyme did efficiently cleave the extended mispaired DNA (but not single nucleotide mismatches) and displayed perfect results in mismatched DNA cleavage when used together with Surveyor nuclease. Simultaneously, Till and coauthors used the same enzyme and another single-stranded DNA-specific nuclease S1 from Aspergillus to detect mutations as small as single nucleotide substitutions (Till et al. 2004). Surprisingly, in suboptimal conditions (higher pH, temperature, and divalent cation concentrations) these nucleases were able to specifically cleave nearly all single-stranded mismatches tested. These intriguing results imply that plenty of other single-stranded DNA-specific nucleases belonging to S1 and mung bean families theoretically might be used for effective mutation detection. Overall, the use of single-stranded specific DNases provides an advantage of cleaving not only single nucleotide mismatches, but also more extended loop regions, which are frequently ignored by resolvases.

Restriction endonucleases can be used for mutation detection as well. An interesting approach for high throughput screening of known SNPs in the analyzing samples was recently published by Che and Chen (2004). This method uses a type II restriction endonuclease to create extendable ends at target polymorphic sites and uses single-base extension (SBE) to discriminate alleles (Figure 6). A restriction site is engineered in one of the two PCR primers so that the restriction enzyme cuts immediately downstream of the targeted SNP site. The digestion of the PCR products generates a 5′-overhang structure at the targeted polymorphic site. This 5′-overhang structure then serves as a template for SBE reaction to generate allele-specific products using fluorescent dye-terminator nucleotides.

Following the SBE, the allele-specific products with different sizes can be resolved by DNA sequencers. Through primer design, one can create a series of PCR products that vary in size and contain only one restriction enzyme recognition site. This allows loading of many PCR products in a single capillary/lane. This method, restriction-enzyme-mediated SBE, was demonstrated by typing multiple SNPs simultaneously for 44 DNA samples. By multiplexing PCR and pooling multiplexed reactions together, this method has the potential to score 50–100 SNPs/capillary/run if the sizes of PCR products are arranged at every 5–10 bases from 100 to 600 base range.

If one is interested in investigating a particular point mutation that disrupts a wild-type restriction site, than a variety of simple, cost efficient, and sensitive techniques utilizing heteroduplex treatment with restriction endonucleases is available (e.g. Zhu et al. 2004). The shortcoming that is shared by all restriction endonuclease-based mutation screening methods is that they cannot identify new, previously unknown mutations. Therefore, their area of application is the studying of already established mutations and polymorphisms.

The last method reviewed in this section will be the novel technique recently proposed by Shagin et al. (2002) for SNP detection. The authors have characterized a novel nuclease from the Kamchatka crab, designated duplex-specific nuclease (DSN). DSN displays a strong preference for cleaving double-stranded DNA and DNA in DNA–RNA hybrid duplexes, compared to single-stranded DNA. Moreover, the cleavage rate of short, perfectly matched DNA duplexes by this enzyme is essentially higher than that for nonperfectly matched duplexes of the same length. Thus, DSN differentiates between one-nucleotide variations in DNA. The authors developed a novel assay for SNP detection based on this unique property, termed "duplex-specific nuclease preference (DSNP)". In this assay, the DNA region containing the SNP site is amplified and the PCR product is mixed with signal probes (mutation or wild-type-specific short oligonucleotides labeled by fluorescent resonance energy transfer (FRET)) and DSN (Figure 7).

During incubation, only perfectly matched duplexes between the DNA template and signal probe are cleaved by DSN to generate sequence-specific fluorescence. The use of FRET-labeled signal probes coupled with the specificity of DSN presents a simple and efficient method for detecting SNPs. In model experiments, the authors have demonstrated the robustness of this assay for the typing of SNPs in methyltetrahydrofolate reductase, prothrombin, and *p53* genes on homozygous and heterozygous genomic DNA. Now mutation detection

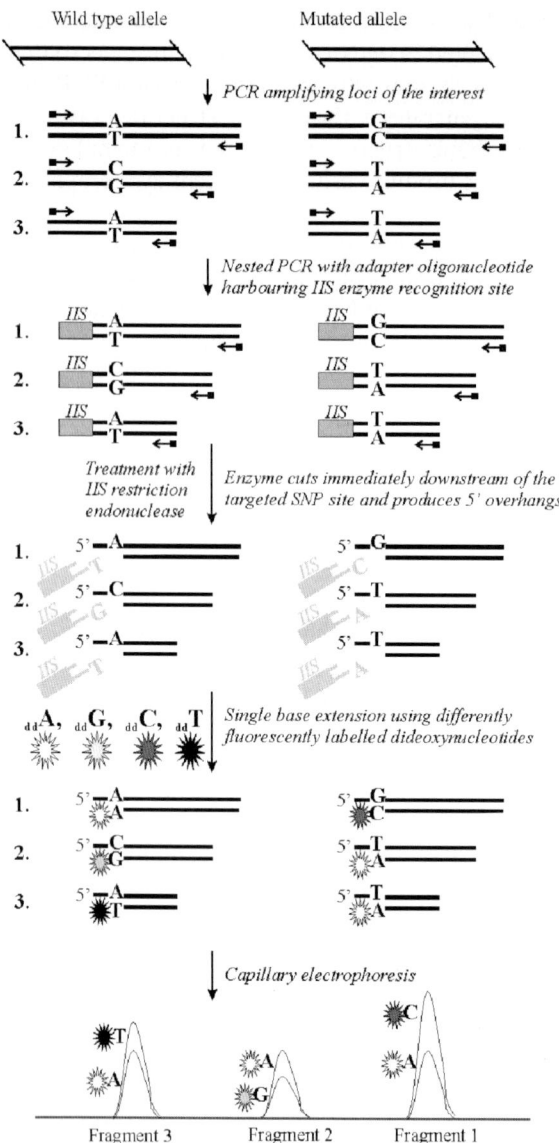

Figure 6. Single-base extension (SBE) technique for multiplex mutation detection. Theoretically, up to 50–100 potent mutation sites may be screened in a single experiment. A restriction site for type II restriction endonuclease is engineered in one of the two PCR primers so that the restriction enzyme cuts immediately downstream of the targeted SNP site. PCR product digestion generates a 5'-overhang structure at the targeted polymorphic site. This 5'-overhang then serves as a template for single base extension reaction to generate allele-specific products using fluorescent dye-terminator nucleotides (e.g. dideoxyribonucleotides). Differentially labeled DNA strands of different lengths can be further resolved in sequencing gels or using capillary electrophoresis.

DNA hybridization in solution for mutation detection 227

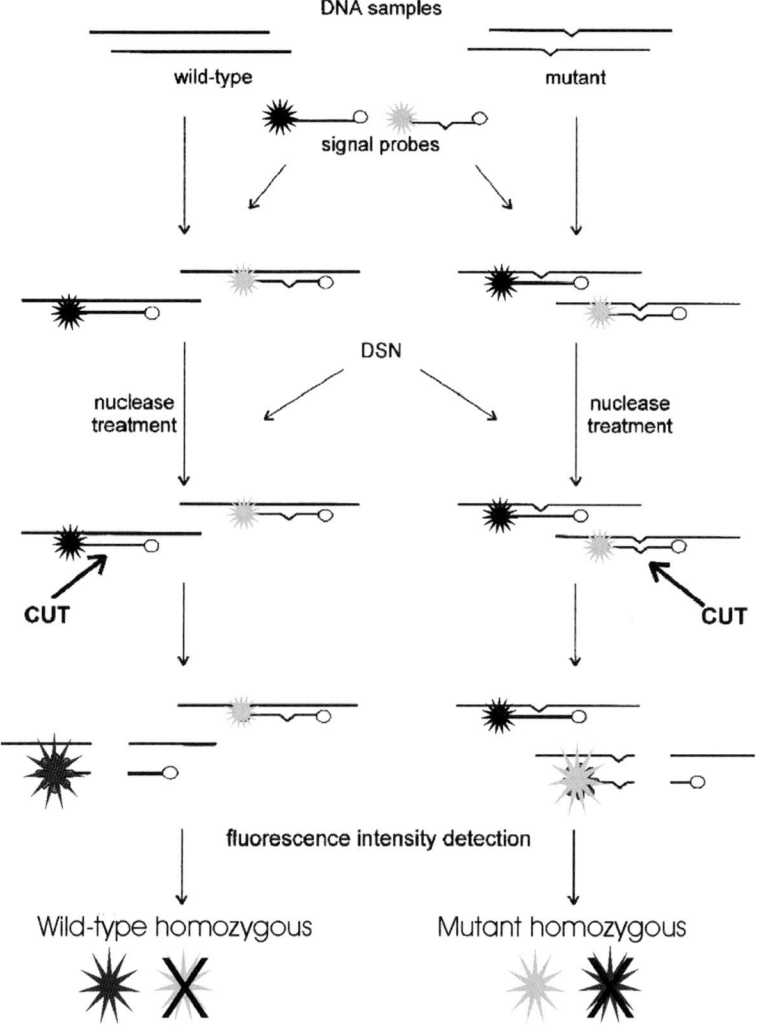

Figure 7. Scheme of the DSNP assay. The black star represents the first fluorescent donor, the gray star the second fluorescent donor, and the open circle the fluorescent quencher.

using this approach is available as a custom service from Eurogene, Inc. The limitation of this promising technique is that no new mutations can be detected.

3.1.3 Artificial endonucleases

The last group of nuclease-based methods for mutation detection involves treatment of the hybridized DNA with artificial nucleases engineered to recognize and to specifically cleave the desired sequence motif. Depending on the further signal detection technique, mutation to be quantified creates or, on the contrary, disrupts such a recognition sequence. Treatment with the nuclease produces a signal evidencing presence or absence of mutation at the position of interest. For example, PCR-amplified genomic locus to be screened for SNPs is hybridized to a signal probe (mutation-specific short oligonucleotide labeled by fluorophore and quencher at the opposite ends). The hybrids are treated with the nuclease, which cleaves duplexes in the presence of mutation, and screened for fluorescence. When quencher and flurophore are localized close to each other, no fluorescence can be detected (as in the case of undigested duplexes), whereas when duplex is cleaved, the distance between fluorophore and quencher is big enough to allow fluorescence emission.

In principle, the rationale of this approach is quite similar to that of restriction endonuclease-based mutation detection, but the current approach theoretically has an advantage of being able to cut DNA at any desired nucleotide position, as nuclease specificity may be engineered. Such artificial nucleases may be proteins or nucleic acid enzymes – ribozymes and DNAzymes (RNA or DNA molecules able to catalyze single- or double-stranded nucleic acid cleavage exactly at the target site). Artificial nucleases of a protein origin are the chimeras comprised of at least two domains, one of which is a nonspecific DNA cleavage domain and another one is a DNA-binding domain. DNA cleavage domain can be engineered from the endonuclease *Fok* I (Lloyd et al. 2005) or can it be an analogue of the metal-binding loop (12 amino acid residues), peptide P1, which has been reported to exhibit a strong binding affinity for a lanthanide ion and DNA cleavage ability in the presence of Ce(IV) (Nakatsukasa et al. 2005).

Specific DNA binding is provided by several Cys2His2 zinc-fingers that are engineered to bind to specific DNA sequences. The Fok I domain must dimerize to cut DNA, and the zinc-finger pairs function most efficiently when their binding sites are separated by precisely 6 bp. When the metal-binding loop is used, no dimerization is needed, as the problem is solved by connecting two distinct zinc finger proteins with this functional linker possessing DNA cleavage activity. To the date, this modern approach of engineering proteins with targeted nuclease activity has been applied only to the tasks of site-specific mutagenesis and *in vivo* gene targeting in eukaryotes like drosophila and higher plants, and the engineering of nuclease specificity is not trivial. However, further progress in this field may be helpful for recruiting these enzymes in mutation detection.

Unlike the former technique, nuclease ribozymes or DNAzymes may be widely used for mutation recovery. These are RNA enzymes with the targeted DNA or RNA cleavage specificity. The latter is provided by a Watson–Crick base pairing, thus making engineering sequence-specific nuclease a much easier task than in the case of artificial nucleases of protein origin. In addition to target binding motif, ribozyme harbors also a DNA/RNA cleavage domain. Being hybridized with the analyzing DNA or RNA, ribozyme binds to the target site and, if it matches perfectly, cuts the target strand, what can be detected as described above in this section. This promising technique works well in the hands of several research groups (Fiammengo and Jaschke 2005) and has bright perspectives for the future.

3.2 Allele-Specific PCR-Based Approaches

Allele-specific PCR is based on polymerase extension from primers that contain a 3'-end base that is complementary to a specific mutation (Figure 8) and inhibition of extension with wild-type DNA due to a 3'-end mismatch. Using a mutant-specific PCR primer with more artificially introduced 3'-terminal mismatches somewhat adds specificity to this amplification of an allele that differs from the wild-type by a single base pair (Parsons et al. 2005). The presence of a PCR product suggests mutation in the analyzing locus at least in one allele, or at least in one sample (when mixed DNA probes are amplified). Taq polymerase is commonly used for this assay, but because of the high rate of nucleotide extension from primer 3'-base mismatches documented for this enzyme, high sensitivity is difficult to achieve and using other polymerases may be advantageous (Gale and Tafoya 2004). If the enzyme having 3' → 5' exonuclease "proofreading" activity is used, than primers must be modified (e.g. by phosphothioate) on the 3'-end to block removal of the critical 3'-mutation-specific base by the polymerase. The protocol can be modified to include a stage of real-time PCR in the procedure.

Following PCR or RT–PCR of a gene segment that may contain allele-specific differences, 100 pg amplified product may be used for a real-time PCR with allele-specific primers and SYBR Green. The use of HEPES buffer at a pH of 6.95 together with AmpliTaq DNA polymerase results in a threshold difference between the correct template and the mismatched template is as many as 20 cycles, depending on the mismatch (Shively et al. 2003). The assay is sensitive, as it permits specific, low-level detection (25 fg DNA) of the SNP, even in the presence of a wild-type allele taken in a 20,000-fold excess (Easterday et al. 2005). Such an approach has been used for many applications, among others, for the

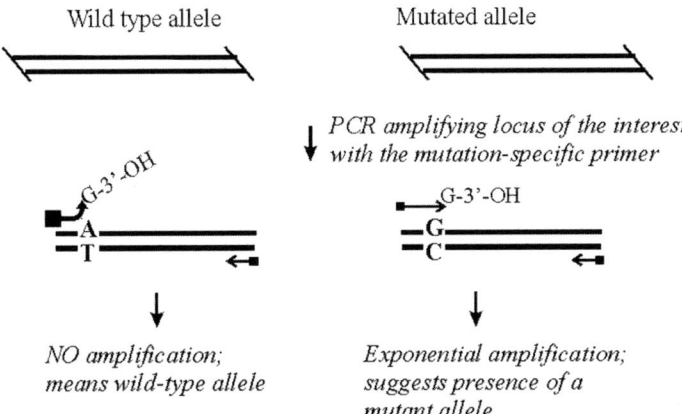

Figure 8. Schematic representation of the allele-specific PCR. PCR primer(s) is (are) generated to selectively amplify one of the two or more alleles, which is provided by 3'-terminal base complementary to a specific mutation or SNP site.

detection of particular strains of viruses and bacteria in environmental samples (Easterday et al. 2005).

The modification of this method termed allele-specific competitive blocker–polymerase chain reaction (ACB–PCR) includes a blocker primer to reduce the amount of background signal generated from the abundant wild-type template. The nonextendable blocker primer preferentially anneals to the wild-type DNA sequence, thereby excluding the annealing of the extendable mutant-specific primer to the wild-type sequence. Inclusion of single-strand DNA binding protein in the ACB–PCR reaction and use of the Stoffel fragment of Taq DNA polymerase both significantly increase allele discrimination. ACB–PCR can detect a base-pair substitution in the presence of a 105-fold excess of wild-type DNA (Parsons et al. 2005). In another variation called LigAmp, two oligonucleotides are hybridized adjacently to a DNA template. One oligonucleotide matches the target sequence and contains adapter sequence for further PCR amplification. If the target sequence is present, the oligonucleotides are ligated together and detected using real-time PCR (Shi et al. 2004).

These methods work well for the qualitative (not quantitative) identification of known SNPs, but are impossible for finding new mutations. In addition, direct PCR assays can be used for the identification of relatively long (more than 15 bp long) insertion/deletion polymorphisms. To this end, primers are designed to flank the site of possible insertion/deletion. PCR fragments of a size greater than expected suggest an insertion in the analyzing locus, whereas shorter PCR products evidence deletion. Using different fluorescent labels, such an approach can be multiplexed, thus allowing analyzing several loci in a single experiment (Diebold et al. 2005).

3.3 Other Enzymatic Approaches for Mutation Scanning

Not only nucleases, but also proteins that specifically bind mispaired nucleotides can be recruited for the mutation analysis. For example, MutS protein is a mismatch binding protein that recognizes mispaired and unpaired base(s) in DNA. Immobilized mismatch binding protein can bind DNA heteroduplexes while allowing homoduplexes to be washed away, thus enriching for rare mutations. In model experiments, unlabeled and fluorescent-labeled oligonucleotides, either perfectly complementary or with single nucleotide mismatches or deletions, were combined to form homo- or heteroduplexes that were further mixed at low ratios of hetero- to homoduplexes and exposed to MutS. Using a capillary DNA sequencer, 29-fold enrichment was detected for oligonucleotides with a single base deletion, whereas a rather modest figure of twofold enrichment was seen for single mismatch oligonucleotides (Baum et al. 2005).

Alternatively, MutS protein chip is available for rapid screening of single-nucleotide polymorphisms. The specific binding of dye-labeled MutS protein with surface-bound DNA or dye-labeled DNA with surface-bound MutS protein is revealed by the obtained fluorescence images (Behrensdorf et al. 2002;

Bi et al. 2003). By measuring the distance from the MutS binding site to DNA ends, one can locate the position of a mutation site. Being a label-free, surface-sensitive technique, quartz crystal microbalance (QCM) (Su et al. 2004) and surface plasmon resonance (SPR) devices have been used to study MutS interactions with the mismatches.

QCM detection principle provides additional information about the structural and viscoelastic properties of adsorbed molecules. The measured motional resistance changes per coupled MutS unit mass (deltaR/deltaf) are found to be indicative of the viscoelastic or structural properties of the bound protein, corresponding to different binding mechanisms. In addition, the deltaR/deltaf values vary remarkably when the MutS protein binds at different distances away from the QCM surface. Thus, these values can be used as a "fingerprint" for MutS mismatch recognition and also used to quantitatively locate the mutation site. However, the robustness of such an approach has not yet been demonstrated in experimental high-throughput screens.

Another mismatch binding protein that can be used for the polymorphism recovery is the cleavage-deficient mutant endonuclease VII that can only attach to mispaired nucleotides (Golz et al. 1998a). The protein can be immobilized on a solid-phase carrier (sepharose) via special tags, thus providing an efficient way to enrich in heteroduplex DNA in several repeated rounds of binding steps (Golz and Kemper 1999). In model experiments, the content of heteroduplex DNAs among the total DNA bound increased from ~10% to 45% after the third cycle of binding with no further increase in additional cycles. This method allows identification of previously unknown mutations, but is not very sensitive, as it allows reliable detection of heteroduplexes in samples with a heteroduplex content starting from 10% (Golz and Kemper 1999).

The technique termed "Glycosylase mediated polymorphism detection" utilizes quite different rationale of using specific glycosylases to detect mimatches (Vaughan and McCarthy 1999; O'Donnell et al. 2001). To this end, target DNA is PCR amplified using three normal dNTPs and a fourth modified dNTP, whose base is a substrate for a specific DNA-glycosylase once incorporated into the DNA (e.g. dUTP). Primers for this amplification are designed so that during extension, the position of the first modified nucleotide (uracil) incorporated into the extended primers differs depending on whether a mutation is present or absent. Subsequent glycosylase excision of the uracil residues followed by chemical or enzymatic cleavage of the apyrimidinic sites allows detection of the mutation in the amplified fragment as a fragment length polymorphism. The technique was shown to be robust and sensitive tool in several applications. However, been published for the first time in 1998 by Patrick Vaughan and Tommie McCarthy (Vaughan and McCarthy 1998), this method never went beyond the use by the same research team.

The last technique described in this section, recently proposed by Li et al., is based on a point mutation detection using high-fidelity DNA ligase (Li et al. 2005). Its protocol has three major steps (Figure 9). First, a hybridization of the target

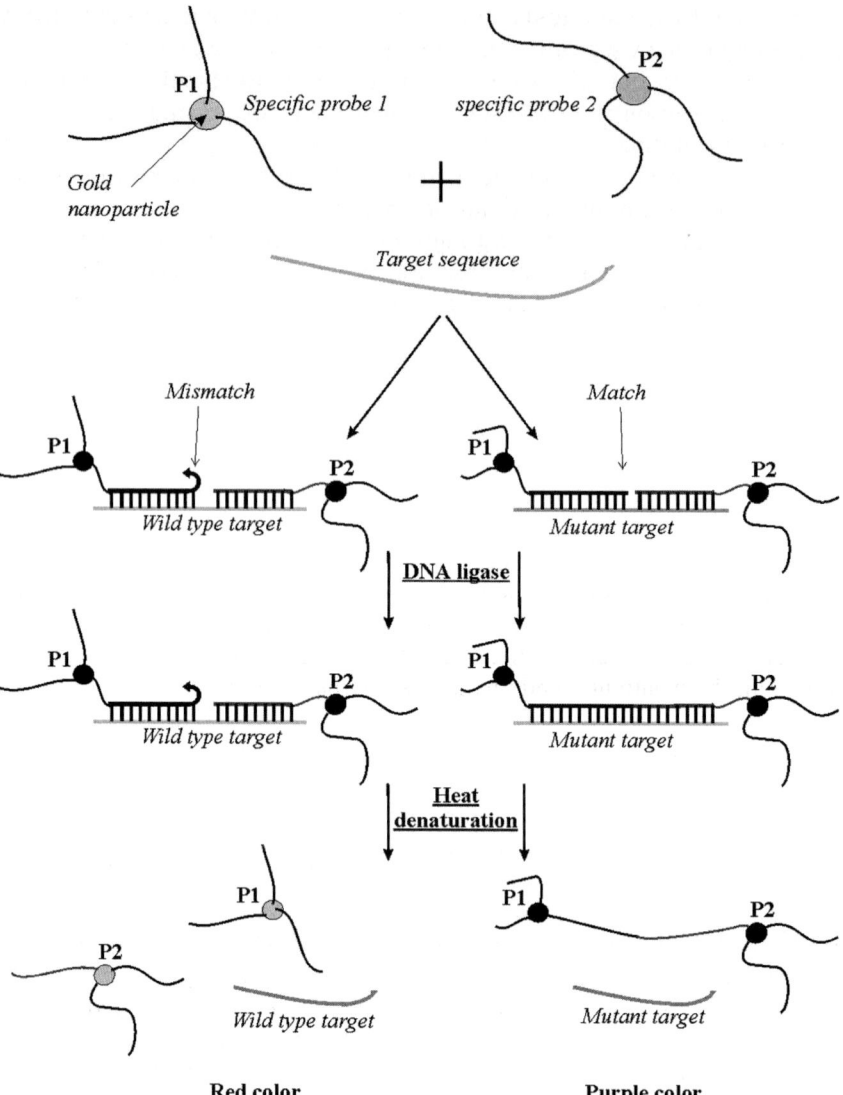

Figure 9. Point mutation detection using high fidelity DNA ligase and two mutation-specific DNA probes, labeled with gold nanoparticles. When the target mutation presents in the analyzing DNA mixture, both probes hybridize perfectly to the target strand, and further ligation step covalently binds both (or more) labeled probes. After heat denaturation the color of covalently attached gold nanoparticle aggregates does not revert to red, in contrast to those corresponding to a wild-type allele.

DNA strand (wild-type or mutant) with two gold nanoparticle-tagged probes occurs. Hybridization results in the formation of an extended polymeric gold nanoparticle–polynucleotide aggregate, turning the solution color from red to purple. At the next stage, high-fidelity ligase is added (Tth DNA ligase in this study) that covalently binds perfectly matched probes while no ligation occurs between mismatched ones. Finally, the mixture is heated to denature double-stranded DNA hybrids. In the case of no ligation occurred, the color reverts to red (as nanoparticles are not bound anymore), whereas in the opposite case the color remains purple. This approach efficacy was demonstrated by scoring single nucleotide mutations in the human oncogene *K-ras* (Li et al. 2005).

4. PHYSICAL APPROACHES

This group of methods look for a physical difference between the mutant strand and wild-type strands of DNA, being based either on physical isolation of imperfectly matched DNA hybrids (like electrophoretic separation), or on finding differences in mismatched versus perfect DNA hybrid physical peculiarities.

One of these approaches is quite similar to well-known protein fingerprinting utilizing the protein molecule digestion with a proteolytic enzyme (usually trypsinization), followed by mass spectrometry analysis of the resulting short peptides. If any protein modification or amino acid substitution occurs, than the position(s) and intensity(s) of the corresponding peaks on the mass specter are changed. Further discovery and development of matrix-assisted laser desorption ionization (MALDI) technique produced revolution in proteomics by making it possible to precisely define peptide amino acid compositions and based on this knowledge to unambiguously identify the whole proteins, even analyzed in complex mixtures. The rationale for this identification is that the number of theoretically possible variants of such peptides for each protein is limited and unique.

Similarly, nucleic acid molecule digested with base-specific enzymes (e.g. uracil DNA glycosylase, RNases A, or T1) or chemicals (see above) yields a set of short oligonucleotides, which can be analyzed using mass spectrometry. If a mutation occurs, then base composition of the respective oligonucleotide will be different from that of the reference sequence, resulting in changes in the resulting mass specter (Bocker 2003). Following analysis with the specialized software, these changes transform to the knowledge regarding (1) type of the mutation (if any) and (2) its localization within the nucleic acid molecule under study. The limitation of such an approach is that not all nucleic acid sequences may be analyzed in such a way: to be informative for mass-specter analysis, small oligonucleotides cannot be too short, but at the same time their length should not exceed 25 nt. To meet these criteria, parallel experiments with treatment using different base-specific reagents are recommended.

Another approach, proposed by Maruyama et al. (2003), is based on the observation that DNA intercalators (including fluorescent dye SYBR Green I used in this study) bind specifically with a duplex DNA. The fluorescence intensity of mismatched oligonucleotides decreases relatively to perfectly matched oligonucleotides. For example, for 40 bp long duplex containing one mismatched pair of nucleotides the fluorescence decreased by more than 13%. Such an assay can distinguish various types of single-base mismatches, the sensitivity being improved in the presence of 20% formamide. This detection method requires only a normal fluorescence spectrophotometer, an inexpensive dye and just 50 pmol of sample DNA.

Undoubtedly, the most widely used physical approach for mutation detection is the *single-strand conformational polymorphism* (SSCP) (e.g. Maekawa et al. 2004). This simple and cost-efficient technique assists in choosing only the DNA fragments of interest with expected mutation. The principle of detection of small changes in DNA sequences is based on the changes in single-strand DNA conformations created by mutations. The electrophoretic mobility of the respective DNAs differs so that SSCP detects such changes, sometimes in a sequence-dependent manner. The limitations faced in SSCP range from the routine polyacrylamide gel electrophoresis problems to the problems of resolving mutant DNA bands. Both could be solved by controlling electrophoresis conditions and by varying physical and environmental conditions like pH, temperature, voltage, gel type and percentage, addition of additives and denaturants (Gupta et al. 2005).

Microchips, which offer an advantage of screening a wide number of polymorphisms in a single experiment (Maekawa et al. 2004) employ hybridization of the sample probe with short oligonucleotides arrayed on a chip. The hybridization temperature and conditions are optimized so that the sequence variants differing from the arrayed oligonucleotides in one nucleotide or more do not hybridize with the chip. This expensive approach has been extensively used for many applications, including SNP detection in the genomes of eukaryotes, bacteria, and mitochondria (Maitra et al. 2004).

Another promising approach, recently developed by Patolsky and colleagues, termed "enzyme-amplified electronic transduction" (Patolsky et al. 2001), made it possible to perform a quantitative analysis of mutations with no PCR preamplification, with the lower limit of sensitivity for the detection of the mutant DNA as low as 1×10^{-14} mol/ml. The authors designed a thiolated sensing oligonucleotide, complementary to the target DNA as far as one base before the mutation site (Figure 10). After hybridizing of the target DNA, normal or mutant, with the sensing oligonucleotide, the resulting assembly is reacted with the biotinylated nucleotide, complementary to the mutation site, in the presence of DNA polymerase. The labeled nucleotide is coupled only to the double-stranded assembly that includes the mutant site. Subsequent binding of avidin–alkaline phosphatase and the biocatalyzed precipitation of an insoluble product on the transducer, provide means to confirm and amplify signal suggesting detection of the mutant allele. Faradaic impedance spectroscopy and microgravimetric QCM analyses were further employed by the

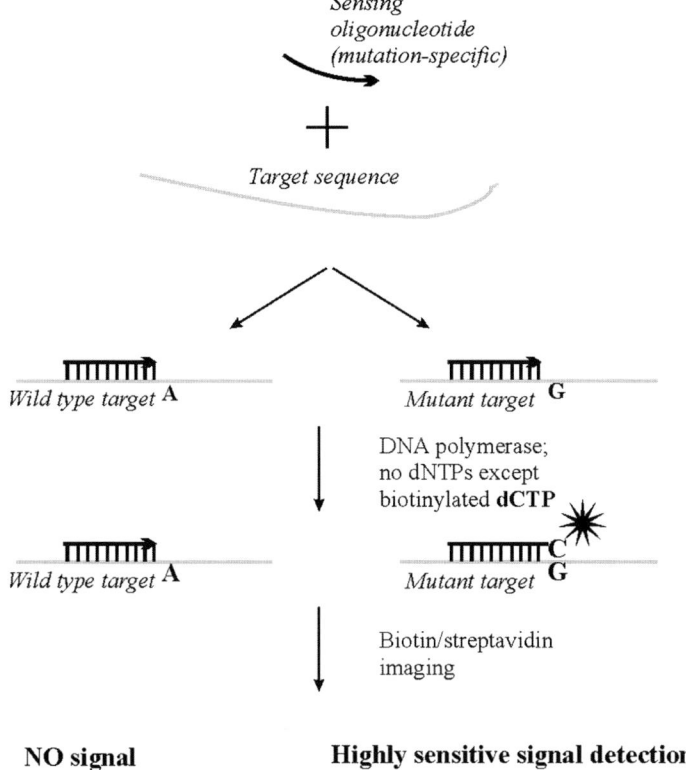

Figure 10. Schematic representation of the enzyme-amplified electronic transduction assay principle (see text).

authors for extremely sensitive and accurate electronic detection of SNPs (Patolsky et al. 2001).

5. BIOINFORMATICAL APPROACHES

These techniques mainly rely on the available sequencing data and deal either with treatment of the information available from databases (Vilella et al. 2005) or with the recognition of polymorphisms from "raw" primary PCR – sequencing data. In the latter case, when more than one allele is PCR-amplified, nucleotide substitutions can be recognized as double peaks on sequencing capillary electrophoregrams (Qiu et al. 2003; Lee and Vega 2004). Another important application is the detection of new bands appeared upon chemical or enzymatic cleavage of the mispaired DNA and the effective discrimination of target against background signaling (Zerr and Henikoff 2005). Overall, bioinformatical techniques provide algorithms or ready-to-use software (Unneberg et al. 2005) to the research community.

REFERENCES

Babon JJ, McKenzie M, Cotton RG (2003) The use of resolvases T4 endonuclease VII and T7 endonuclease I in mutation detection. Mol Biotechnol 23:73–81

Bannwarth S, Procaccio V, Paquis-Flucklinger V (2005) Surveyor nuclease: a new strategy for a rapid identification of heteroplasmic mitochondrial DNA mutations in patients with respiratory chain defects. Hum Mutat 25:575–582

Baum L, Ng A, Leung WK (2005) Developing the use of mismatch binding proteins for discovering rare somatic mutations. Mol Cell Probes 19:163–168

Behrensdorf HA, Pignot M, Windhab N, Kappel A (2002) Rapid parallel mutation scanning of gene fragments using a microelectronic protein-DNA chip format. Nucleic Acids Res 30:e64

Bi LJ, Zhou YF, Zhang XE, Deng JY, Zhang ZP, Xie B, Zhang CG (2003) A MutS-based protein chip for detection of DNA mutations. Anal Chem 75:4113–4119

Birkenkamp K, Kemper B (1995) *In vitro* processing of heteroduplex loops and mismatches by endonuclease VII. DNA Res 2:9–14

Bocker S (2003) SNP and mutation discovery using base-specific cleavage and MALDI-TOF mass spectrometry. Bioinformatics 19 (Suppl 1):i44–53

Bui CT, Lambrinakos A, Babon JJ, Cotton RG (2003) Chemical cleavage reactions of DNA on solid support: application in mutation detection. BMC Chem Biol 3:1

Chalaya T, Gogvadze E, Buzdin A, Kovalskaya E, Sverdlov ED (2004) Improving specificity of DNA hybridization-based methods. Nucleic Acids Res 32:e130

Che Y, Chen X (2004) A multiplexing single nucleotide polymorphism typing method based on restriction-enzyme-mediated single-base extension and capillary electrophoresis. Anal Biochem 329:220–229

Comai L, Young K, Till BJ, Reynolds SH, Greene EA, Codomo CA, Enns LC, Johnson JE, Burtner C, Odden AR, Henikoff S (2004) Efficient discovery of DNA polymorphisms in natural populations by Ecotilling. Plant J 37:778–786

Cotton RG (1999) Mutation detection by chemical cleavage. Genet Anal 14:165–168

Cotton RG, Rodrigues NR, Campbell RD (1988) Reactivity of cytosine and thymine in single-base-pair mismatches with hydroxylamine and osmium tetroxide and its application to the study of mutations. Proc Natl Acad Sci USA 85:4397–4401

Deeble VJ, Roberts E, Robinson MD, Woods CG, Bishop DT, Taylor GR (1997) Comparison of enzyme mismatch cleavage and chemical cleavage of mismatch on a defined set of heteroduplexes. Genet Test 1:253–259

Del Tito BJ, Jr., Poff HE, III, Novotny MA, Cartledge DM, Walker RI, II, Earl CD, Bailey AL (1998) Automated fluorescent analysis procedure for enzymatic mutation detection. Clin Chem 44:731–739

Diebold R, Bartelt-Kirbach B, Evans DG, Kaufmann D, Hanemann CO (2005) Sensitive detection of deletions of one or more exons in the neurofibromatosis type 2 (NF2) gene by multiplexed gene dosage polymerase chain reaction. J Mol Diagn 7:97–104

Easterday WR, Van Ert MN, Zanecki S, Keim P (2005) Specific detection of *bacillus anthracis* using a TaqMan mismatch amplification mutation assay. Biotechniques 38:731–735

Fiammengo R, Jaschke A (2005) Nucleic acid enzymes. Curr Opin Biotechnol 16:614–621

Ford ME, Whitcomb DC (1999) Analysis of the hereditary pancreatitis-associated cationic trypsinogen gene mutations in exons 2 and 3 by enzymatic mutation detection from a single 2.2-kb polymerase chain reaction product. Mol Diagn 4:211–218

Fuhrmann M, Oertel W, Berthold P, Hegemann P (2005) Removal of mismatched bases from synthetic genes by enzymatic mismatch cleavage. Nucleic Acids Res 33:e58

Gale JM, Tafoya GB (2004) Evaluation of 15 polymerases and phosphorothioate primer modification for detection of UV-induced C:G to T:A mutations by allele-specific PCR. Photochem Photobiol 79:461–469

Goldrick MM (2001) RNase cleavage-based methods for mutation/SNP detection, past and present. Hum Mutat 18:190–204

Golz S, Birkenkamp-Demtroder K, Kemper B (1998a) Enzymatic mutation detection. Procedure for screening and mapping of mutations by immobilised endonuclease VII. Nucleic Acids Res 26:1132–1133

Golz S, Greger B, Kemper B (1998b) Enzymatic mutation detection. Phosphate ions increase incision efficiency of endonuclease VII at a variety of damage sites in DNA. Mutat Res 382:85–92

Golz S, Kemper B (1999) Enzymatic mutation detection: enrichment of heteroduplexes from hybrid DNA mixtures by cleavage-deficient GST-tagged endonuclease VII. Nucleic Acids Res 27:e7

Guo DC, Qi Y, He R, Gupta P, Milewicz DM (2003) High throughput detection of small genomic insertions or deletions by Pyrosequencing. Biotechnol Lett 25:1703–1707

Gupta V, Arora R, Ranjan A, Bairwa NK, Malhotra DK, Udhayasuriyan PT, Saha A, Bamezai R (2005) Gel-based nonradioactive single-strand conformational polymorphism and mutation detection: limitations and solutions. Methods Mol Biol 291:247–261

Huang J, Kirk B, Favis R, Soussi T, Paty P, Cao W, Barany F (2002) An endonuclease/ligase based mutation scanning method especially suited for analysis of neoplastic tissue. Oncogene 21:1909–1921

Inganas M, Byding S, Eckersten A, Eriksson S, Hultman T, Jorsback A, Lofman E, Sabounchi F, Kressner U, Lindmark G, Tooke N (2000) Enzymatic mutation detection in the P53 gene. Clin Chem 46:1562–1573

Kosak HG, Kemper BW (1990) Large-scale preparation of T4 endonuclease VII from over-expressing bacteria. Eur J Biochem 194:779–784

Lambrinakos A, Yakubovskaya M, Babon JJ, Neschastnova AA, Vishnevskaya YV, Belitsky GA, D'Cunha G, Horaitis O, Cotton RG (2004) Novel TP53 gene mutations in tumors of Russian patients with breast cancer detected using a new solid phase chemical cleavage of mismatch method and identified by sequencing. Hum Mutat 23:186–192

Lee WH, Vega VB (2004) Heterogeneity detector: finding heterogeneous positions in Phred/Phrap assemblies. Bioinformatics 20:2863–2864

Li J, Chu X, Liu Y, Jiang JH, He Z, Zhang Z, Shen G, Yu RQ (2005) A colorimetric method for point mutation detection using high-fidelity DNA ligase. Nucleic Acids Res 33:e168

Lloyd A, Plaisier CL, Carroll D, Drews GN (2005) Targeted mutagenesis using zinc-finger nucleases in Arabidopsis. Proc Natl Acad Sci USA 102:2232–2237

Maekawa M, Nagaoka T, Taniguchi T, Higashi H, Sugimura H, Sugano K, Yonekawa H, Satoh T, Horii T, Shirai N, Takeshita A, Kanno T (2004) Three-dimensional microarray compared with PCR-single-strand conformation polymorphism analysis/DNA sequencing for mutation analysis of K-ras codons 12 and 13. Clin Chem 50:1322–1327

Maitra A, Cohen Y, Gillespie SE, Mambo E, Fukushima N, Hoque MO, Shah N, Goggins M, Califano J, Sidransky D, Chakravarti A (2004) The human MitoChip: a high-throughput sequencing microarray for mitochondrial mutation detection. Genome Res 14:812–819

Maruyama T, Takata T, Ichinose H, Park LC, Kamaiya N, Goto M (2003) Simple detection of point mutations in DNA oligonucleotides using SYBR Green I. Biotechnol Lett 25:1637–1641

Mashal RD, Koontz J, Sklar J (1995) Detection of mutations by cleavage of DNA heteroduplexes with bacteriophage resolvases. Nat Genet 9:177–183

Mitani N, Tanaka S, Okamoto Y (2006) Surveyor nuclease-based genotyping of SNPs. Clin Lab 52:385–386

Nakatsukasa T, Shiraishi Y, Negi S, Imanishi M, Futaki S, Sugiura Y (2005) Site-specific DNA cleavage by artificial zinc finger-type nuclease with cerium-binding peptide. Biochem Biophys Res Commun 330:247–252

O'Donnell KA, Tighe O, O'Neill C, Naughten E, Mayne PD, McCarthy TV, Vaughan P, Croke DT (2001) Rapid detection of the R408W and I65T mutations in phenylketonuria by glycosylase mediated polymorphism detection. Hum Mutat 17:432

Oleykowski CA, Bronson Mullins CR, Godwin AK, Yeung AT (1998) Mutation detection using a novel plant endonuclease. Nucleic Acids Res 26:4597–4602

Parsons BL, McKinzie PB, Heflich RH (2005) Allele-specific competitive blocker-PCR detection of rare base substitution. Methods Mol Biol 291:235–45

Patolsky F, Lichtenstein A, Willner I (2001) Detection of single-base DNA mutations by enzyme-amplified electronic transduction. Nat Biotechnol 19:253–257

Qiu P, Shandilya H, D'Alessio JM, O'Connor K, Durocher J, Gerard GF (2004) Mutation detection using Surveyor nuclease. Biotechniques 36:702–707

Qiu P, Shandilya H, Gerard GF (2005) A method for clone sequence confirmation using a mismatch-specific DNA endonuclease. Mol Biotechnol 29:11–18

Qiu P, Soder GJ, Sanfiorenzo VJ, Wang L, Greene JR, Fritz MA, Cai XY (2003) Quantification of single nucleotide polymorphisms by automated DNA sequencing. Biochem Biophys Res Commun 309:331–338

Roberts E, Deeble VJ, Woods CG, Taylor GR (1997) Potassium permanganate and tetraethylammonium chloride are a safe and effective substitute for osmium tetroxide in solid-phase fluorescent chemical cleavage of mismatch. Nucleic Acids Res 25:3377–3378

Rowley G, Saad S, Giannelli F, Green PM (1995) Ultrarapid mutation detection by multiplex, solid-phase chemical cleavage. Genomics 30:574–582

Schmalzing D, Belenky A, Novotny MA, Koutny L, Salas-Solano O, El-Difrawy S, Adourian A, Matsudaira P, Ehrlich D (2000) Microchip electrophoresis: a method for high-speed SNP detection. Nucleic Acids Res 28:e43

Shagin DA, Rebrikov DV, Kozhemyako VB, Altshuler IM, Shcheglov AS, Zhulidov PA, Bogdanova EA, Staroverov DB, Rasskazov VA, Lukyanov S (2002) A novel method for SNP detection using a new duplex-specific nuclease from crab hepatopancreas. Genome Res 12:1935–1942

Shi C, Eshleman SH, Jones D, Fukushima N, Hua L, Parker AR, Yeo CJ, Hruban RH, Goggins MG, Eshleman JR (2004) LigAmp for sensitive detection of single-nucleotide differences. Nat Methods 1:141–147

Shi R, Otomo K, Yamada H, Tatsumi T, Sugawara I (2006) Temperature-mediated heteroduplex analysis for the detection of drug-resistant gene mutations in clinical isolates of Mycobacterium tuberculosis by denaturing HPLC, Surveyor nuclease. Microbes Infect 8:128–135

Shively L, Chang L, LeBon JM, Liu Q, Riggs AD, Singer-Sam J (2003) Real-time PCR assay for quantitative mismatch detection. Biotechniques 34:498–502, 504

Smith MJ, Humphrey KE, Cappai R, Beyreuther K, Masters CL, Cotton RG (2000) Correct heteroduplex formation for mutation detection analysis. Mol Diagn 5:67–73

Sokurenko EV, Tchesnokova V, Yeung AT, Oleykowski CA, Trintchina E, Hughes KT, Rashid RA, Brint JM, Moseley SL, Lory S (2001) Detection of simple mutations and polymorphisms in large genomic regions. Nucleic Acids Res 29:e111

Su X, Robelek R, Wu Y, Wang G, Knoll W (2004) Detection of point mutation and insertion mutations in DNA using a quartz crystal microbalance and MutS, a mismatch binding protein. Anal Chem 76:489–494

Sun H, Yabuki A, Maizels N (2001) A human nuclease specific for G4 DNA. Proc Natl Acad Sci USA 98:12444–12449

Surdi GA, Yaar R, Smith CL (1999) Discrimination of DNA duplexes with matched and mismatched tandem repeats by T4 endonuclease VII. Genet Anal 14:177–179

Taylor GR (1999) Enzymatic and chemical cleavage methods. Electrophoresis 20:1125–1230

Tessitore A, Toniato E, Gulino A, Frati L, Ricevuto E, Vadala M, Vingolo E, Martinotti S (2002) Prenatal diagnosis of a rhodopsin mutation using chemical cleavage of the mismatch. Prenat Diagn 22:380–384

Till BJ, Burtner C, Comai L, Henikoff S (2004) Mismatch cleavage by single-strand specific nucleases. Nucleic Acids Res 32:2632–2641

Unneberg P, Stromberg M, Sterky F (2005) SNP discovery using advanced algorithms and neural networks. Bioinformatics 21:2528–2530

Vaughan P, McCarthy TV (1998) A novel process for mutation detection using uracil DNA-glycosylase. Nucleic Acids Res 26:810–815

Vaughan P, McCarthy TV (1999) Glycosylase mediated polymorphism detection (GMPD)–a novel process for genetic analysis. Genet Anal 14:169–175

Vilella AJ, Blanco-Garcia A, Hutter S, Rozas J (2005) VariScan: analysis of evolutionary patterns from large-scale DNA sequence polymorphism data. Bioinformatics 21:2791–2793

Yeung AT, Hattangadi D, Blakesley L, Nicolas E (2005) Enzymatic mutation detection technologies. Biotechniques 38:749–758

Youil R, Kemper BW, Cotton RG (1995) Screening for mutations by enzyme mismatch cleavage with T4 endonuclease VII. Proc Natl Acad Sci USA 92:87–91

Youil R, Toner TJ, Bull E, Bailey AL, Earl CD, Dietz HC, Montgomery RA (2000) Enzymatic mutation detection (EMD) of novel mutations (R565X and R1523X) in the FBN1 gene of patients with Marfan syndrome using T4 endonuclease VII. Hum Mutat 16:92–93

Zerr T, Henikoff S (2005) Automated band mapping in electrophoretic gel images using background information. Nucleic Acids Res 33:2806–2812

Zhu D, Xing D, Shen X, Liu J (2004) A method to quantitatively detect H-ras point mutation based on electrochemiluminescence. Biochem Biophys Res Commun 324:964–969

CHAPTER 10

CURRENT ATTEMPTS TO IMPROVE THE SPECIFICITY OF NUCLEIC ACIDS HYBRIDIZATION

ANTON A. BUZDIN

Shemyakin-Ovchinnikov Institute of Bioorganic Chemistry, Russian Academy of Sciences, 16/10 Miklukho-Maklaya, 117997 Moscow, Russia
Phone: + (7495) 3306329; Fax: + (7495) 3306538; E-mail: anton@humgen.siobc.ras.ru

Abstract:	Being very useful and informative, many techniques based on nucleic acids hybridization suffer from the cross-annealing of repetitive DNA, presenting in reassociating samples. This "wrong" annealing causes "nonspecific" hybridization of nonorthologous DNA fragments, thus producing chimeric sequences and at the final stage significantly hampering the analysis of the resulting cDNA or genomic libraries. Such chimeras may constitute up to 40–60% of DNA libraries. Importantly, the number of chimerical clones positively correlates with the complexity of hybridizing genomic or cDNA mixtures. The hybridization specificity is a crucial factor determining both the fidelity the efficiency of all hybridization-based analytical techniques. In this chapter, I review the current attempts to increase the specificity of hybridization at both stages: during nucleic acids reassociation and at the stage of selection of proper hybrids. To this end, approaches based on chemical modifications, improving hybridization kinetics, and improving selection of perfectly matched duplexes, have been developed.
Keywords:	Hybridization specificity, PCR selection effect, targeted genomic difference analysis (TGDA), thermodynamically stable duplexes, phenol emulsion reassociation technique (PERT), repetitive element, genomic repeat, chimerical clone, hybridization temperature, melting point, perfectly matched hybrids, mispaired DNA rejection (MDR), TILLING, mismatch-sensitive nuclease, mung bean nuclease, $C_o tA$ fraction.
Abbreviations:	MDR, mispaired DNA rejection; PCR, polymerase chain reaction; PERT, phenol emulsion reassociation technique; RDA, representative differential analysis; RE, repetitive element; SH, subtractive hybridization; SSH, suppression subtractive hybridization; TGDA, targeted genomic difference analysis.

TABLE OF CONTENTS

1. Introduction 242
2. Improving Hybridization Kinetics 244
   2.1 Simplification of Hybridizing Mixtures 246
   2.2 Chemical Modifications 248
3. Improving Selection of Perfectly Matched Hybrids 249
4. Protocols 255
   4.1 Targeted Genomic Difference Analysis 255
   4.2 Using Competitor DNA to Decrease
       the Background of Genomic Repeats 258
   4.3 Mispaired DNA Rejection 260
References 262

1. INTRODUCTION

Many popular (approximately 300 PubMed citations per year) experimental techniques for genome and transcriptome analysis, such as coincidence cloning (Chapter 8) and subtractive hybridization (SH) (Chapters 3, 6, 7), including representative differential analysis (RDA) (Lisitsyn and Wigler 1995) and suppression subtractive hybridization (SSH) (Diatchenko et al. 1996), are based on DNA hybridization in solution, followed by polymerase chain reaction (PCR) amplification of certain hybridized fractions (Sasaki et al. 1994; Nagayama et al. 2001).

The hybridization specificity is a crucial factor determining both the fidelity of the natural biological processes and the efficiency of hybridization-based analytical techniques. Hybridization specificity (f) is determined as a relative factor for match versus mismatch discrimination: $f = \exp\text{-}|\Delta G_{m\text{-}mm}/RT|$, where $G_{m\text{-}mm}$ is the free energy penalty for binding to sites that differ from the perfectly complementary sequences by a single base-pair substitution. If $G_{m\text{-}mm}$ is ~4 kcal/mol (Roberts and Crothers 1991), hybridization specificity will be ~1/100–1/1000, and theoretically it is possible to find a range of conditions (so-called stringency conditions) where perfect complexes will be substantially more stable than the complexes containing mismatches (Broude 2002).

However, being very useful and informative, the techniques based on nucleic acids hybridization are not free from some imperfections. The well-known disadvantage of complex DNA mixture hybridization is the cross-annealing of repetitive DNA, presenting in reassociating samples (Hames and Higgins 1985). This "wrong" annealing causes "nonspecific" hybridization of nonorthologous DNA fragments, thus producing chimeric sequences and at the final stage significantly hampering the analysis of the resulting cDNA or genomic libraries (see Figure 1). Such chimeras may constitute up to 40–60% of DNA libraries.

Importantly, the number of chimerical clones positively correlates with the complexity of hybridizing genomic or cDNA mixtures. This is probably the

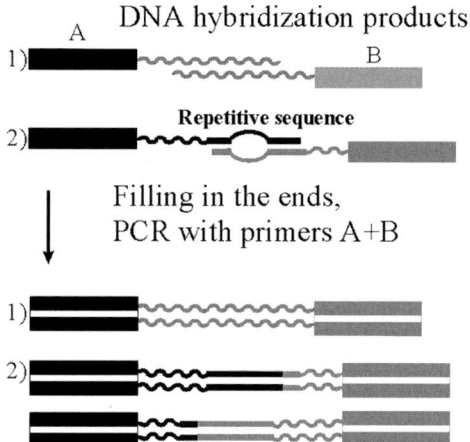

Figure 1. Along with the normal duplex formation, chimeric hybrids between the repetitive sequences may occur, especially when complex genomic mixtures are hybridized. In many techniques, these "wrong" hybrids are not properly recognized and are further amplified and included in the final clone libraries, thus forming up to 60% of the overall sequence information.

major reason that limits SH applications to the comparison of cDNA samples (representing only the modest part of genomic sequences) and of small genomes (such as prokaryotic, yeast or planarian DNA; see Chapters 1 and 3). SH-based approaches in their present form are hardly applicable to the recovery of differences between the complex genomic DNAs like mammalian ones. Two major related techniques dealing with complex genomic DNA subtractions, namely RDA (Lisitsyn and Wigler 1993; Lisitsyn and Wigler 1995) and targeted genomic difference analysis (TGDA) (Buzdin et al. 2002; Buzdin et al. 2003a), all employ dramatic genomic DNA simplification prior to hybridization step.

In the first case, this simplification is achieved by means of the so-called PCR selection effect, when the total pool of fragmented genomic DNA with ligated adapters is PCR preamplified for 50–100 cycles. This results in a great bias in different DNA fragment concentrations in the resulting amplicons: most of the sequences turn to be underrepresented or completely lost due to rather inefficient PCR amplification with *Taq* DNA polymerase, whereas the others, forming relatively small (~10% or less) fraction of the initial pool, are overrepresented because of an optimal length/GC-content ratio making them preferable targets for *Taq* polymerase. Therefore, RDA utilizes a random genomic DNA simplification based on the fragment size and GC-content. Consequently, the resulting pool of differential sequences, which appear after the subtraction stage, lacks most of the differential sequences presented in the genome originally. RDA is, thus, worth applicable for the recovery of some marker sequences, but cannot be used for comprehensive genome or cDNA analyses. The second approach TGDA is based on a specific PCR amplification of a group of genomic sequences of interest (e.g. sequences flanking insertions of human retroelements amplified with primer(s) specific to the retroelement 5′- or 3′-terminus (Buzdin et al. 2002; Buzdin et al. 2003a; Mamedov et al. 2005)). These sequences are selectively amplified with the reasonable number of PCR cycles (25–40 depending on the requirement of nested PCR amplification), and are further subtracted resulting in a comprehensive library enriched in DNAs presented in one of the comparing samples (tracer) but absent from the others (driver). The strength of TGDA is the complete rather than random (as for RDA) recovery of differential sequences of the interest (different groups of repetitive elements (REs), multigene family members, pseudogenes, duplicated, or multiplicated genomic loci). At present, this technique has been successfully applied to the recovery of human-specific endogenous retroviruses (Buzdin et al. 2002), L1 retrotransposons (Buzdin et al. 2003a), and for the experimental identification of polymorphisms created in human populations by the insertions of *Alu* repeats (Mamedov et al. 2005).

However, both TGDA and RDA approaches will be inefficient for the complete comparisons of the whole genomes. Of course, such a comparison may be done by means of complete genome sequencing (which is extremely expensive and time consuming; note, even now multiple gaps in human genome assembly are not filled [http://genome.ucsc.edu/cgi-bin/hgGateway]), but it is low probable that in the nearest future it will be possible to compare, say, 100 individual human genomes through the complete genome sequencing. Theoretically, for many applications SH could become an alternative to shotgun genome sequencing (except for identification of "fine" differences like single nucleotide substitutions or microsatellite length polymorphisms). "For many applications" means screening for relatively large deletions, duplications, translocations, insertions of pseudogenes, and transposable elements, as well as of the exogenous sequences like retroviruses. To improve the existing in-solution nucleic acids hybridization techniques and, in particular, subtraction-based methods, two major approaches seem reasonable:

– To improve hybridization kinetics (in order to insure that only the most thermodynamically stable duplexes [i.e. those lacking mispaired nucleotides] are formed during the stage of nucleic acids hybridization)
– To improve the recognition of perfectly matched duplexes (e.g. by selectively PCR amplifying them).

Both approaches appear to be fruitful in many cases.

2. IMPROVING HYBRIDIZATION KINETICS

Coincidence cloning, SH, and other techniques described in this book, all follow the uniform rules of DNA hybridization in solution. Although hybridization kinetics was better studied for SH (Sverdlov and Ermolaeva 1994; Ermolaeva and Wagner 1995; Milner et al. 1995; Ermolaeva et al. 1996), all major conclusions and theoretical considerations defined for SH, will be true for other techniques based on nucleic acids hybridization in solution as well. SH has

Current attempts to improve the specificity

become the practice since 1984, when Palmer and Lamar (1984) proposed a simple general idea of specific separation of enriched tracer–tracer homoduplexes from other components of the reannealed mixture: tracer DNA fragments (those containing differential sequences to be found) should have termini different from those of driver fragments (those serving as the background for differential tracer fragments; see Chapters 1, 3, 6, and 7). The authors prepared a mouse recombinant library enriched in Y chromosome fragments. The female mouse DNA (driver) was randomly cut into fragments, whereas the male DNA (tracer) was cleaved with *Mbo* I restriction endonuclease. Both DNAs were mixed at a ratio of 100:1, respectively, denatured, and reannealed. Only reassociated tracer homoduplexes contained sticky *Mbo* I ends at both termini and therefore could be selectively ligated to a *Bam* HI-digested pBR322 vector. This principle has been successfully applied to the isolation of DNA probes corresponding to a deletion spanning Duchenne muscular dystrophy locus (Kunkel et al. 1985).

The expected enrichment of the subtracted DNA with a sequence difference (E^d (t) value) is expressed by a formula (Ermolaeva and Sverdlov 1996):

$$E^d (t) = (1 + RD_0t)/(1 + RT_0t)$$

where R [M^{-1}s^{-1}] is the reassociation rate constant, and D_0 and T_0 are initial molar concentrations of driver and tracer, respectively. The maximum enrichment at $t \to \infty$ is D_0/T_0. For finite t values, like 14 h (overnight incubation), the enrichment value increases as RD_0 increases. Thus, to reach better results, one should increase the values of R, D_0, or both. As an example, Lamar (Lamar and Palmer 1984) and Kunkel (Kunkel et al. 1985), with coauthors, clearly realizing that the rate of hybridization is crucial to achieve substantial enrichment, increased the R value by addition of chemical accelerators, such as phenol, to reannealing mixtures.

However, mammalian genomes are too complex to reach sufficiently high D_0 values, and only major differences (like presence/absence of Y chromosome or extended deletions) can be isolated in such a way. As genome size increases beyond 5×10^8 bp (complexity comparable with that of arabidopsis or drosophila genomes), the kinetics of hybridization start to become an increasingly important factor limiting enrichment of the target (Milner et al. 1995). To enhance the kinetics of hybridization, increased hybridization times, higher driver concentrations, greater driver/tracer ratios, longer DNA fragments, and the use of techniques that enhance the rate of reassociation, e.g. phenol emulsion reassociation technique (PERT) (Kohne et al. 1977) or solvent exclusion (Barr and Emanuel 1990), may be effective. At a driver DNA concentration of 3.125 mg/ml the effective enhancement under PERT is only 2.2-fold (Kohne et al. 1977). However, compared to the rate in 1.0 M NaCl, the relative enhancement in the presence of 11% dextran sulphate and 1.5 M NaCl is 11.9-fold (Kohne et al. 1977; Barr and Emanuel 1990), although driver DNA concentration is limited to 1.0 mg/ml, reducing the achievable enhancement (Kohne et al. 1977; Barr and Emanuel 1990) to about fourfold.

2.1 Simplification of Hybridizing Mixtures

One of the widely used genomic subtraction schemes, called representational difference analysis (RDA) (Lisitsyn and Wigler 1995), involves selective PCR amplification of reannealed double-stranded tracer fragments, with all other types of hybrids not amplified. To ensure the selectivity, authors used special PCR adapters that functioned in the same way as *Mbo* I ends of the renatured tracer in previous works (Lamar and Palmer 1984; Kunkel et al. 1985). To increase the enrichment, RDA uses complexity reduction of driver and tracer DNA before subtraction. To this end, driver and tracer DNA fragments are repeatedly amplified so that the resulting driver and tracer represent a depleted pool of initial fragments as a result of the size-bias of PCR amplification. These simplified, fragmented genomes (or amplicons) are then used for subtraction. Thus, in the RDA technique (Lisitsyn and Wigler 1993), D_0 is increased by means of random genome simplification that allows molar concentrations of driver and tracer to be increased at the same mass concentrations. The simplification and repetitive cycles of subtraction make it possible to obtain very high enrichment of tracer with target fragments (Lisitsyn et al. 1994a). RDA was successfully used to clone DNA losses and amplifications in tumors (Carulli et al. 1998) and to generate specific genetic markers linked to a trait of interest (Lisitsyn et al. 1994b). An alternative to great PCR-cycle-number–based simplification of the hybridizing mixtures is the use of infrequent-cutter endonucleases providing very limited sets of fragments, which would not be too long for PCR amplification (see more detailed insight in Chapter 6, Section 3).

However, an evident RDA drawback is that due to random genome simplification only a minor fraction of the genome (2–10%) is actually compared, while 90–98% remains beyond the analysis. Therefore, isolation of genomic differences using this technique is, in a sense, a matter of luck. When TGDA is used (Buzdin et al. 2002), the genome simplification is targeted, the simplified fraction is not random but contains a fairly definite portion of the genome. The complexity C of the simplified portion depends on the repetitive target content in the genome. With N target repeats in the genome and an average size of the amplified fragments of ~256 bp (an average fragment produced by a frequent-cutter restriction endonuclease), $C = 256 N$. In the case of human endogenous retroviral family HERV-K (which was the first target for TGDA application), which amount to about 2000 in number of representatives (Buzdin et al. 2003b), C is as low as $~5 \times 10^5$, which is only 0.017% of the whole human genome complexity. This results in a dramatic (3.6×10^7) increase in the hybridization rate of the simplified versus the original genomic DNA, providing that mass concentrations during the subtraction are the same (Ermolaeva et al. 1996). The mass concentrations used by the authors (Buzdin et al. 2002; Buzdin et al. 2003a) (150 ng of driver DNA and 1.5 ng of tracer DNA per 1 μl) correspond to the driver and tracer molar concentrations of 5×10^{-10} and 5×10^{-12}. At $R = 10^6$ (Hames and Higgins 1985), one could expect ~20-fold enrichment after 14 h of hybridization. If the DNA is not simplified, the enrichment will be just negligible. The enrichment value 16, experimentally found by the authors, was in good agreement with the theory.

The successful application of TGDA partly depends on the divergence between the members of the repetitive element group under comparison (transposons, pseudogenes, etc.). If this divergence is high, oligonucleotide primers designed using the group consensus sequence may fail to prime PCR with the group members diverged too far from the consensus. However, the technique is aimed at the comparison of highly homologous REs forming evolutionarily young groups (sequence divergence less than 10%), with integrations polymorphic between closely related species or even within one species. The detailed TGDA protocol is given in the Section 4.1.

If sequences of interest are unique, then TGDA cannot be applied and a very serious problem appears when working with complex genomes: low reassociation rate of the nonrepetitive DNA. Indeed, DNA reassociation rate for each particular fragment is proportional to the square of its concentration; therefore, REs presented in a genome by ~10 (some pseudogenes), ~1000 (several mammalian endogenous retroviral families), or ~1,000,000 (human *Alu* retrotransposons) copies will hybridize, respectively, 10^2-, 10^6-, and 10^{12}-fold faster than fragments representing unique genomic sequences. As the latter's reassociation rate is incomparably lower, so that ~99% reassociation may take months or years, an enormous background of repetitive sequences appears when reasonable (~days) hybridization time is used. In this case, a great majority of double-stranded molecules in solution are reassociated repeats, whereas unique sequences mostly still in a single-stranded form. Therefore, the true genomic sequence representations will be enormously biased in such clone libraries.

A rather efficient attempt to improve the situation is the addition of competitor DNA fractions containing genomic repeats (Sambrook and Russell 2001) into hybridization mixture. Such competitors are mostly fractions of quickly reassociating double-stranded DNA, purified from single-stranded DNA using hydroxyapatite column chromatography. Such fractions, for example, commercially available "$C_o t$ A" and "$C_o t$ B" DNAs from Gibco BRL (USA) are greatly enriched in genomic repeats and may be used to decrease the background of repetitive sequences in cloned libraries. To this end, the initial genomic DNAs to be hybridized must be fragmented (either by sonication or digestion with restriction endonucleases), tagged (through the ligation of adapter sequences, by incorporation of biotin or other signal molecules), denatured, and allowed to hybridize in the presence of competitor DNA, taken in a 100–1000-fold weight excess.

The major part of genomic repeats presenting in the sample DNA will hybridize to competitor DNA. At the next stage, it is crucial to isolate the "proper" hybrids (those formed by original genomic DNA fragments) from the hybridization mixture while discarding genomic–competitor DNA duplexes and single-stranded DNAs. This can be done, for example, by using selective PCR amplification of the proper hybrids (see protocol in Section 4.2), or using biotin–streptavidin systems. To our experience, the use of $C_o t$ A DNA taken in 100-fold weight excess results in a decrease of genomic repeat-containing clones from ~93% to 76–78% of the libraries, when nonsimplified frequent-cutter endonuclease digested human genomic DNA is hybridized (Chalaya et al. 2004), which is significantly closer to the natural genomic occurrence of REs, occupying approximately two thirds of human DNA (Lander et al. 2001; Venter et al. 2001). Nevertheless, the number of chimerical clones representing improperly matched duplexes remained high in such libraries (~56%). To achieve better results with this approach, one has to titer genomic and competitor DNA concentrations.

Another important parameter, hybridization temperature, is a well-known regulator of hybridization specificity: the lower the temperature, the more is the background. However, after the value of 65°C the temperature increase has little or no effect on the hybridization specificity (Chalaya et al. 2004). For instance, exactly the same proportion of chimerical clones (~56%) was produced when frequent-cutter enzyme-digested human genomic DNA was hybridized at 65°C and 85°C.

2.2 Chemical Modifications

As mentioned above, the hybridization temperature increase from 65°C to 85°C has absolutely no effect on the hybridization specificity (~100–300 bp long fragments of human genomic DNA hybridized). This means that both perfectly and imperfectly matched duplexes of such lengths are stable enough to be formed at 85°C. Therefore, using temperature as an instrument of hybridization specificity control requires shorter hybridizing fragments. However, it is frequently puzzling to unambiguously map fragments shorter than 100 bp in the genomic sequence. Titering hybridization temperature conditions from 85°C (when ~56% background duplexes are formed) to 94°C (when DNA is denatured) will be probably helpful for creating higher-hybridization-fidelity systems; however, this temperature interval is rather small, and the data obtained are frequently hardly reproducible.

The approach based on DNA chemical modifications, or synthesis of DNA analogs, makes it possible to increase the stability of complementary nucleotide interactions, thus to increase hybridization temperatures and, therefore, to significantly enlarge the operational temperature interval between the point where hybridization specificity is not sufficient, and the melting point (see, e.g. Mouritzen et al. 2003). The further "fine-tuning" of the temperature conditions will make it possible to find out the conditions ensuring the highest specificity of hybridization for genomic DNA fragments of a given length.

Another group of methods utilizes chemical modifications for discriminating mispaired versus perfectly matched DNA (or RNA) duplexes (Cotton et al. 1988). Under the special conditions, some chemicals do preferentially modify mismatched nucleotides due to their higher availability, and the following specific glycoside bond cleavage results in a degradation of "wrong" hybrids.

For mispaired C bases modification, hydroxylamine (NH2OH) is used, whereas mispaired T was originally modifying with osmium tetroxide (OsO4) (Cotton 1999), which is now replaced by potassium permanganate (KMSnO4) solution with a coadditive triethylammonium chloride due to high toxicity of the former (Roberts et al. 1997). Modified DNAs are further simultaneously cleaved by piperidine and purified. However, these products cannot be efficiently cloned or PCR-amplified, most probably, due to some background DNA modifications.

The main advantage of this approach includes nearly 100% efficiency in cleaving hybrids having mispaired C or T nucleotides, whereas its important shortcomings are multiple manipulations and the fact that toxic chemicals are required. Finally, the most serious drawback of all chemical modifications-based methods for improving hybridization selectivity is that the products cannot be further PCR-amplified and cloned.

3. IMPROVING SELECTION OF PERFECTLY MATCHED HYBRIDS

Improving selection of perfectly matched duplexes in hybridization mixtures is an alternative promising approach aimed to remove background chimerical sequences from the resulting clone libraries. This approach does not deal with the improvement of hybridization conditions, but, instead, it is focused on the exclusive amplification of the "proper" hybrids (Figure 1). To this end, hybridized DNA may be treated with some chemical reagents specifically modifying mispaired nucleotides or producing double- or single-strand DNA breaks there. However, the PCR amplification and cloning of such chemicals-treated DNA is problematic, and the second approach comprising hybridized DNA exposure to the nucleases specifically recognizing improperly matched DNA, seems to be advantageous. The method called mispaired DNA rejection (MDR), recently published by Chalaya and coauthors (Chalaya et al. 2004) makes it possible to almost completely exclude the chimerical sequences from analyzing DNA subsets (Figure 2). The technique is based on the observation that overwhelming majority of cross-hybridizingREs, although sharing considerable sequence similarity, are not entirely identical to each other. Their DNA heteroduplexes are therefore imperfectly matched, having quite a number of mispaired bases. These latter can form single nucleotide mismatches or even extended single-stranded DNA loop regions. All such structural deviations from normal properly paired DNA duplexes can be recognized and cut by certain enzymes, termed here mismatch-specific nucleases. Mispaired DNA sensitive nucleases, serving *in vivo* as reparation or viral life cycle machinery units, are now successfully employed by investigators for mutation detection. Such approaches are both simple and rather efficient, such as, for example, TILLING technique for large-scale mutation screening (Till et al. 2004), Surveyor mutation detection system (Qiu et al. 2004), and elegant high-fidelity technique for endonuclease/ligase-based mutation scanning by Huang and others (Huang et al. 2002). The most commonly used mismatch-specific nucleases are phage T7 endonuclease I (Babon

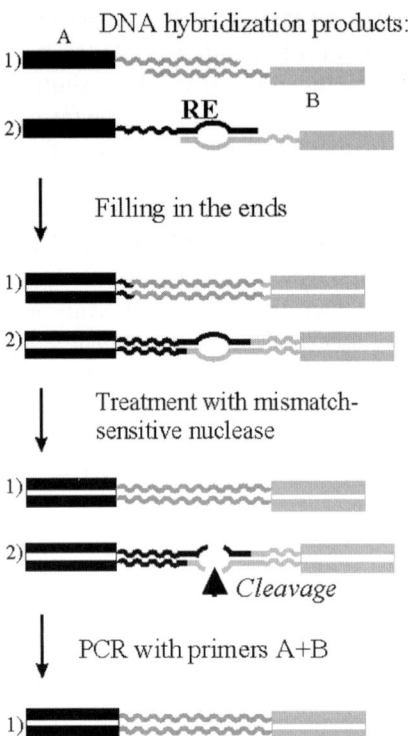

Figure 2. Outline of the mispaired DNA rejection (MDR) approach. Nonperfectly matched DNA hybrids, which constitute most of chimeric sequences, are recognized and cleaved by mismatch-sensitive nucleases, thus preventing their exponential amplification in the final PCR(s) and strongly increasing the occurrence of target nonchimeric sequences in the resulting clone libraries.

et al. 2003), T4 endonuclease VII (Mikhailov and Rohrmann 2002), modified bacterial endonuclease V (Huang et al. 2002), plant CEL I and Surveyor nucleases (Kulinski et al. 2000; Qiu et al. 2004). The authors demonstrated that these enzymes, cleaving DNA at mispaired base positions, can be used for eliminating chimerical hybrids from DNA hybridization mixtures, thus strongly reducing the number of background chimerical clones from 44–60% to 0–4%. MDR can be applied to both cDNA and genomic DNA subtractions of very complex DNA mixtures. This technique was also useful for the genome-wide recovery of highly conserved DNA sequences, as demonstrated by comparing human and pygmy marmoset genomes (Chalaya et al. 2004).

In order to investigate MDR efficiency, Chalaya et al. used the testing system (see Figure 3) comprising (1) digestion of mammalian genomic DNA with frequent-cutter enzyme, (2) ligation of different oligonucleotide suppression adapters (required for the PCR-suppression effect described in the Chapter 2) to digested DNA, (3) melt and annealing of two DNA portions harboring different

Current attempts to improve the specificity

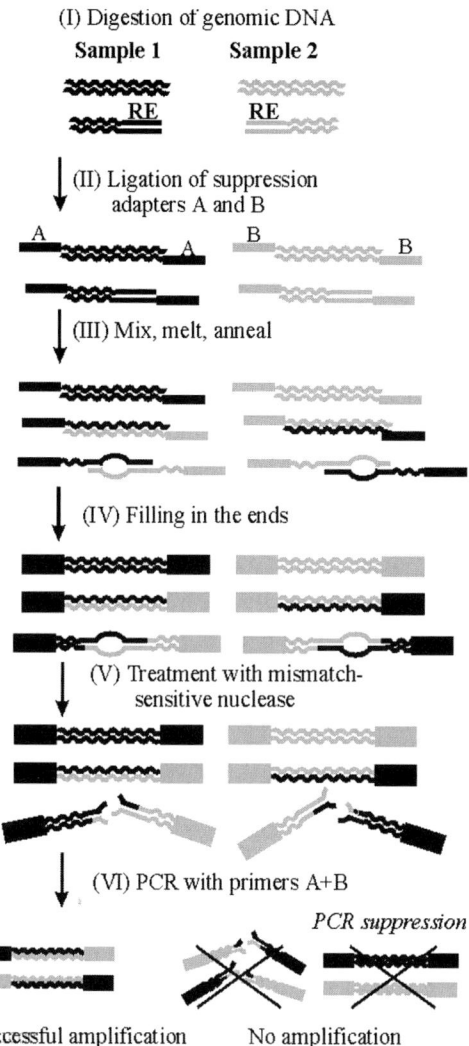

Figure 3. The testing system used to investigate MDR efficiency (see text). The use of MDR reduced the background chimeric clone proportion from 44–60% to 0–4%.

adapters, (4) filling-in the ends of DNA duplexes with DNA polymerase, (5) treatment with mismatch-sensitive nuclease, and (6) PCR amplification of heteroduplexes, that were not cleaved at the previous stage, with primers specific to both adapters using PCR-suppression effect, described in details earlier (Gurskaya et al. 1996).

Briefly, it includes the ligation of restriction fragments to a panhandle-like structure-forming adapter. The authors used standard adapters (Lavrentieva et al. 1999) forming after ligation to restriction fragments ~40 bp long GC-rich inverted repeats at their termini. Therefore, such single-stranded DNA fragments contained self-complementary termini capable of forming strong intramolecular stem-loop structures. PCR of the DNA fragments with such termini is therefore suppressed in homoduplexes when primers targeted at the 5'-ends of the ligated adapters are used. In contrast, heteroduplex molecules have different termini unable to form stem-loop structures, and can be further efficiently PCR amplified in this system. Nested PCR with primers A2 and B2 increases the specificity of the amplification. This procedure thus ensures exclusive amplification of only the heteroduplex DNA. The control experiments had all of the stages mentioned above, except the (5) step, i.e. treatment of hybridized DNA with nucleases.

Two mismatched DNA sensitive nucleases were used: Surveyor nuclease that recognizes and cleaves mispaired DNA structures within DNA duplexes and mung bean nuclease, which degrades single-stranded DNA and, therefore, is able to attack loop structures in chimeric hybrids. Mammalian DNAs were chosen for model experiments because they stand among most complex eukaryotic genomes, thus producing very complex hybridization mixtures, far more complex than those of cDNAs. Thus, by solving the challenge of unwanted chimera formation for complex mammalian genome libraries, one may be assured that this obstacle will be surmounted for lower complexity libraries too (such as those of cDNAs or of less complex genomes).

The resulting DNA libraries were cloned into *Escherichia coli*, and random transformants from each library were sequenced. The authors applied the following criteria for the chimera detection: such sequences did not match genomic databases entirely, but their separate 5'- and 3'-terminal fragments did match the databases. Figure 4 depicts the results of the analysis of six DNA libraries. It is clear that the addition of C_0t A fraction and the hybridization temperature increase from 65°C to 85°C has essentially no effect on the number of chimerical clones, in contrast to the addition of mismatch sensitive nucleases. Both mung bean and Surveyor nucleases display the strong effect on the chimera formation, greatly reducing their number from 44–60% clones to 0–4%. Many sequenced inserts contained genomic REs, which is not surprising, as they constitute a major part of mammalian DNA (Lander et al. 2001). Such RE sequences even if they correspond to correct genomic loci may match different positions on the genomic DNA, thus making their exact mapping problematic. Therefore it is desirable to minimize the portion of such kind of sequences in the libraries. Interestingly, the proportion of REs containing inserts differed considerably among the libraries: C_0t A-libraries contained high number of REs independently on the addition of the nuclease (87–93% of the sequenced clones), C_0t A + /N-libraries – slightly smaller proportion of REs (76–78%), and finally C_0t A + /N + library (H6, mung bean nuclease added) had only 44% of RE-containing inserts. These data show that the best results in the library construction can be achieved with both (1) addition of RE-containing competitor

Current attempts to improve the specificity

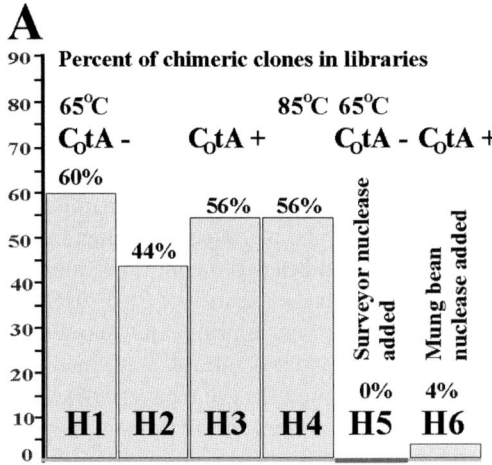

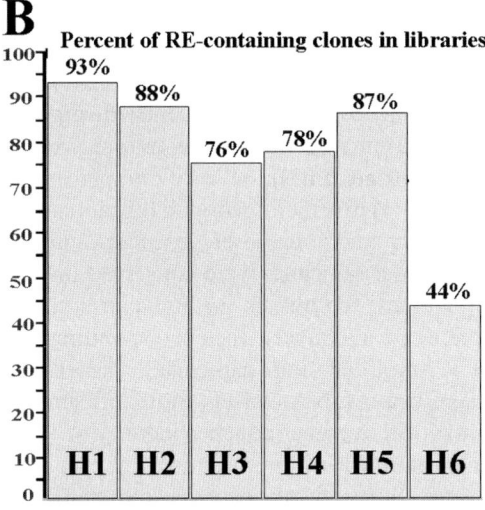

Figure 4. Comparison of six DNA libraries, created under different hybridization conditions with or without the use of MDR. H1, human–human DNA hybridization at 65°C (T65), without competitor C_otA DNA (C_otA-), no mismatch sensitive nucleases added (N-); H2, human–chimpanzee DNA, T65, C_otA-, N-; H3, human–human DNA, T65, C_otA added (C_otA+), N-; H4, human–human DNA, T85, $C_otA +$, N-; H5, human–chimpanzee DNA, T65, C_otA -, Surveyor nuclease added; H6, human–human DNA, T65, $C_otA +$, mung bean nuclease added. (A) Column height reflects the proportion of chimeric clones in analyzed libraries. The number of chimeric sequences is dramatically decreased in libraries, treated with mismatch sensitive nucleases. (B) Column height reflects the proportion of clone inserts, containing repetitive element (RE) sequences. It can be seen that the addition of C_otA competitor DNA alone slightly decreases the number of RE-containing clones, but the combination of both C_otA addition and nuclease digestion yields the best result in library construction.

DNA into hybridization mixture and (2) treatment of hybridized DNA with mismatch sensitive nucleases.

Also, MDR was applied to interspecies DNA hybridizations (analogous to coincidence cloning approach, Chapter 8). To this end in two hybridization experiments the authors hybridized human and chimpanzee DNA. Human and chimpanzee genomes are closely related, displaying ~98% sequence identity (Lander et al. 2001). The results suggest that MDR reduces the number of chimerical sequences from 44% even to the absence of detected chimeras. All sequenced inserts from Surveyor nuclease-treated library did contain sequences, highly conservative between the two genomes (average identity of 98.3%). Some inserts contained regions, evolutionary conserved among the sequenced mammalian genomes – these of human, chimpanzee, mouse, and rat. This observation suggests that MDR could also be applied for the recovery of evolutionary conserved sequences between different genomes. To investigate this, the authors performed another interspecies hybridization, between human and new world monkey *Callithrix pygmaea* genomes, followed by the subsequent digestion with Surveyor nuclease. The pygmy marmoset *C. pygmaea* genome is more divergent from human DNA than that of chimpanzee (human and new world monkey ancestor lineages separated roughly 45 million years ago (Sverdlov 1998; Sverdlov 2000), thus showing about 20% DNA sequence divergence (assuming the average nucleotide substitution rate in primate genomes to be 2.2×10^{-9} bases per million years (Consortium 2002)). Seventy-one percent of cloned inserts represented moderately (~14%) divergent genomic repeats, which are believed to be present in both human and marmoset genomes, and the remaining 29% were unique sequences, most of which were conserved among human, chimpanzee, mouse, and rat genomes. To confirm the high conservation value of these sequences among human and marmoset, the corresponding loci from *C. pygmaea* genome were PCR-amplified and sequenced. Indeed, all sequenced marmoset loci displayed significant DNA conservation and similarity to the corresponding human loci with the average sequence identity of 95%, thus showing about fourfold slower mutation rate for these loci than neutral base substitution rate. Interesting enough, Surveyor nuclease was more efficient for DNA library refinement than mung bean nuclease, probably because of the ability to recognize and to cleave the DNA at the one-nucleotide mismatches, in contrast to mung bean nuclease, which is specific to more extended single-strand DNA loop regions (Figure 5).

The results presented above strongly suggest that MDR technique may provide a useful tool for the refinement of various DNA libraries obtained with the use of DNA reassociation, including subtractive and normalized genomic and cDNA libraries. The technique may also considerably improve genome wide recovery of evolutionary conserved sequences. The experimental techniques for identification of evolutionary conserved regions are required for the comparison of sequenced and/or unsequenced genomes, thus making MDR a universal method. Whenever the case the technique application will hopefully diminish

Control G/C heteroduplex

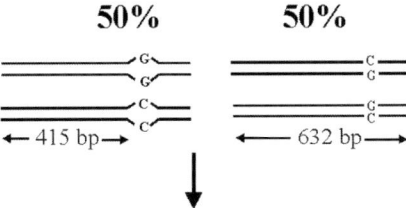

Surveyor nuclease

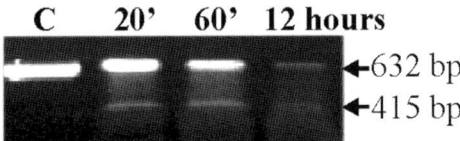

Mung bean nuclease

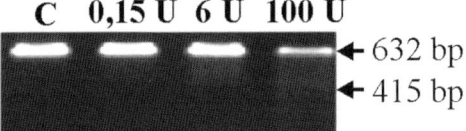

Figure 5. Comparison of Surveyor and mung bean nuclease digestion of control DNA. Control DNA, supplied with Surveyor mutation detection kit, contains 50% of perfectly matched 632 bp long DNA duplexes, and 50% of duplexes having single mismatched base pair (25% of G-G and 25% of C-C mispaired nucleotides). The DNA cleavage at the exact mismatch site gives 415 bp long products. Surveyor nuclease digestion: the 415 bp long fragment band is clearly seen in all experiments, thus showing specific mismatched DNA cleavage. Mung bean nuclease digestion: no detectable band at 415 bp could be detected, slight smearing appeared when large amounts of nuclease were used, thus suggesting the lack of efficient specific cleavage at single mispaired nucleotides by mung bean nuclease.

the confusions caused by cross-hybridization of closely related but however different paralogous sequences. MDR protocol is given in the Section 4.3.

4. PROTOCOLS

4.1 Targeted Genomic Difference Analysis

TGDA is schematically presented in Figure 6. The method includes two steps: whole-genome selective amplification of the flanks adjacent to interspersed REs (in our case human L1 retrotransposons and HERV-K endogenous retroviruses) in both genomic DNAs under comparison (Figure 6A), and then SH of the selected amplicons (Figure 6B). The first step is based on the PCR-suppression effect (Lukyanov et al. 1995; Lukyanov et al. 1997; see Chapter 2).

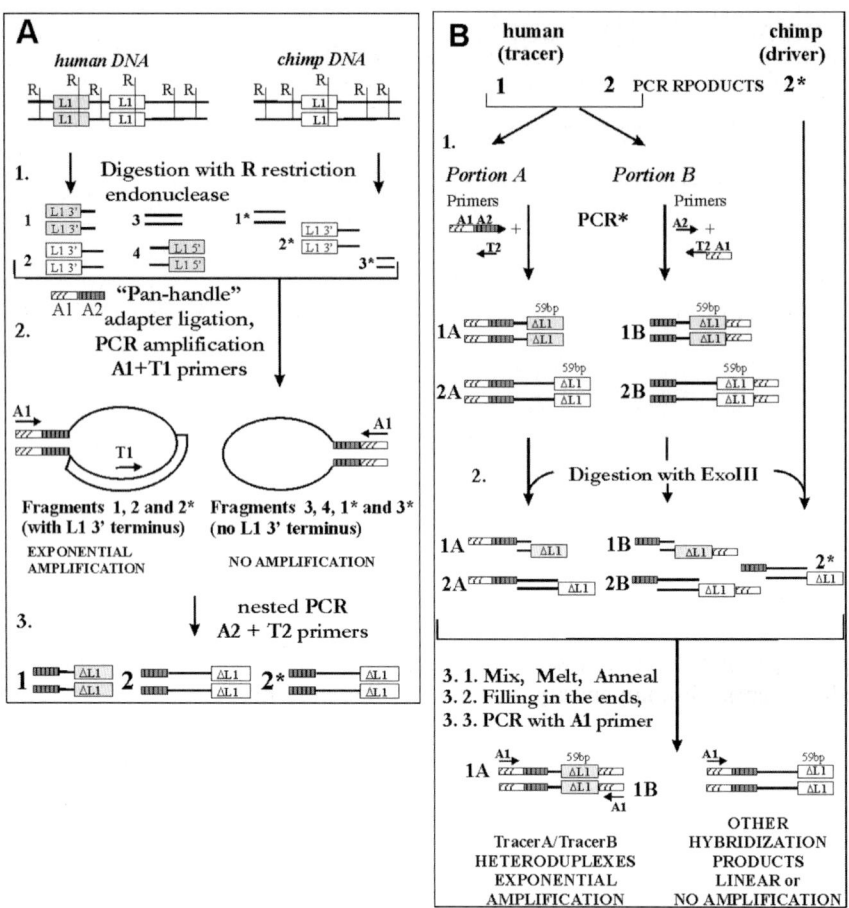

Figure 6. TGDA schematic representation. (A) Stages of selective amplification of L1-flanking genomic regions. Gray and open boxes denote human-specific and other L1s, respectively. R, positions of restriction sites. Hatched boxes designate suppression adapters used. Different types of restriction fragments are enumerated (with asterisks for chimp DNA). (B) Stages of subtractive hybridization (SH). Stage 1: PCR* carried out in accordance with the step-out PCR technique using A1 + A1A2 + T2, or A1 + A1T2 + A2 sets of primers.

Briefly, it includes digestion of the genomic DNAs with a frequent-cutter restriction enzyme, R (Figure 6A, stage 1, using *Alu* I), and ligation of the resulting restriction fragments to a stem-loop structure forming oligonucleotide adapter (Figure 6A, stage 2, see Chapter 2, Section 6). As a result, all DNA restriction fragments had inverted repeats at their termini. Therefore, the single-stranded fragments contained self-complementary termini capable of forming strong intramolecular stem-loop structures (panhandle-like structures, Figure 6A). PCR of the DNA fragments with such termini is suppressed when only one primer targeted at the 5'-ends of the ligated adapter (Figure 6A, stage 2) is used. In contrast, a pair of A1- and T1-primers targeted at the single-stranded part of the stem-loop structure (Figure 6A, stage 2) can initiate DNA synthesis by DNA polymerase. The amplified DNA in this case will have different termini unable to form stem-loop structures, and can be further efficiently amplified with A1 + T1 primers. Nested PCR with A2 and T2 primers increases the specificity of the amplification. This procedure ensures efficient, nearly exclusive amplification of only the fragments that contain the target sequence, HERV-K LTR or L1 in this example. We used it to prepare amplicons containing the DNA flanks of the LTRs 5'-parts and L1 3'-termini both for human and chimpanzee DNAs. T1 and T2 primers directed the DNA synthesis to the outside of the target repeat.

The subtraction in shown schematically on Figure 6B. This allows direct isolation of sequences present solely in one of two related genomes (we will define them as targets), without any preliminary knowledge of the genome sequences. SH is based on reassociation of both genomic DNAs under comparison. After digestion and mixing at a large excess of one DNA (defined as driver DNA) over the other one (defined as tracer DNA), the resulting short fragments were denatured and cooled to reanneal. During the reassociation most of the tracer DNA hybridizes to the excess driver DNA, except for the targets, which form homoduplexes. The self-reassociated tracer is enriched in these reassociated target fragments, compared with the original tracer genome. In this example we tried to identify human specific targets, and therefore human and chimpanzee (our closest relative) DNAs were used as tracer and driver, respectively. We prepared two separate portions of the tracer DNA (Figure 6B, left, stage 1) by reamplification of the human amplicons obtained at the previous stage (Figure 6A, stage 3). We used the step-out PCR (Matz et al. 1999) with primers A1A2, A1, and T2, or A1T2, A1, and A2 for amplifications of portions A and B, respectively. The resulting portion A DNA fragments contained A1A2 sequence at one end and T2 sequence at the other, whereas the corresponding terminal sequences of the portion B fragments were A2 and A1T2.

The following 5'-protruding single-stranded termini formation (Figure 6B, stage 2) is a critical stage for the whole procedure: it prevents cross-hybridization of the repetitive parts common for all the amplicons, and ensures subsequent specific amplification of the double-stranded tracer A/B heteroduplexes formed during the subtraction process. To form 5'-protruding single-stranded termini, we digested A and B tracers with the nuclease *Exo* III until ~60 nt were removed from each 3'-end. The driver DNA was digested similarly to remove ~40 terminal nt.

Tracers A and B and a 100-fold excess of the driver (Figure 6B, stage 3) were mixed, melted, and allowed to reanneal. The resulting mixture contained single-stranded fragments of the both tracers and the driver, double-stranded hybrids formed between the tracers and the driver, homoduplexes formed as a result of self-reassociation of tracers A and B, and heteroduplexes formed by cross-reassociation of the tracers A and B complementary strands (tracer A/B fraction). Once the protruding ends of the latter heteroduplexes have been filled-in with DNA polymerase, the heteroduplexes acquired targets for primer A1 at both termini and were the only fragments that could be exponentially amplified with this primer.

The PCR products were cloned in *E. coli*, ~500 random transformants were arrayed and further analyzed for each library.

4.1.1 DNA Samples and oligonucleotides

Genomic DNA was extracted from 20 samples of individual human placentas, human blood samples, or blood samples of chimpanzees using a genomic DNA purification kit (Promega). Suppression adapters are listed in Chapter 2, Section 6.1.

4.1.2 Preparation of tracer and driver DNAs

Digestion of human and chimpanzee DNAs, adapter ligation, and PCR amplification of LTR-flanking regions was all done as described (Lavrentieva et al. 1999). We amplified 1 ng aliquots of human amplicons according to step-out PCR procedure (Matz et al. 1999) with set A (0.01 μM A1A2, 0.2 μM A2, 0.2 μM T2) or set B (0.01 μM A2T2, 0.2 μM A2, 0.2 μM A1) of primers using 15 cycles at 95°C for 15 s, 57°C for 10 s, 72°C for 90 s.

We digested 150 ng each of the resulting tracer A and B samples, and 3000 ng of the initial chimp amplicon (driver) with *Exo* III nuclease separately at 16°C using the following conditions: tracer A, 20 units of the *Exo* III, 11 min (40 terminal nt to be removed); tracer B, 20 units, 14 min (60 nt to be removed); driver, 400 units, 11 min (40 residues to be removed). We mixed 15 ng each of the digested tracer A and B samples with 1500 ng of digested driver. DNA samples were purified by phenol/chloroform extraction, precipitated with ethanol, and dissolved in 5 μl of sterile water.

4.1.3 Subtractive hybridization

We mixed both tracer A/driver and tracer B/driver samples, transferred them into a hybridization buffer (0.5 M NaCl, 50 mM Hepes, pH 8.3, 0.2 mM EDTA), denatured at 95°C for 10 min, and hybridized at 65°C for 14 hs. The final 15 μl mixture was diluted with 185 μl dilution buffer (50 mM NaCl, 5 mM Hepes, pH 8.3, 0.2 mM EDTA). We PCR-amplified 1 μl of this diluted mixture with 0.4 μM A1 primer. The PCR conditions were as follows: (1) 72°C for 6 min to fill in the ends of DNA duplexes; (2) 95°C for 15 s, 65°C for 10 s, 72°C for 90 s, 15 cycles. PCR products obtained were further cloned in *E.coli* using a TA-cloning system (Promega), and ~500 individual clones were sequenced for each library.

4.2 Using Competitor DNA to Decrease the Background of Genomic Repeats

4.2.1 Starting material

DNA samples. In our experiments, we extracted DNA from four mixed human blood samples and from blood sample of chimpanzee *Pan paniscus* using a genomic DNA purification kit (Promega) according to the manufacturers' recommendations.

Oligonucleotides. We used the standard suppression adapters A1A2 (5'-*GTAAT-ACGACTCACTATAGGGCAGCGTGGTCGCGGCCGAGGT*-3') and B1B2 (5'-*CGACGTGGACTATCCATGAACGCATCGAGCGGCCGCCCGGGCA-GGT*-3'). For nested PCR amplifications, the following primers specific for the suppression adapter set were used: A1, 5'-*GTAATACGACTCACTATAGGGC*-3', and B1, 5'-*CGACGTGGACTATCCATGAACGCA*-3'. A2, 5'-*AGCGTG-GTCGCGGCCGAGGT*-3', and B2, 5'-*TCGAGCGGCCGCCCGGGCAGGT*-3'. Oligonucleotides were synthesized using an ASM-102U DNA synthesizer (Biosan, Novosibirsk, Russia).

4.2.2 DNA preparation for hybridization

Digestion of genomic DNA. About 1μg of genomic DNA was digested with 10 units of frequent-cutter blunt end-producing restriction endonuclease *Alu* I (Fermentas) at 37°C, for 2 h. DNA was phenol–chloroform extracted, ethanol precipitated and dissolved in 25 μl of sterile water.

Ligation of the suppression adapters. The suppression adapter ligation was done as described previously in this book (Lavrentieva et al. 1999). We used T4 DNA ligase (Promega) and suppression adapters A1A2 and B1B2 (see above), annealed to 10 nt long oligonucleotide complementary to the adapter 3'-terminal part, A3 and B3, respectively). Ligated DNA was purified using Quiagen PCR product purification kit, ethanol precipitated and dissolved in 5 μl of hybridization buffer (0.5 M NaCl, 50 mM Hepes, pH 8.3, 0.2 mM EDTA).

4.2.3 DNA hybridization

We mixed 800 ng of each of both DNA samples assigned for hybridization in a volume of 8 μl of 1x hybridization buffer, denatured at 95°C for 10 min, and hybridized at 65°C or 85°C for 50 h. The final 8 μl mixture was diluted with 72 μl of dilution buffer (50 mM NaCl, 5 mM Hepes, pH 8.3, 0.2 mM EDTA). C_0tA fraction competitor DNA (Gibco BRL, USA) was added in 100x weight excess to the hybridization mixture. In control experiments, no C_0tA DNA was added.

Filling in the termini of hybridized DNA. We used AmpliTaq DNA polymerase (1 unit/1 μg of hybridized DNA) to fill in the ends of DNA duplexes at 72°C for 20 min.

4.2.4 PCR amplification of hybridization products and library construction

Nested PCR amplification. DNA samples were dissolved in 100 μl of water and 1 μl was PCR amplified with 0.2 μM primers specific for the used suppression adapter set: A1 and B1. The PCR conditions were as follows: 95°C for 15″, 65°C for 10″, 72°C for 90″, 15 cycles. To increase the amplification specificity, we used an additional round of nested PCR for 500-fold dissolved products of the latter PCR with 0.2 μM primers A2 and B2, under the same cycling conditions. The number of nested PCR cycles varied substantially depending on the particular hybridization.

Clone library construction. The PCR products obtained were cloned in *E.coli* strain DH5α using a TA-cloning system (Promega). We sequenced positive clones by the dye termination method using an Applied Biosystems 373 automatic DNA sequencer.

DNA sequence analysis. We used BLAT search (http://genome.ucsc.edu/cgi-bin/hgBLAT) to map clone inserts within human and chimpanzee genomes. Homology searches against GenBank were done using the BLAST web server at National Center for Biotechnology Information (NCBI; http://www.ncbi.nlm.nih.gov/BLAST) (Altschul et al. 1990). For multiple alignments the ClustalW program (Thompson et al. 1994) was used.

4.3 Mispaired DNA Rejection

4.3.1 Starting material

DNA samples. In our experiments, we extracted DNA from four mixed human blood samples, or from blood samples of chimpanzee *P. paniscus* and marmoset *C. pigmaea* using a genomic DNA purification kit (Promega) according to the manufacturers' recommendations.

Oligonucleotides. We used the standard suppression adapters A1A2 (5'-*GTAAT-ACGACTCACTATAGGGCAGCGTGGTCGCGGCCGAGGT*-3') and B1B2 (5'-*CGACGTGGACTATCCATGAACGCATCGAGCGGCCGCCCGGGCAGGT*-3'). For nested PCR amplifications, the following primers specific for the suppression adapter set were used: A1, 5'-*GTAATACGACTCACTATAGGGC*-3', and B1, 5'-*CGACGTGGACTATCCATGAACGCA*-3'. A2, 5'-*AGCGTGGTCGCGGCC-GAGGT*-3', and B2, 5'-*TCGAGCGGCCGCCCGGGCAGGT*-3'. Oligonu-cleotides were synthesized using an ASM-102U DNA synthesizer (Biosan, Novosibirsk, Russia).

4.3.2 DNA preparation for hybridization

Digestion of genomic DNA. About 1 µg of genomic DNA was digested with 10 units of frequent-cutter blunt end-producing restriction endonuclease *Alu* I (Fermentas) at 37°C, for 2 h. DNA was phenol–chloroform extracted, ethanol precipitated, and dissolved in 25 µl of sterile water.

Ligation of the suppression adapters. The suppression adapter ligation was done as described previously in this book (Lavrentieva et al. 1999). We used T4 DNA ligase (Promega) and suppression adapters A1A2 and B1B2 (see above), annealed to 10 nt long oligonucleotide complementary to the adapter 3'-terminal part, A3 and B3, respectively). Ligated DNA was purified using Quiagen PCR product purification kit, ethanol precipitated and dissolved in

5 µl of hybridization buffer (0.5 M NaCl, 50 mM Hepes, pH 8.3, 0.2 mM EDTA).

4.3.3 DNA hybridization

We mixed 800 ng of each of both DNA samples assigned for hybridization in a volume of 8 µl of 1x hybridization buffer, denatured at 95°C for 10 min, and hybridized at 65°C or 85°C for 50 h. The final 8 µl mixture was diluted with 72 µl of dilution buffer (50 mM NaCl, 5 mM Hepes, pH 8.3, 0.2 mM EDTA). In some experiments, C_0tA fraction competitor DNA (Gibco BRL, USA) was added in 100x weight excess to the hybridization mixture.

Filling in the termini of hybridized DNA. We used AmpliTaq DNA polymerase (1 unit/1 µg of hybridized DNA) to fill in the ends of DNA duplexes at 72°C for 20 min.

4.3.4 Hybridized DNA treatment with mismatch sensitive nucleases

About 100 ng aliquots of hybridized DNA were digested with 1 µl Surveyor nuclease (Transgenomic, USA) in 20 µl of 1× buffer supplied by the manufacturer, overnight incubation at 42°C, or treated with 0.1 unit of mung bean nuclease (Promega) at 37°C for 15 min. DNA samples were phenol–chloroform extracted and ethanol precipitated.

4.3.5 PCR amplification of hybridization products and library construction

Nested PCR amplification. DNA samples were dissolved in 100 µl of water and 1 µl was PCR amplified with 0.2 µM primers specific for the used suppression adapter set: A1 and B1. The PCR conditions were as follows: 95°C for 15″, 65°C for 10″, 72°C for 90″, 15 cycles. To increase the amplification specificity, we used an additional round of nested PCR for 500-fold dissolved products of the latter PCR with 0.2 µM primers A2 and B2, under the same cycling conditions. The number of nested PCR cycles varied substantially depending on the particular hybridization.

Clone library construction. The PCR products obtained were cloned in *E.coli* strain DH5α using a TA-cloning system (Promega). We sequenced positive clones by the dye termination method using an Applied Biosystems 373 automatic DNA sequencer.

DNA sequence analysis. We used BLAT search (http://genome.ucsc.edu/cgi-bin/hgBLAT) to map clone inserts within human and chimpanzee genomes. Homology searches against GenBank were done using the BLAST web server at NCBI (http://www.ncbi.nlm.nih.gov/BLAST) (Altschul et al. 1990). For multiple alignments the ClustalW program (Thompson et al. 1994) was used.

4.3.6 PCR amplification of evolutionary conserved sequences

A DNA sample of 40 ng of old world monkey *C. pigmaea* blood were PCR amplified using multiple sets of 0.2 µM unique genomic primers flanking the presumable conserved genomic loci. The resulting PCR products were analyzed on 1.2% agarose gels and sequenced.

REFERENCES

Altschul SF, Gish W, Miller W, Myers EW, Lipman DJ (1990) Basic local alignment search tool. J Mol Biol 215:403–410

Babon JJ, McKenzie M, Cotton RG (2003) The use of resolvases T4 endonuclease VII and T7 endonuclease I in mutation detection. Mol Biotechnol 23:73–81

Barr FG, Emanuel BS (1990) Application of a subtraction hybridization technique involving photoactivatable biotin and organic extraction to solution hybridization analysis of genomic DNA. Anal Biochem 186:369–373

Broude NE (2002) Stem-loop oligonucleotides: a robust tool for molecular biology and biotechnology. Trends Biotechnol 20:249–256

Buzdin A, Khodosevich K, Mamedov I, Vinogradova T, Lebedev Y, Hunsmann G, Sverdlov E (2002) A technique for genome-wide identification of differences in the interspersed repeats integrations between closely related genomes and its application to detection of human-specific integrations of HERV-K LTRs. Genomics 79:413–422

Buzdin A, Ustyugova S, Gogvadze E, Lebedev Y, Hunsmann G, Sverdlov E (2003a) Genome-wide targeted search for human specific and polymorphic L1 integrations. Hum Genet 112:527–533

Buzdin A, Ustyugova S, Khodosevich K, Mamedov I, Lebedev Y, Hunsmann G, Sverdlov E (2003b) Human-specific subfamilies of HERV-K (HML-2) long terminal repeats: three master genes were active simultaneously during branching of hominoid lineages (small star, filled). Genomics 81:149–156

Carulli JP, Artinger M, Swain PM, Root CD, Chee L, Tulig C, Guerin J, Osborne M, Stein G, Lian J, Lomedico PT (1998) High throughput analysis of differential gene expression. J Cell Biochem 30–31:286–296

Chalaya T, Gogvadze E, Buzdin A, Kovalskaya E, Sverdlov ED (2004) Improving specificity of DNA hybridization-based methods. Nucleic Acids Res 32:e130

Consortium MGS (2002) Initial sequencing and comparative analysis of the mouse genome. Nature 420:520–562

Cotton RG (1999) Mutation detection by chemical cleavage. Genet Anal 14:165–168

Cotton RG, Rodrigues NR, Campbell RD (1988) Reactivity of cytosine and thymine in single-base-pair mismatches with hydroxylamine and osmium tetroxide and its application to the study of mutations. Proc Natl Acad Sci USA 85:4397–4401

Diatchenko L, Lau YF, Campbell AP, Chenchik A, Moqadam F, Huang B, Lukyanov S, Lukyanov K, Gurskaya N, Sverdlov ED, Siebert PD (1996) Suppression subtractive hybridization: a method for generating differentially regulated or tissue-specific cDNA probes and libraries. Proc Natl Acad Sci USA 93:6025–6030

Ermolaeva OD, Sverdlov ED (1996) Subtractive hybridization, a technique for extraction of DNA sequences distinguishing two closely related genomes: critical analysis. Genet Anal 13:49–58

Ermolaeva OD, Wagner MC (1995) SUBTRACT: a computer program for modeling the process of subtractive hybridization. Comput Appl Biosci 11:457–462

Ermolaeva OD, Lukyanov SA, Sverdlov ED (1996) The mathematical model of subtractive hybridization and its practical application. Proc Int Conf Intell Syst Mol Biol 4:52–58

Gurskaya NG, Diatchenko L, Chenchik A, Siebert PD, Khaspekov GL, Lukyanov KA, Vagner LL, Ermolaeva OD, Lukyanov SA, Sverdlov ED (1996) Equalizing cDNA subtraction based on selective suppression of polymerase chain reaction: cloning of Jurkat cell transcripts induced by phytohemaglutinin and phorbol 12-myristate 13-acetate. Anal Biochem 240:90–97

Hames BD, Higgins SJ (eds) (1985) Nucleic acid hybridization: a practical approach. IRL Press, Oxford, Washington DC

Huang J, Kirk B, Favis R, Soussi T, Paty P, Cao W, Barany F (2002) An endonuclease/ligase based mutation scanning method especially suited for analysis of neoplastic tissue. Oncogene 21:1909–1921

Kohne DE, Levison SA, Byers MJ (1977) Room temperature method for increasing the rate of DNA reassociation by many thousandfold: the phenol emulsion reassociation technique. Biochemistry 16:5329–5341

Kulinski J, Besack D, Oleykowski CA, Godwin AK, Yeung AT (2000) CEL I enzymatic mutation detection assay. Biotechniques 29:44–46, 48

Kunkel LM, Monaco AP, Middlesworth W, Ochs HD, Latt SA (1985) Specific cloning of DNA fragments absent from the DNA of a male patient with an X chromosome deletion. Proc Natl Acad Sci USA 82:4778–4782

Lamar EE, Palmer E (1984) Y-encoded, species-specific DNA in mice: evidence that the Y chromosome exists in two polymorphic forms in inbred strains. Cell 37:171–177

Lander ES, Linton LM, Birren B, Nusbaum C, Zody MC, Baldwin J, Devon K, et al. (2001) Initial sequencing and analysis of the human genome. Nature 409:860–921

Lavrentieva I, Broude NE, Lebedev Y, Gottesman, II, Lukyanov SA, Smith CL, Sverdlov ED (1999) High polymorphism level of genomic sequences flanking insertion sites of human endogenous retroviral long terminal repeats. FEBS Lett 443:341–347.

Lisitsyn N, Wigler M (1993) Cloning the differences between two complex genomes. Science 259:946–951

Lisitsyn N, Wigler M (1995) Representational difference analysis in detection of genetic lesions in cancer. Methods Enzymol 254:291–304

Lisitsyn NA, Leach FS, Vogelstein B, Wigler MH (1994a) Detection of genetic loss in tumors by representational difference analysis. Cold Spring Harb Symp Quant Biol 59:585–587

Lisitsyn NA, Segre JA, Kusumi K, Lisitsyn NM, Nadeau JH, Frankel WN, Wigler MH, Lander ES (1994b) Direct isolation of polymorphic markers linked to a trait by genetically directed representational difference analysis. Nat Genet 6:57–63

Lukyanov K, Diatchenko L, Chenchik A, Nanisetti A, Siebert P, Usman N, Matz M, Lukyanov S (1997) Construction of cDNA libraries from small amounts of total RNA using the suppression PCR effect. Biochem Biophys Res Commun 230:285–288

Lukyanov KA, Launer GA, Tarabykin VS, Zaraisky AG, Lukyanov SA (1995) Inverted terminal repeats permit the average length of amplified DNA fragments to be regulated during preparation of cDNA libraries by polymerase chain reaction. Anal Biochem 229:198–202

Mamedov IZ, Arzumanyan ES, Amosova AL, Lebedev YB, Sverdlov ED (2005) Whole-genome experimental identification of insertion/deletion polymorphisms of interspersed repeats by a new general approach. Nucleic Acids Res 33:e16

Matz M, Shagin D, Bogdanova E, Britanova O, Lukyanov S, Diatchenko L, Chenchik A (1999) Amplification of cDNA ends based on template-switching effect and step-out PCR. Nucleic Acids Res 27:1558–1560

Mikhailov VS, Rohrmann GF (2002) Binding of the baculovirus very late expression factor 1 (VLF-1) to different DNA structures. BMC Mol Biol 3:14

Milner JJ, Cecchini E, Dominy PJ (1995) A kinetic model for subtractive hybridization. Nucleic Acids Res 23:176–187

Mouritzen P, Nielsen AT, Pfundheller HM, Choleva Y, Kongsbak L, Moller S (2003) Single nucleotide polymorphism genotyping using locked nucleic acid (LNA). Expert Rev Mol Diagn 3:27–38

Nagayama K, Enomoto N, Miyasaka Y, Kurosaki M, Chen CH, Sakamoto N, Nakagawa M, Sato C, Tazawa J, Ikeda T, Izumi N, Watanabe M (2001) Overexpression of interferon gamma-inducible protein 10 in the liver of patients with type I autoimmune hepatitis identified by suppression subtractive hybridization. Am J Gastroenterol 96:2211–2217

Qiu P, Shandilya H, D'Alessio JM, O'Connor K, Durocher J, Gerard GF (2004) Mutation detection using Surveyor nuclease. Biotechniques 36:702–707

Roberts E, Deeble VJ, Woods CG, Taylor GR (1997) Potassium permanganate and tetraethylammonium chloride are a safe and effective substitute for osmium tetroxide in solid-phase fluorescent chemical cleavage of mismatch. Nucleic Acids Res 25:3377–3378

Roberts RW, Crothers DM (1991) Specificity and stringency in DNA triplex formation. Proc Natl Acad Sci USA 88:9397–9401

Sambrook J, Russell DW (2001) Molecular cloning: a practical manual. CSHL Press, Cold Spring Harbour, NY

Sasaki H, Nomura S, Akiyama N, Takahashi A, Sugimura T, Oishi M, Terada M (1994) Highly efficient method for obtaining a subtracted genomic DNA library by the modified in-gel competitive reassociation method. Cancer Res 54:5821–5823

Sverdlov ED (1998) Perpetually mobile footprints of ancient infections in human genome. FEBS Lett 428:1–6

Sverdlov ED (2000) Retroviruses and primate evolution. Bioessays 22:161–171

Sverdlov ED, Ermolaeva OD (1994) [Kinetic analysis for subtractive hybridization of transcripts]. Bioorg Khim 20:506–514

Thompson JD, Higgins DG, Gibson TJ (1994) CLUSTAL W: improving the sensitivity of progressive multiple sequence alignment through sequence weighting, position-specific gap penalties and weight matrix choice. Nucleic Acids Res 22:4673–4680

Till BJ, Burtner C, Comai L, Henikoff S (2004) Mismatch cleavage by single-strand specific nucleases. Nucleic Acids Res 32:2632–2641

Venter JC, Adams MD, Myers EW, Li PW, Mural RJ, Sutton GG, Smith HO, et al. (2001) The sequence of the human genome. Science 291:1304–1351

CHAPTER 11

CONCEPTS ON MICROARRAY DESIGN FOR GENOME AND TRANSCRIPTOME ANALYSES

HELDER I. NAKAYA, EDUARDO M. REIS, SERGIO VERJOVSKI-ALMEIDA

Departamento de Bioquimica, Instituto de Quimica, Universidade de São Paulo, 05508900 São Paulo, SP, Brazil.
Phone: +55-1130912173; Fax: +551130912186; E-mail:verjo@iq.usp.br

Abstract: Microarray technology has revolutionized molecular biology by permitting many hybridization experiments to be performed in parallel. With the size of a glass microscope slide, this tool can carry thousands of DNA fragments in an area smaller than a postage stamp. In this chapter, we will describe microarray chips that host nucleic acid probes, which are the most commonly used type of microarrays. Science in this field is mostly data-driven, where biological hypothesis are generated upon analysis and comparison of a huge amount of potentially meaningful differential data derived from microarray hybridizations. DNA microarray technology is under a constant and rapid evolution. The first paper reporting DNA microarray as a tool for transcript-level analyses has been published in 1995, and that chip had about 1000 *Arabidopsis* genes. Almost 11 years have passed and advances in miniaturization, robotic, and informatics, as well as the development of alternative approaches to microarray construction have permitted to put more than 250,000 different spots into a single square centimeter. This rapid advance in the microarray field, combined with the falling price of technology and the acquisition of whole-genome sequence information for hundreds of organisms has caused biologists to abandon their home-made equipment in favor of one of an expanding range of commercial platforms now available on the market. However, we are still not able to represent the entire genome of any eukaryotic organism in a unique chip or even to analyze the great complexity of the human transcriptome. Therefore, how to choose and design the best probes to construct DNA microarray chips is a crucial step to the appropriate use of this powerful technique (Figure 1).

Keywords: Microarray chip, DNA microarray, spotted DNA microarray, cDNA microarray, antisense transcription, genomic microarray, comparative genomic hybridization, CGH, oligoarray, *in situ* synthesis, Affymetrix, NimbleGen, Agilent, microarray probe, gene-oriented array, oligonucleotide probe, epigenomic microarray, tilling array, CpG island, expression profiling, gene atlas, alternative splicing, transcript variants, pre-mRNA, transcriptome annotation, whole-genome oligonucleotide array, microRNAs (miRNAs), methylation, bisulfite oligonucleotide array, methylated DNA immunoprecipitation, chromatin immunoprecipitation (ChIP), single nucleotide polymorphisms (SNPs), intronic transcription, noncoding transcripts, linkage disequilibrium.

Abbreviations: ASO, allele-specific oligonucleotide hybridization; ASPE, allele-specific primer extension; BAC, bacterial artificial chromosome; BLAST, basic local sequence alignment tool; CAGE, cap analysis of gene expression; CGH, comparative genomic hybridization; ChIP, chromatin immunoprecipitation; Cy3, cyanine-3; Cy5, cyanine-5; DMD, Digital Micromirror Device; dNTP, deoxyribonucleotidetriphosphate; EST, expressed sequence tag; MAS, Maskless Array Synthesizer; MeDIP, methylated DNA immunoprecipitation; MiRNA, microRNA; MM, mismatch; MPSS, massively parallel signature sequencing; Oligo, oligonucleotide; PAP, poly(A)-polymerase; PCR, polymerase chain reaction; PM, perfect match; RACE, rapid amplification of cDNA ends; SAGE, serial analysis of gene expression; SBE, single-base extension; SNP, single nucleotide polymorphism; UV, ultraviolet; WGA, whole-genome amplification.

TABLE OF CONTENTS

1. Building a Microarray Chip 266
   1.1 Spotted DNA Microarrays 268
   1.2 *In situ* Synthesis 271
2. Selecting the Probes 277
   2.1 Gene-Oriented Arrays 278
   2.2 Epigenomic Microarrays 279
   2.3 Tiling Arrays 279
3. Specific Question, Specific Chip 280
   3.1 Transcriptional Profiling 280
   3.2 Comparative Genome Hybridization 282
   3.3 Alternative Splicing 284
   3.4 Transcriptome Annotation 285
   3.5 Small MicroRNA Profiling 289
   3.6 Methylation Pattern 289
   3.7 ChIP-Chip 293
   3.8 Genotyping 295
   3.9 Intronic Transcription 297
4. Conclusions 299
References 300

1. BUILDING A MICROARRAY CHIP

Various types of microarrays with different probe materials can be produced, including DNA/RNA and oligonucleotides [1], soluble proteins [2], membrane proteins [3], peptides [4], carbohydrates [5], small molecules [6], tissue [7], and live cells [8], with each technology possessing distinctive characteristics while

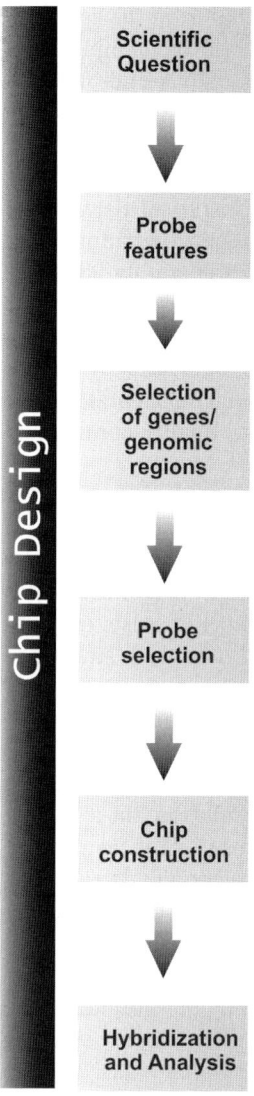

Figure 1. Steps comprised on chip design.

providing unique opportunities to increase our understanding of how a living being operates. In this review we will concentrate on the construction and use of DNA microarrays.

The principle behind microarray chips composed of nucleic acid probes is simple: DNA or oligonucleotide probes representing genes or genomic regions of an organism capture, by preferential binding of complementary single-stranded nucleic-acid sequences, the labeled RNA, DNA, or cDNA molecules (targets) applied to the chip. The intensity of the label signal from the captured targets reflects the abundance of that target within the hybridized sample.

Single-stranded DNA probes, in the form of DNA (e.g. cDNA and bacterial artificial chromosome – BACs) or oligonucleotides are placed on a substrate made by glass or silicon. Based on the principle of whether or not there is direct contact between the sample probe and the support substrate, a robot arrayer uses contact (using printing pins) or noncontact (using piezoelectrical deposition) printing methods. An alternative method of printing is the semiconductor-based technology, which consists in synthesizing oligonucleotides *in situ*, building up nucleotide by nucleotide each element of the array and using ink-jet printing or photolithographic methods, similar to those used in the semiconductor industry. This technology offers the advantage of higher density and consistency [9, 10]. Therefore, the key trends have been a shift from cDNA- to oligonucleotide-based microarrays and from "in-house or home-brew" to higher quality commercial platforms [11].

Oligonucleotide microarrays have several advantages in comparison to DNA microarrays. Today, microarray companies such as Agilent Technologies and NimbleGen offer custom oligo arrays with up to hundreds of thousands features, resulting in platforms with a very flexible custom design. Oligonucleotides contained in these chips have great sensitivity as discussed below. Also, overlapping sense and antisense transcription (RNAs transcribed from both strands of DNA in the same genomic *locus*), which is being recognized as a common event in the eukaryotic cells, can be discriminated by these oligonucleotide arrays.

The type of microarray chip depends on the scientific question behind it. Chips comprised either by short (200–500 bp) PCR products of cDNA sequences or by oligonucleotides (17–70 bp) have high resolution but a limited genomic coverage (Figure 2). This kind of chip is largely used for measurement of mRNA transcript levels from annotated genes. Large genomic deletions or duplications are better detected by low-resolution microarray chips carrying fragments of several kilobases (kb) in length, such as BACs and cosmids (Figure 2). Although oligonucleotides can be either spotted or *in situ* synthesized onto a microarray chip, large DNA molecules can only be spotted.

1.1 Spotted DNA Microarrays

Selected DNA fragments spotted onto microarray chips can be derived from genomic regions, cDNA libraries, BAC clones, or synthetic oligonucleotides (Figure 3). In order to obtain the required concentration of a specific DNA for spotting, amplification and purification steps must be performed. For oligonucleotide microarrays, this is obtained by an appropriate dilution of the purchased oligo set. Printing of DNA fragments is then performed by a robot arrayer using printing pins or piezoelectrical deposition. In general, the latter printing method generates small, homogenous spots, whereas the results of contact printing depend largely on the quality of the printing pins. Successful printing also

Concepts on microarray design for genome 269

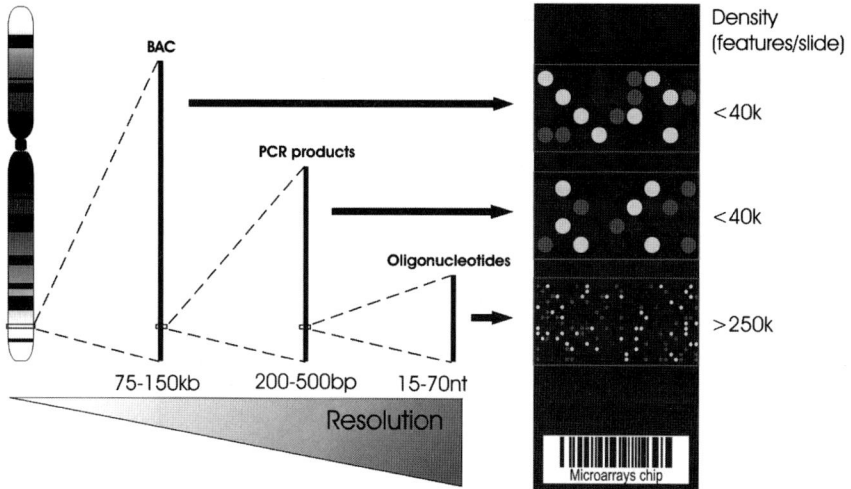

Figure 2. Array density and length of different DNA probes. Bacterial artificial chromosome (BAC) clones can represent large spans of genomic DNA at the expense of low-tiling resolution. Spots comprised by PCR products provide better resolution than BACs using relatively few array features in comparison to spots comprised by oligonucleotides, which have the highest resolution.

requires controlled environmental conditions, such as optimized air humidity, temperature, and the absence of dust and dirt particles. The DNA within the created spots is fixed onto the array surface by covalent bonds randomly formed by cross-linking the DNA backbone of spotted probes to the chemically coated surface, using heat or ultraviolet radiation (UV) [12].

The advantages of being highly customizable and having a low manufacturing cost per array is balanced by the fact that substrate properties or pen-tip diameter of this type of microarray chip limit the density to less than 1000 features per cm^2. Another limitation of cDNA microarrays is the possibilities of cross hybridization between mRNAs and other nonspecific elements of the cDNA clone, and the painstaking effort of maintaining accurate and viable cDNA libraries. Also, several reports show a widespread occurrence of antisense transcription in the human genome [13, 14] and double-stranded cDNA microarrays are unable to discriminate between sense and antisense overlapping messages. These problems are largely circumvented by the use of oligonucleotide arrays.

1.1.1 cDNA microarray

This common variety of microarray chip uses cDNA molecules immobilized to a glass slide to assay parallel expression of RNAs transcribed from particular genes. A cDNA is a nucleic acid molecule reversely transcribed from mRNA. To immobilize these molecules, a PCR amplification of cDNA libraries and purification of PCR products is needed. Double-strand DNA-amplified fragments

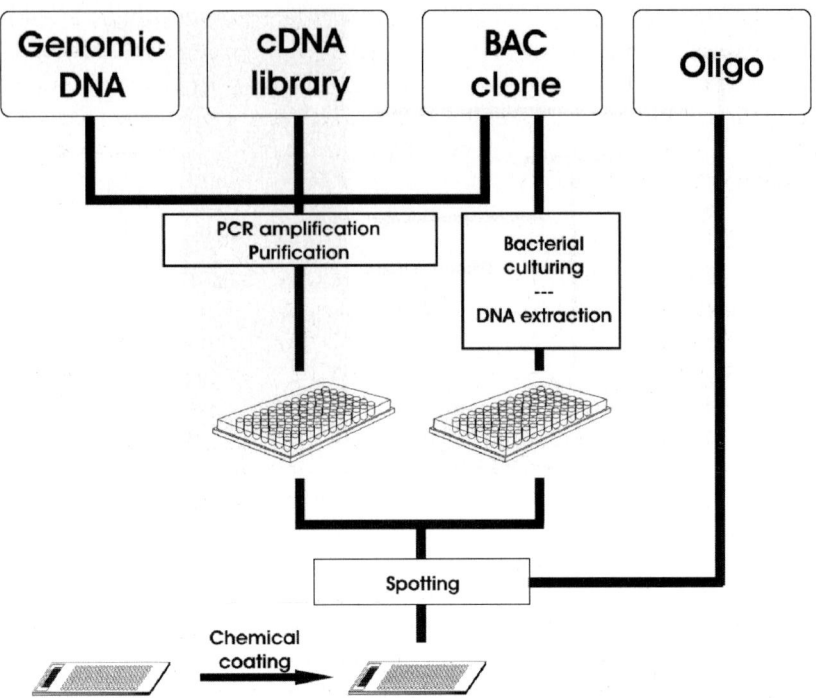

Figure 3. Types of spotted DNA microarrays. PCR reactions are performed in order to amplify the DNA for spotting. For PCR reactions the templates are either genomic DNA, cDNA from a library, or BAC clones. Alternatively, BAC clones can be directly spotted following amplification with bacterial culturing and DNA extraction and purification. An appropriate dilution of the purchased oligonucleotide set is required to obtain the DNA amount necessary for spotting.

are deposited onto coated glass slides by pen tips of a spotter robot. Following spotting, cDNAs are fixed to the slide surface by UV cross-linking. The glass surfaces of cDNA microarrays can be chemically modified in various ways to immobilize DNA and some studies show that the spotting cDNA concentration, surface chemistries, and blocking strategies affect the performance and quality of cDNA microarray data [15, 16].

1.1.2 Genomic microarray

Spots on a DNA microarray can represent large spans of genomic DNA (gDNA) for comparative genomic hybridization (CGH) analysis. This chip uses BAC clones, and facilitates global experimentation using relatively few array features, at the expense of low-tiling resolution [17]. Cloned gDNA for probes is isolated from bacterial cultures by standard DNA extraction protocols [18]. To avoid large-scale bacterial culturing, DNA fragments may be obtained by PCR amplification of BAC DNA using degenerate oligonucleotide primers [19] or linkers [20]. While oligonucleotide and small PCR fragments facilitate a more detailed investigation at selected genomic regions, the large insert BAC clone arrays (typically ~150 kb in size) efficiently capture signals from samples of low DNA quantity and quality for genome-wide analysis, since BAC arrays require 200–400 ng of DNA, whereas oligonucleotide and cDNA platforms typically require microgram amounts [21].

1.1.3 Oligoarray

Microarray chips carrying spotted longmer oligonucleotides have recently become more widely used. Single-stranded probes with 50–70 bases representing exons of genes combine the advantages of flexible and controlled probe design with the higher probe specificity as compared to double-stranded cDNA. Presynthesis of oligos or cDNAs has the important advantage that the sequences eventually placed on the array can be exactly those desired; on the other hand presynthesis significantly increases the fixed cost attached to a multiprobe array and thus in practice limits the number of features spotted per array to a few thousands. Pen-tip spotting methods [1, 22] will continue to be a relatively low-tech but robust and affordable method for small laboratories to generate their own arrays, with a moderate number of features. Ink-jet methods also can be used to print presynthesized oligos [23]. With the recent and significant improvements in spotting technology and acquisition of the genomic sequence from many organisms, whole-genome longmer oligonucleotide sets for printing are now available for many species.

For example, a chip containing thousands of oligonucleotides has been commercialized by the GE Healthcare division of General Electric. This platform named "CodeLink Bioarray Platform" (Figure 4) is constructed by piezoelectrical deposition of presynthesized and functionally validated 30 mer oligonucleotide probes onto a proprietary 3D aqueous gel matrix [24]. The CodeLink platform offers several bioarrays for both expression and single nucleotide polymorphism (SNP) studies in humans, mice, and rats [25].

1.2 *In situ* Synthesis

Another approach to manufacture DNA arrays employs the *in situ* synthesis of oligonucleotides. Production of these microarrays requires more sophisticated and costly equipment, and these arrays are generally produced commercially [26]. The larger vendor corporations, such as Affymetrix, NimbleGen Systems, and Agilent Technologies (Figure 5) provide suites of components, reagents, and services. The main features of each platform are presented below.

Comprehensive comparative studies of data generated from the most widely used commercial platforms have been carried out by different laboratories [27–29]. In each study, gene expression measurements from the platforms being compared were generated from a common source of biologically different RNAs. Correlations in expression levels and comparisons for significant expression

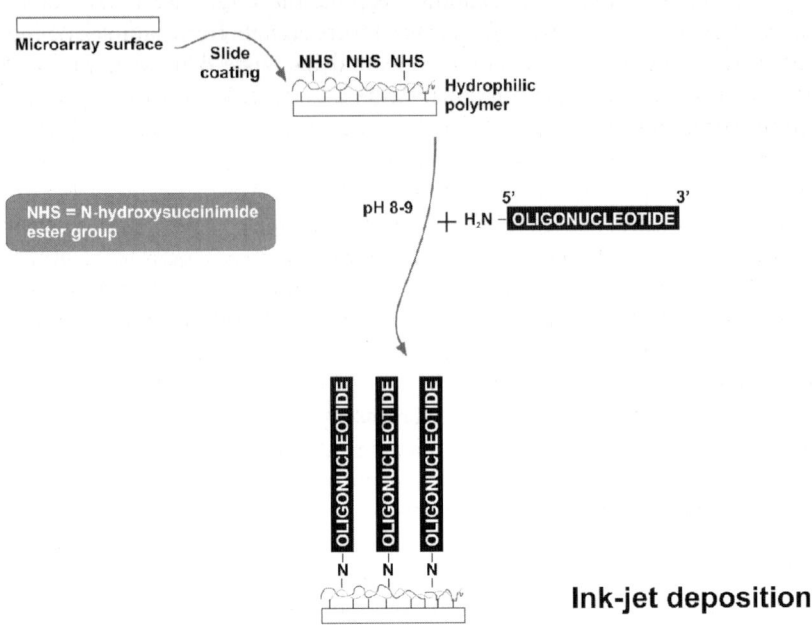

Figure 4. CodeLink bioarray platform. A unique and well characterized 30 mer oligonucleotide for each gene is deposited on a proprietary 3D gel matrix. Attachment is accomplished through covalent interaction between the amine-modified group present on the 5′-end of the oligonucleotide and the activated functional group present in the gel matrix. The 3D gel matrix provides an aqueous environment, allowing for maximal interaction between probe and target.

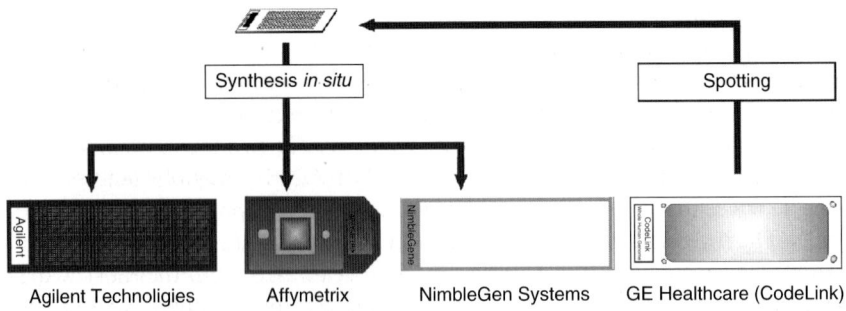

Figure 5. The four major vendors of microarray chips. For the three commercial microarrays on the left the oligonucleotides are synthesized *in situ*, whereas for the array on the right they are pre-synthesized and spotted.

Concepts on microarray design for genome 273

changes in genes present on all platforms, revealed considerable divergence across platforms [27, 29]. Unsupervised clustering and principle component analysis suggested that the largest variation in measurements from the commercial platforms was attributable to the platforms themselves. Although gene sets did overlap to some extent across these platforms, the majority of genes that were identified as differentially expressed were exclusively identified with each technology [27]. Other sources of divergence across platforms can be attributed to the detection of distinct types or sets of alternatively spliced transcript variants, represented in each array [28] and by the labeling/hybridization protocols of each technology – one-color based microarrays (Affymetrix and CodeLink) compared to the two-color arrays from Agilent [29]. However, later analyses under more controlled conditions have demonstrated that good reproducibility can be achieved across laboratories and platforms [30–32]. The conclusion of these latter studies is that the main factors that influence variation are the biological samples and human factors, rather than technical diversity. Specific attention can be given to these negative factors in order to minimize inconsistencies; nevertheless a small degree of variability is probably unavoidable with such a sensitive and complex technology.

1.2.1 Affymetrix

By applying photolithographic technologies derived from the semiconductor industry to the fabrication of high-density microarrays, Affymetrix of Santa Clara, California, pioneered this field and has dominated for many years. High-density Affymetrix oligonucleotide arrays, also called GeneChips have become the pharmaceutical industry standard owing to its extensive genetic content, high levels of reproducibility, and minimal start up time [11]. A major advantage of GeneChips is that they are designed *in silico*, thereby eliminating management of DNA clone libraries or oligonucleotide sets, and the possibility of misidentified tubes, clones, or features [33]. The disadvantage of this platform is that it demands a dedicated scanner and utilizes short 25 mer oligonucleotides, which are less sensitive than the longer 60 mers utilized in other technologies. Additionally, to increase sensitivity multiple oligonucleotides are required for transcript detection.

Affymetrix focused on light-directed synthesis for the construction of high-density DNA probe arrays using two techniques: photolithography and solid-phase DNA synthesis (Figure 6). The glass substrate, or chip, is first covalently modified with a silane reagent to provide hydroxyalkyl groups, which serve as the initial synthesis sites. Synthetic linkers modified with photosensitive protecting groups are attached to a glass surface. Using a photolithographic mask, light is then directed to specific areas on the surface to remove the protection groups from the exposed linkers. The first of a series of chemical building blocks, hydroxyl-protected deoxynucleosides, is incubated with the surface, and chemical coupling occurs at those sites that have been illuminated in the preceding step. Another mask is used to deprotect and direct light to, other sites.

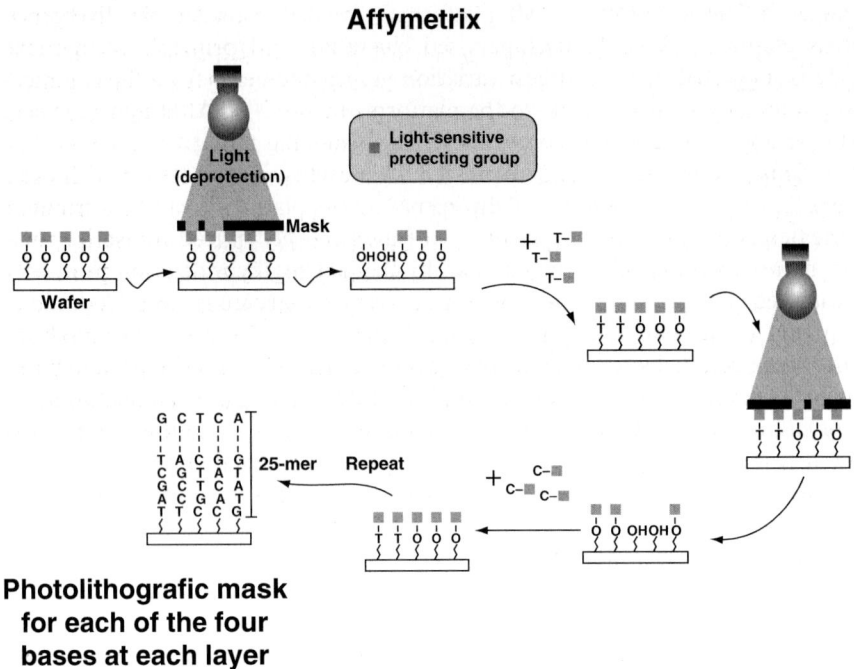

Figure 6. Construction of an Affymetrix chip by light-directed oligonucleotide synthesis. Light is directed through a mask to deprotect and activate selected sites, and new protected nucleotides (in this example, thymidine derivatives) couple to the activated sites. The process is repeated, activating different sets of sites and coupling different bases (in this example, cytosine derivatives) allowing arbitrary DNA probes to be constructed at each site.

New deoxynucleosides are added and the process is repeated until the desired length of oligonucleotide is synthesized. The amount of nucleic acid information encoded on the array in the form of different probes is limited only by the physical size of the array and the achievable lithographic resolution. A 1.28 × 1.28 cm array can include over a million different oligonucleotide sequences.

For gene expression purposes, the oligonucleotides are generally 25 bases long and each transcript is represented by 11–20 such probes. Probe sequences are ideally spread throughout the gene sequence, and are generally more concentrated at the 3′-end. In addition, each perfect match (PM) probe is paired with a mismatch (MM) probe, an identical probe except for a single base difference in a central position. The MM probes act as specificity controls and enable subtraction of background and cross-hybridization. The use of multiple independent probes for each gene greatly improves signal-to-noise ratios, improves the accuracy of RNA quantization (averaging and outlier rejection), increases the dynamic range and reduces the rates of false positive and miscalls [9].

Concepts on microarray design for genome

1.2.2 NimbleGen systems

NimbleGen manufactures custom, high-density DNA arrays based on its proprietary Maskless Array Synthesizer (MAS) technology. The MAS system is a solid-state, high-density DNA array fabrication instrument comprised of a maskless light projector, a reaction chamber, a personal computer, and a DNA synthesizer. NimbleGen builds its arrays using photo-mediated synthesis chemistry with its MAS system.

A digital micromirror device (DMD) employs a solid-state array of miniature aluminum mirrors to pattern up to 786,000 individual pixels of light (Figure 7). The DMD creates "virtual masks" that replace the physical chromium masks used in traditional arrays. These "virtual masks" reflect the desired pattern of UV light with individually addressable aluminum mirrors controlled by the computer. The DMD controls the pattern of UV light projected on the microscope slide in the reaction chamber, which is coupled to the DNA synthesizer. The UV light selectively cleaves a UV-labile protecting group at the precise location where the next nucleotide will be coupled. The patterns are coordinated with the DNA synthesis chemistry in a parallel, combinatorial manner such that up to 786,000 unique probe features are synthesized in a single array [34].

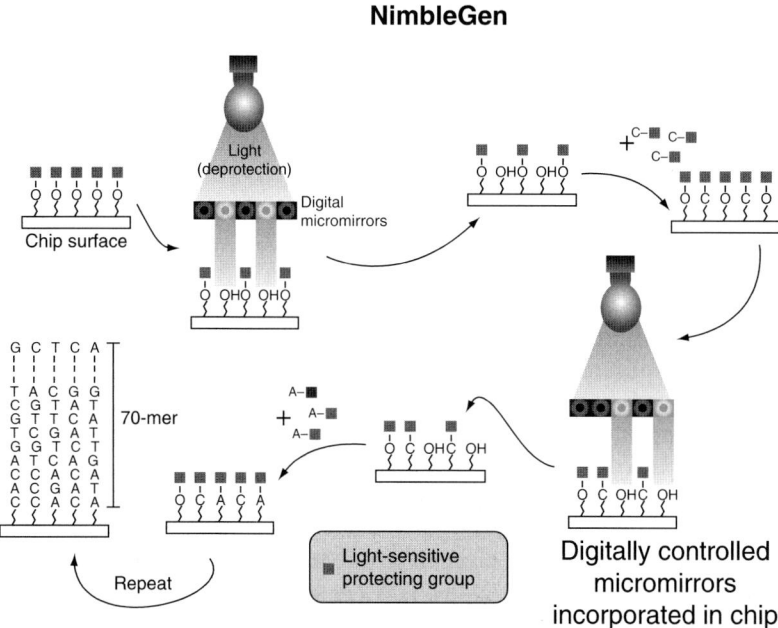

Figure 7. Construction of microarrays using NimbleGen System's MAS technology. Digital micromirrors reflect a pattern of UV light, which deprotects the nascent oligonucleotide and allows addition of the next base.

The light-directed synthesis methods, both photolithographic [35] and digital micromirror-based [34], have the potential to achieve feature sizes not much larger than a wavelength of light. This should enable substantial further reductions in cost and in hybridization volume with consequent reduction in the amount of biological sample required.

1.2.3 Agilent technologies

Agilent Technologies uses proprietary SurePrint ink-jet technology and offers a flexible microarray platform. Oligonucleotides (60 mer long) are synthesized *in situ* and are built up a base at a time on standard glass slides, resulting in arrays with more than 230,000 unique features. The iterative oligonucleotide synthesis loop begins when the first nucleotide of each oligo is printed onto the activated glass surface of the microarrays. In phosphoramidite synthesis reactions, the reactive sites on the nucleotides are blocked with chemical groups that can be removed selectively. This allows the bases to be added to the oligo chain one base at a time in a very controlled manner. After the first base is printed, the trityl group that protects the 5′-hydroxyl group on the nucleotide is removed and oxidized to activate it, enabling it to react with the 3′-group on the next nucleotide. In between each step, the excess reagents are washed away so that they will not randomly react later in the synthesis. The process of printing a nucleotide followed by detritylation, oxidation, and washing is repeated 60 times (Figure 8). After the last base in the oligo chain is printed, the microarrays undergo a final deprotection step, before moving on to quality control testing [36]. Ink-jet synthesis yields are ~98% per stage with chemical deprotection, as opposed to ~95% for photodeprotection, allowing the ink-jet technology to be optimized with longer oligos and higher stringency hybridization conditions. *In situ* ink-jet synthesis should have a valuable niche for rapid turnaround of custom arrays in small lots, unless it is overtaken by the micromirror technologies.

This 60 mer oligonucleotide platform contrasts with the short 25 mers probes employed by Affymetrix. Although short oligonucleotides should in theory provide the greatest discrimination between related sequences, they often have poor hybridization properties. The 60 mers provide enhancements in sensitivity over 25 mers in part due to the larger area available for hybridization. In light-directed synthesis, failure of photodeprotection at any stage terminates the oligo. The yields per stage in the Affymetrix synthesis process are such that attempts to make 60 mers would result in very few of them running to even half that length; Affymetrix settled on 25 mers partly for this reason. Another advantage of Agilent chips is that only one 60 mer per gene or transcript is required [36]. The reason why Affymetrix uses multiple probe pairs to estimate the abundance of each target transcript is partly by the need to make up for the performance limitations of 25 mers.

Concepts on microarray design for genome

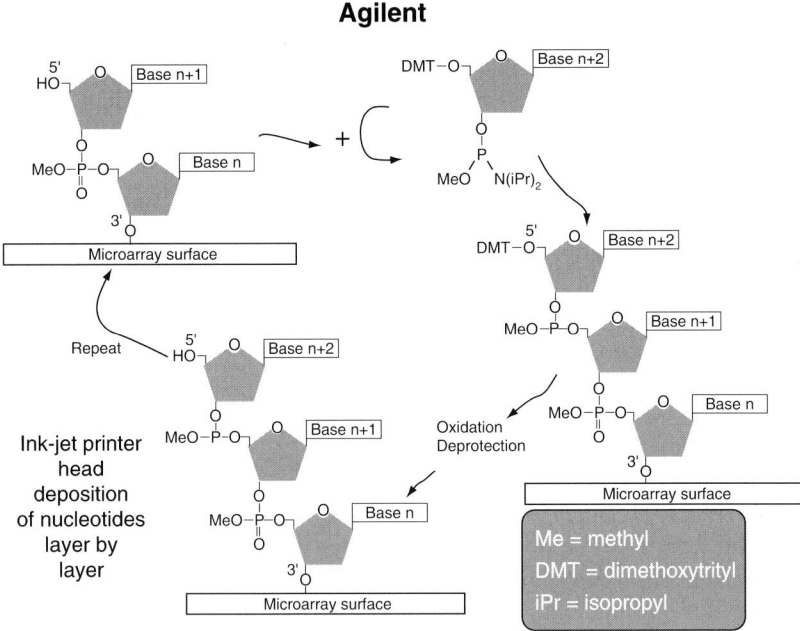

Figure 8. Ink-jet synthesis of probes in the Agilent microarray chips. This figure shows the general cycle of oligo synthesis via phosphoramidite chemistry.

2. SELECTING THE PROBES

For a given organism under study, DNA microarray probes can be designed as soon as a sequence of genomic region or transcript from that species becomes available. With several whole-genomes already sequenced and millions of expressed sequence tags (ESTs) deposited into public databases, microarrays are able to increase our understanding of basic biological processes if the investigator selects a set of probes that are suitable to answer specific question, as discussed in the following sections.

The probe sequences used in gene-oriented arrays are selected on the basis of gene and EST data from public databases according to a number of criteria; most importantly, they should be unique for the gene (avoiding, e.g. characteristic sequences of gene families). To minimize the probability of unspecific target cross-hybridization, sequence comparison *in silico* tests can be performed between each probe and all known transcripts from that species using basic local sequence alignment tool (BLAST). Probes with a unique exact-match target are ideal; for 50–70 mer oligonucleotide probes, mismatched target sequences with no more than 35% identity with no gaps are usually desirable, to decrease the probability of cross-hybridization. In addition, probes should be relatively uniform in their hybridization properties, which are determined by a similar overall gyanine–cytosine (GC) content, melting temperature (T_m), and tendency to form secondary structure.

Microarrays aim to provide accurate measurements of true expression values of the phenomenon under study. This is achieved by a high specificity (reduced false–positive rate) and a high sensitivity (reduced false–negative rate) of microarray probes. cDNA probes or longer oligonucleotide probes provide greater sensitivity at the expense of reduced specificity.

In general, the specificity of oligonucleotide probes is evaluated by experiments using target RNAs that share various degrees of sequence similarity [36, 37]. For a given hybridization stringency condition and protocol, these experiments determine the maximal degree of sequence similarity for which no cross-hybridization is detected, thus revealing the probe parameters for good specificity. These parameters can be applied for the design of novel probes that should work well with the predefined hybridization protocol.

Probe sensitivity is generally defined as the lowest target concentration at which an acceptable accuracy is obtained [36, 38]. For a given organism, additional control probes containing DNA sequences with no homology to any known transcript or to the genome sequence are often used to estimate the cutoff detection limit parameters. Moreover, tiling arrays can use signal intensity information of consecutive probes in a predefined transcriptional unit in order to determine a detection cutoff that can be applied to identify novel transcripts in nonannotated genomic regions [13, 14].

2.1 Gene-Oriented Arrays

Microarrays designed for measuring gene expression levels are generally biased toward known and predicted protein-coding genes. These genes can be determined using several approaches, such as large-scale sequencing of ESTs, comparative genomic annotation, or full-length cDNA cloning experiments. Once the gene sequence is obtained, cDNA or oligonucleotide probes can be designed and placed onto a microarray chip. Then, expression levels of genes can be assessed by relative hybridization between these probes and labeled targets derived from different cell conditions or types. Although independent experiments are required to validate selected probes in terms of specificity and sensitivity (discussed above), signal intensity comparison of a given probe under different controlled conditions can be used to estimate cutoff detection limit parameters that increase the specificity of measurements.

Compared to tiling arrays, the gene-oriented platform is a relatively easy-to-handle tool since it uses relatively few probes for each gene. A single chip is capable of measuring the expression levels of all known messages of specific types of transcripts, being these messages protein-coding genes [39] or, for example, intronic noncoding RNAs [40] and micro-RNAs [41]. Pre-mRNA splicing at every exon–exon junction [42] or SNPs [43] of thousands of genes can also be

Concepts on microarray design for genome

monitored using this platform. Therefore, this tool has the advantage that a rapid evaluation of the differences between two or more transcriptomes can be made by hybridizing the different cDNA preparations to identical chips and comparing the hybridization patterns.

2.2 Epigenomic Microarrays

Only a minor fraction of eukaryotic genomes is occupied by genes; however, histone and nonhistone chromosomal proteins and methylated DNA bases are distributed over both genic and intergenic regions. Once mapped, the microarray platform can be used to obtain the profiling patterns of these widespread epigenomic features, such as DNA methylation [44], DNA replication [45], DNA binding, and chromatin-associated proteins and histone modifications [46]. Alternatively to the already-mapped sites, microarray-based strategies are able to identify novel DNA binding sites or novel DNA methylation regions by probing upstream and downstream regions of genes [47], or by probing predicted CpG islands of a genome [44]. Certainly, epigenomic microarrays will become a standard research tool for understanding chromatin structure and gene expression during development [46].

Similar to gene-oriented arrays, epigenomic microarrays are easy-to-handle tools in comparison to tiling arrays and permit that many different experiments be performed at a low cost and lower labor analysis. However, for identification of the complete set of epigenomic features of an organism, a tiling-array platform is the best tool since it covers long contiguous genomic regions.

2.3 Tiling Arrays

With the completion of sequencing of many genomes, attention has shifted to determining the complete set of transcribed sequences and regulatory elements. This recent trend in genomics has involved the development of tiling arrays: microarrays that represent a complete non-repetitive tile path over a *locus*, chromosome or whole-genome, irrespective of any genes that may be annotated in that region [17]. Potential uses for such unbiased representation of gDNA include empirical annotation of the transcriptome [48], chromatin-immunoprecipitation-chip studies [49], characterization of the methylation state of CpG islands [50], analysis of alternative splicing [48, 51], and CGH [52].

Numerous options exist for tiling genomic sequences with oligonucleotides or PCR products, leading to microarray designs of different sequence resolutions and feature densities (Figure 9). Oligonucleotide arrays comprise 25–70 bp probes, which are synthesized directly on the slides or prepared in solution and then deposited. The second type of tiling array is constructed using PCR products typically of ~1 kb in length, or BAC arrays – typically at 1 Mb resolution (see Section 1). One caveat of PCR-based tiling arrays is that their construction is labor intensive and therefore they are not readily scalable to the study of large

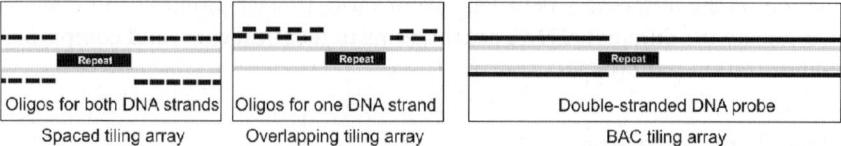

Figure 9. Comparison of different whole-genome array designs. Probes can be overlapping or spaced at regular intervals; comprised by oligonucleotides, PCR products or BAC clones; single- or double-stranded; and designed to interrogate one or both DNA strands of genomic regions without annotation bias. This figure shows three different combinations of whole-genome array design.

genomes at a high resolution. As an example, tiling of the entire human genome using this platform would require approximately 2 million PCR reactions at 1 kb resolution and necessitate extensive informatics infrastructure to support the effort [53].

3. SPECIFIC QUESTION, SPECIFIC CHIP

The many applications of microarrays chip are being used to answer important questions in biology and medicine. Beside transcriptome analysis, microarrays chip is currently useful to determine the methylation status of CpG islands, to identify DNA binding sites of transcription factors and to discover novel genes or alternative isoforms of genes. Also, different types of DNA probes can detect from large chromosome deletions of millions of nucleotides that are associated to cancer as well as single nucleotide substitutions that may affect important proteins of different metabolic pathways. Thus, it is clear that the appropriate application is no longer determined by technical improvements but by the efficient chip design, derived from the specific aim of the assay. Here in this chapter, we focus on specific chip designs applied to answer specific biological questions.

3.1 Transcriptional Profiling

The power of microarray technology lies in its ability to simultaneously measure the expression of thousands of genes, thus providing a snapshot of the transcriptome in different states of tissues and cells. The most common application of microarray chips is still the expression profiling of mRNAs. Comparison of mRNA expressions in a high-throughput way raises a number of hypothesis and points to important biological functions of genes inside cells under different situations, such as disease states, tissues from a given organism, drug treatments, and gene disruptions. Large-scale EST sequencing [54], serial analysis of gene expression (SAGE) [55], and massively parallel signature sequencing (MPSS) [56] technologies also provide a transcriptional profiling of tissues and cell

Concepts on microarray design for genome 281

types. However, these techniques are relatively more expensive and less flexible than microarray technology.

In 1995, microarray technology was used for the first time to assess the transcriptional profiling of ~1000 *Arabidopsis* genes [1]. This number represents only 4% of the 26,330 annotated genes from *Arabidopsis*. Since then, microarray technology has evolved fast and cDNA/oligo microarray platforms containing all genes from this organism are now available (Figure 10). In 11 years of research, more than 60 papers were published by different labs across the world, reporting changes on transcriptional levels of *Arabidopsis* genes using microarray chips (Figure 10). Such simultaneous measurements of *Arabidopsis* gene expression helped scientists to gain comprehensive insights into the response of *Arabidopsis* to several environmental conditions.

Microarray probes can only be designed based on previous information of sequences of known or predicted genes. The exon structure of a gene defines where probes can be designed. In general, probes are designed close to the 3′-end of the transcript. The reason is that most of labeling protocols use the poly-A tail of target mRNAs for priming the labeling reaction.

Microarray chips are valuable tools for functional genomic studies and could accelerate the annotation of novel genes. Thanks to the large number of EST

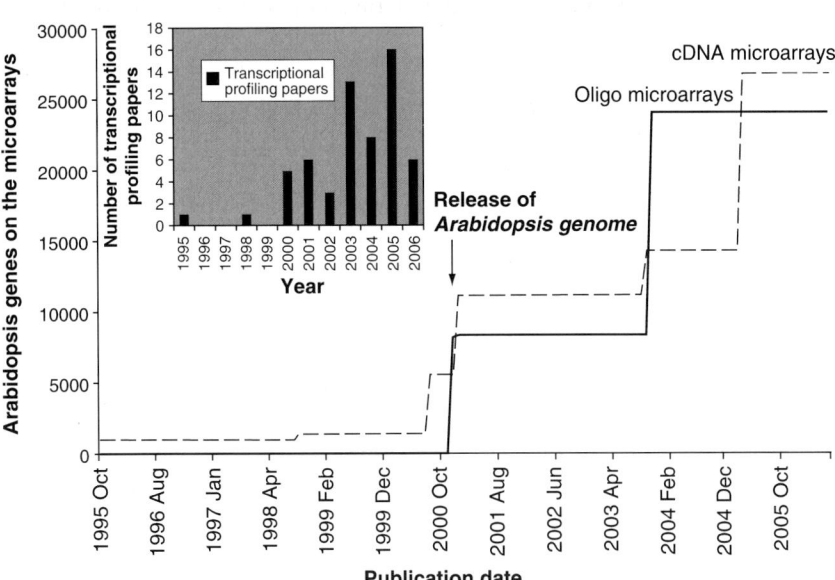

Figure 10. Use of microarray technology to assess transcriptional profiling. The line graph shows the increasing number of *Arabidopsis* genes represented in oligo and cDNA microarrays that were used in the *Arabidopsis* transcriptional profiling papers, which were published between 1995 and 2006. The number of such published papers is shown on the bar graph.

sequencing projects and the complete sequencing of many genomes, bioinformatics analysis can predict thousands of genes. Tissue-specific pattern of mRNA expression of known and predicted genes can confirm their expression and give important clues about gene function [39]. The gene atlas of the mouse and human protein-encoding transcriptomes, described by Su et al. [39], identifies hundreds of regions of correlated transcription and show that some genes are subject to both tissue and parental allele-specific expression, suggesting a link between spatial expression and imprinting. Also, hypotheses about the biological roles of genes with unknown function can be raised by comparison of their expression levels with possible coregulated known genes [57]. In addition, the identification of groups of genes with similar expression profiles can uncover gene families or metabolic pathways that are affected in a specific condition [58].

The use of arrays as tools for gene expression profiling on a genomic scale has some limitations. One is that this technology is only able to measure relative levels of mRNA expression, and not absolute amounts. Another current limitation is that it is not reliable to compare the levels of different mRNAs from the same sample, due to differences on labeling and hybridization of each probe and target. Recently, these limitations started to be addressed by different approaches that allow quantitative estimation of absolute endogenous transcript abundances in cells, that are based on a common oligonucleotide reference [59] or on a set of exogenous RNA controls [60]. Moreover, most hybridization arrays are not designed to differentiate between alternatively spliced transcripts of the same gene and, in some cases, between highly homologous members of a gene family. Finally, a change in messenger RNA does not necessarily correlate with a change in protein expression [61], and the translated protein often requires further modifications to attain its full activity. These latter two points are a common and legitimate criticism of array technology because it measures an intermediate step (mRNA levels) and not a functional product (active protein). However, until sensitive and reproducible proteomic technologies become universally accessible to the research community, hybridization arrays will continue to be the best opportunity for studying gene expression on a genomic scale [62].

3.2 Comparative Genome Hybridization

The CGH array technique allows the detection of chromosomal copy number changes on a genome-wide and with a high-resolution scale. It is used in human genetics and oncology, with great promise for clinical application. In typical CGH experiments, test and control DNA samples (e.g. tumor and normal cells) are isolated and used to create fluorescently labeled probes, typically cyanine-3 (Cy3) and cyanine-5 (Cy5). The probes are pooled and competitively cohybridized to a glass slide spotted with a known array of mapped genomic clones, cDNAs, or oligonucleotides. Log ratios of the Cy5–Cy3 intensities are measured for each clone. Next, a log ratio profile is assembled to determine relative copy number changes between the test and control samples, which may comprise loss or gain/amplification of specific genomic regions (Figure 11).

Different CGH platforms differ in the spatial resolution (e.g. the number of genomic bases) for the detection of copy number changes. This can vary from megabase to kilobase resolution. The main factors affecting the resolution of CGH arrays are the number and size of elements on the arrays, the chromosomal distribution of printed elements, as well as the amplitude of a chromosomal copy number change. Until recently, PCR-amplified BACs have been the main source of DNA for the assembly of CGH arrays covering large genomic regions, entire chromosomes, and eventually the whole human genome [52, 63, 64]. The large insert size of BAC clones (~150 kbp) provide multiple sites for target binding, giving a good sensitivity for detection of small changes in copy number. Genome-wide CGH arrays based on cDNA clones have also been developed. While cDNA CGH arrays provide a direct link to RNA expression measurements, they preclude the analysis of chromosomal gain or loss in nontranscribed regions. Furthermore, the smaller probe size requires larger amounts of gDNA for target generation and often result in a lower signal-to-noise ratio as compared to large-insert CGH arrays [64]. The large-scale operation required for DNA isolation or PCR amplification of large-insert clones necessary for manufacturing the arrays are elaborate, time consuming, and has to deal with the possibility of clone contamination along the process of array fabrication

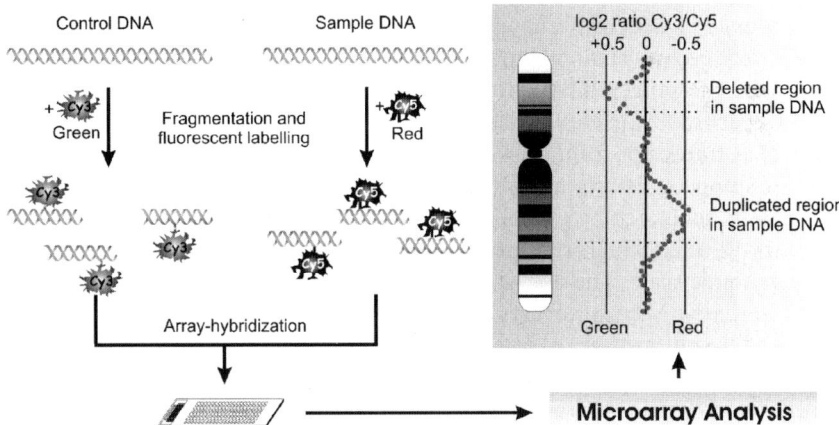

Figure 11. Principles of array comparative genomic hybridization (CGH). Sample and control DNA are fragmented and labeled with fluorescent dyes, combined, and cohybridized to a microarray containing spots of genomic material (tiling array). The sample and reference competitively bind to the spots and the resulting fluorescence intensity ratios are reflected by their relative quantities, as shown by the computer generated CGH fluorescence ratio profile (*right*). The center line in the CGH profile represents the balanced state of the chromosomal copy number ($\log_2$ ratio = 0). Gains are viewed to the right ($\log_2$ ratio = –0.5) and losses ($\log_2$ ratio = +0.5) to the left of the centre line.

[65]. CGH arrays comprised of synthetic oligonucleotides are emerging as an alternative technology to eliminate the need for clone management and lessen probe identity errors. Assembly of high-density arrays composed of small (25-60 nt) oligonucleotide probes are likely to provide a better resolution than BAC and cDNA arrays [65]. Oligonucleotide CGH arrays are readily available through academic institutions (Sanger Center, UCSF, DKFZ) as well as through commercial suppliers (Agilent, Affymetrix, NimbleGen). In practice, current resolution of oligonucleotide CGH arrays is limited by the lower signal-to-noise ratio from individual probes, which requires that measured intensities from several adjacent probes are combined to calculate a moving average of signal intensities. Also, methods for reduction of genome complexity are often applied to limit nonspecific target binding to short oligonucleotide probes such as those present in Affymetrix platforms [66]. Future developments in gDNA target amplification and labeling will be required to expand the use of whole-genome tiling oligoarrays for CGH analysis [66].

3.3 Alternative Splicing

Almost all protein-coding genes of humans have a split structure with several exons and introns. Intronic sequences are removed from the primary transcript by the process of pre-mRNA splicing, an essential step in eukaryotic gene expression. Alternative splicing is the differential processing of exon junctions to produce a new transcript variant from one gene, and is a major determinant of the protein functional diversity underlying human physiology, development, and behavior [67]. Much of the available genomic information on alternative splicing is derived by the alignment and conservation analysis of large numbers of ESTs and messenger RNAs to genome sequences of different organisms [68, 69]. In general, exons are called "constitutive" when are presented in every example of a transcript from a given locus and called "alternative" if they are sometimes skipped. Efforts are now being directed at studying relevant transcript variants generated by alternative splicing at a global level.

Microarrays offer a high-resolution means for monitoring pre-mRNA splicing on a genomic scale. The use of this technology has permitted the discovery of new alternative splicing events not previously detected in cDNA or EST sequences [42] and large-scale detection of cell- and tissue-specific alternative splicing events involving exons that were initially identified using EST/cDNA sequence data [70, 71]. Moreover, alternative splicing microarrays have facilitated the global analysis of alternative exons regulated by specific splicing factors [72, 73] and have led to the discovery of sequence motifs that correlate with tissue-specific alternative splicing [74].

The splice array is based on the design of probes located on constitutive exons, alternative exons, as well as on the constitutive and alternative splice junctions. Frey et al. [75] designed a platform containing probes for all 1.14 million putative exons of the mouse genome and Johnson et al. [42] for every

Concepts on microarray design for genome 285

exon–exon junction in more than 10,000 multiexon human genes. In addition, other groups [70, 76–78] used splicing-sensitive microarray containing both exon and splice junction oligonucleotide probes to assay splicing of a large number of human genes. This platform permits the detection of all different types of splice events: exon skipping, novel exon, internal exon deletion, intron retention, or alternative usage of splice donor or acceptor sites. Another microarray format employing a fiber-optic-based system for the detection of specific splice variants has been described, and this approach has been used to monitor splice variants in different transformed cell lines and tumors [79–81].

Probes should be designed with homogeneous T_m and similar lengths to obtain a common thermodynamic profile and junction probes being preferably centered on the splice site. This positional constraint for junction probes may complicate probe design, making probe composition not suitable to get the desired thermodynamic parameters. However, junction probes can be designed with a sequence up to two nucleotides off-centre, which maintained the expected specificity [82]. Cross-hybridization tests of probe sequences can be performed by BLAST analyses against the human EST databases using parameters for short, nearly exact matches.

Although longer exon probes are better for detecting exons, longer splice junction probes present a unique problem. Since about half of a splice junction probe will be derived from one exon and about half from another, each junction probe has perfect complementarity over about half of its length to other RNA forms that contain a different exon [83]. Fehlbaum et al. [82] evaluated the specificity of probes of splice arrays using three different probe lengths (24, 30, and 40 mer) and labeled targets from only two variants of a gene (long and short isoforms). Junction probes were designed to detect specifically each type of isoform. Their results showed that the junction probes with 30 and 40 bases long detect both isoforms. Due to potential hybridization of half of the junction probes to a single exon, only the 24 mer seems to have the specificity required for isoform-specific detection of alternatively spliced events [82].

Alternative splicing analysis can use signals derived from the hybridization of labeled targets to the constitutive exon oligonucleotides relative to exon–exon junction probes [77]. In theory, constitutive exon probes measure the total amount of RNA from the particular gene, whereas hybridization signals from an exon–exon junction oligonucleotide would reflect the amount of RNA containing that particular junction (Figure 12). Therefore, the ratio of hybridization intensity from a probe spanning a specific exon–exon junction to that from a constitutive exon probe would provide exon-skipping or -inclusion indexes, reflecting the level of that alternatively spliced RNA in the two comparison samples (Figure 13) [77].

3.4 Transcriptome Annotation

Even with a finished genome sequence, computational gene prediction or traditional molecular methodologies are not able to identify all of the transcription units. These approaches – sequencing randomly selected cDNA clones, aligning

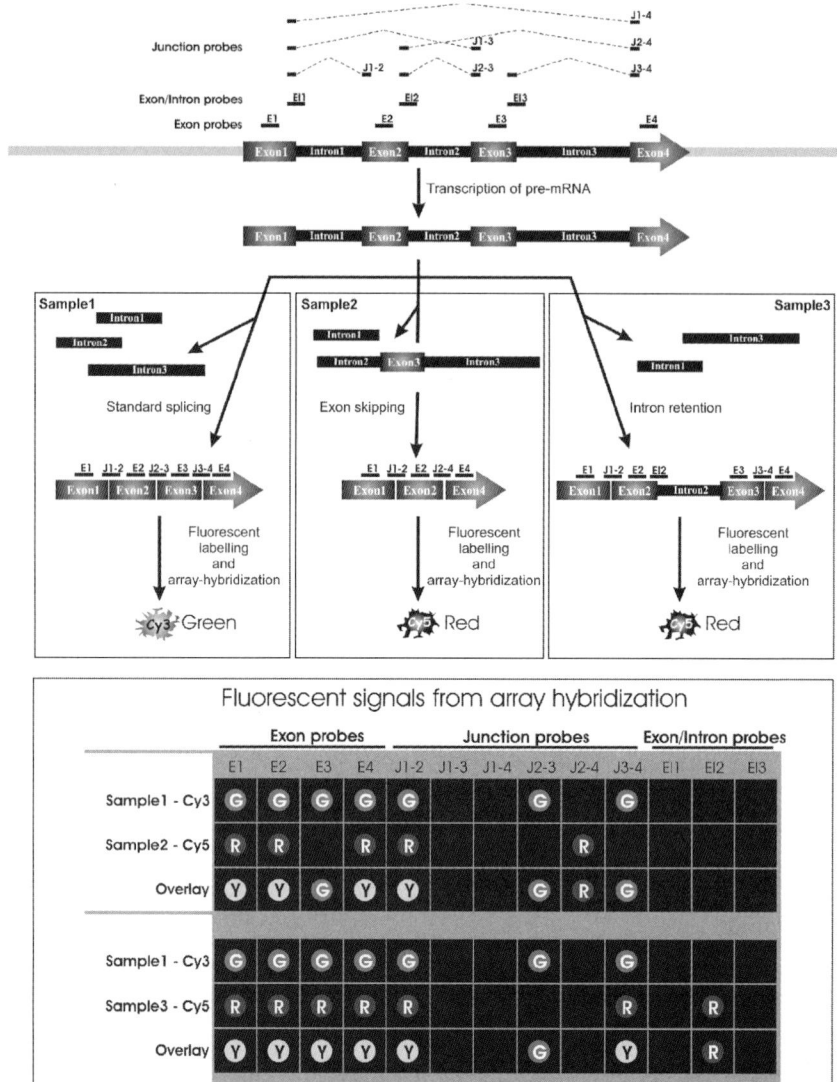

Figure 12. Principles of alternative splicing detected by splicing-sensitive microarray. Probes spanning all exons and all exon–exon and exon–intron junctions are designed and placed onto microarrays (*upper panel*). The standard splicing (sample 1) and two types of alternative splicing (samples 2 and 3) of a pre-mRNA are represented as three different samples (*middle panel*). Sample 1 is labeled with Cy3 (green) and sample 2 and 3 with Cy5 (red). Targets are mixed in two different combinations (sample 1 with sample 2, and sample 1 with sample 3) and each combination is hybridized to a microarray. The lower panel shows a schematic view of scanned images of these two microarray slides. Alternative splicing is detected by different hybridization signals of exon and junction probes.

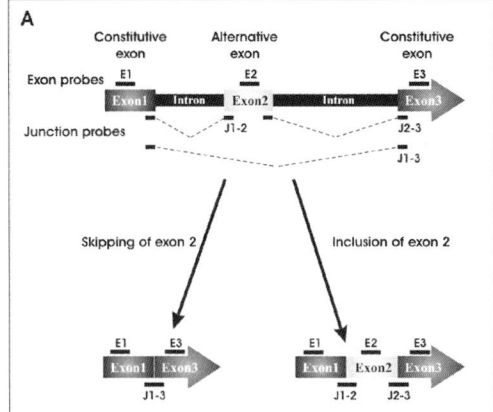 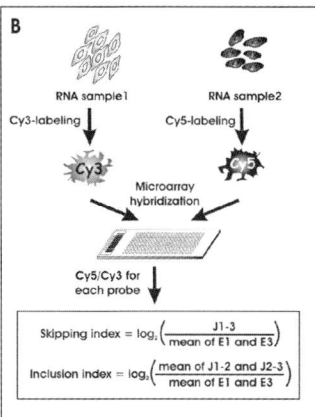

Figure 13. Detection of alternative splicing by microarray. Panel A: Design of oligonucleotide probes. The microarray probes contain oligonucleotides that target all possible exon-exon junction sequences (J1–2, J1–3, and J2–3). Probes E1 and E3 are complementary to flanking constitutive exons and probe E2 is complementary to alternative exon2. Panel B: Data collection and analysis. RNA samples 1 and 2 are isolated and labeled separately with Cy5 or Cy3 fluorescent dye, mixed, and hybridized to oligonucleotides in microarray. Red (Cy5) and green (Cy3) fluorescence are measured and the ratio of the two values is calculated for each oligonucleotide. To assess differences in splicing pattern between the two samples, skipping indexes and inclusion indexes are calculated. The skipping index of alternative exon2 is $\log_2$ of Cy5/Cy3 from the exon1–exon3 junction oligonucleotide (probe J1–3 in Panel A) divided by the mean of Cy5/Cy3 from the constitutive exons 1 and 3 (probes E1 and E3 in Panel A). The inclusion index of exon2 is $\log_2$ of the mean of Cy5/Cy3 from exon1–exon2 and exon2–exon3 (probes J1–2 and J2–3 in Panel A) divided by the mean of Cy5/Cy3 from exon1 and exon3 (probes E1 and E3 in Panel A).

protein sequences identified in other organisms, sequencing more genomes, and manual curation – successfully identified expressed transcripts for tens of thousands of genes, but they eventually reach a point of greatly diminished returns. These methods fail in detecting transcripts that are low abundance or expressed in rare cell types or in response to specific stimuli. Tiling microarrays can be used to circumvent some of these problems, allowing confirmation of the predicted gene models as well as being a tool for new exon and gene discovery (Figure 14) [84].

Microarray technology has permitted a refined high-throughput mapping of the transcriptional activity in the human genome. A pioneering study from Kapranov and colleagues [48] revealed a tenfold excess of transcriptionally active regions along chromosome 21 and 22 than originally predicted by mapping of known genes. Later, this study was extended to ten human chromosomes where sites of transcription of polyadenylated and nonpolyadenylated RNAs were mapped at 5 bp resolution in eight cell lines [14]. Interestingly, the major proportion of the transcriptional output of the human genome was comprised by unannotated, nonpolyadenylated transcripts [14]. In another study, Bertone et al. [13] constructed a set of 134 high-density oligonucleotide microarrays to

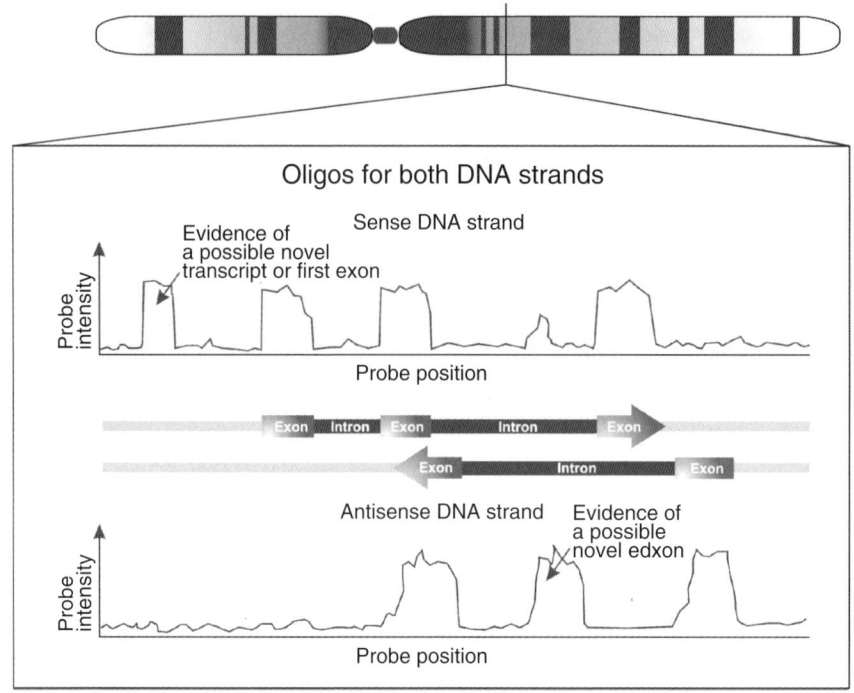

Figure 14. Mapping of the transcriptional activity in the genome using a tiling array. Tiling microarrays are designed to assay transcription at intervals of the genome using regularly spaced probes that can be overlapping or separated. For each DNA strand, transcription within a genomic region is represented by regions of greater fluorescent intensity. Annotated genes aligned with these microarray fluorescence intensities can reveal unannotated exons or novel transcripts.

represent ~1.5 Gb of nonrepetitive gDNA from each strand of the human genome. This approach identified thousands of new transcribed regions and confirmed the transcription of predicted genes on a global scale. Also, conservation between many of the novel transcribed sequences and well-characterized mouse proteins provides strong evidence that a large number of them are likely to encode functional transcripts [13].

Whole-genome oligonucleotide arrays have also been useful for studying another poorly understood aspect of the transcriptome, natural antisense transcription, because they can simultaneously monitor gene expression on both strands of a genome (Figure 14). Natural antisense RNAs are endogenous coding or noncoding transcripts that exhibit complementary sequences to transcripts of already known function, named sense transcripts [85]. These antisense messages might be involved in several biological processes, such as alternative splicing, alteration of methylation pattern, and competitive transcriptional interference (for RNA polymerase II) within the same *locus* [86]. In the oligonucleotide tiling array study of the human genome, a significant proportion of exonic sequence, represented by known exons, mRNAs, and ESTs, was found to exhibit antisense transcription [13]. This result demonstrates the utility of tiling arrays for helping to unravel the high-complexity of eukaryotic transcriptomes.

3.5 Small MicroRNA Profiling

MicroRNAs (miRNAs) represent a class of small noncoding RNAs encoded in the genomes of plants and animals that are thought to regulate gene expression of target mRNAs. Mature miRNAs are about 22 nucleotides long and typically excised from 60- to 80-nucleotide foldback RNA precursor structures [87]. In animals, most miRNAs function through the inhibition of effective mRNA translation of target genes through imperfect base pairing with the 3'-untranslated region (3'UTR) of target mRNAs [88]. Some miRNA functions include control of cell proliferation, cell death, and fat metabolism in flies, neuronal patterning in nematodes, modulation of hematopoietic lineage differentiation in mammals, and control of leaf and flower development in plants [88]. Also, altered expression of a few miRNAs has been found in some tumor types [89–92].

Several DNA chips have been designed to expression profile miRNAs or their hairpin precursors across several human and mouse tissues [41, 93–97], during mouse brain development [98] or in human B cell chronic lymphocytic leukemia [99]. Oligonucleotide probes with sequences complementary to miRNAs can be spotted [41, 93, 96–99] or *in situ* synthesized [94, 95] onto a microarray platform and used to capture labeled miRNAs (Figure 15). In general, miRNAs should be first isolated from total RNA by excision from poly-acrylamide gel or by size-fractioning using commercial column-purification kits. Then, methods involving PCR-based amplification or ligation strategies can be used to label mature and active miRNAs [94–99]. A direct tiling/labeling procedure and hybridization approach was also developed by others [41]. Essentially, polynucleotide tails 20–50 nt long are appended to the 3'-ends of all miRNAs by the poly(A)-polymerase (PAP) enzyme (Figure 15). The 3'-tail is a mixture of standard and amine-modified nucleotides, and tailed miRNAs can subsequently be labeled with any monoreactive NHS–ester dyes, such as Cy3 and Cy5 [41].

The application of DNA microarray technology to parallel expression measuring of the entire endogenous set of miRNAs may offer higher sensitivity, high throughput, and higher comparative capabilities over the other methods used to detect miRNAs, e.g. Northern blot analysis, cloning, and membrane arrays using radioactive detection methods [95]. However, the detection of miRNAs with microarrays still meet significant difficulties, mostly due to the short size of miRNAs and the sequence similarity between miRNA family members.

3.6 Methylation Pattern

DNA methylation in CpG dinucleotides is an epigenetic mark crucial in regulation of gene expression. DNA methylation is required to complete embryonic development, and has been directly implicated in genomic imprinting and X chromosome inactivation in mammals. Cytosine methylation is also important for silencing of repetitive elements such as transposons and retroviruses, and for epigenetic

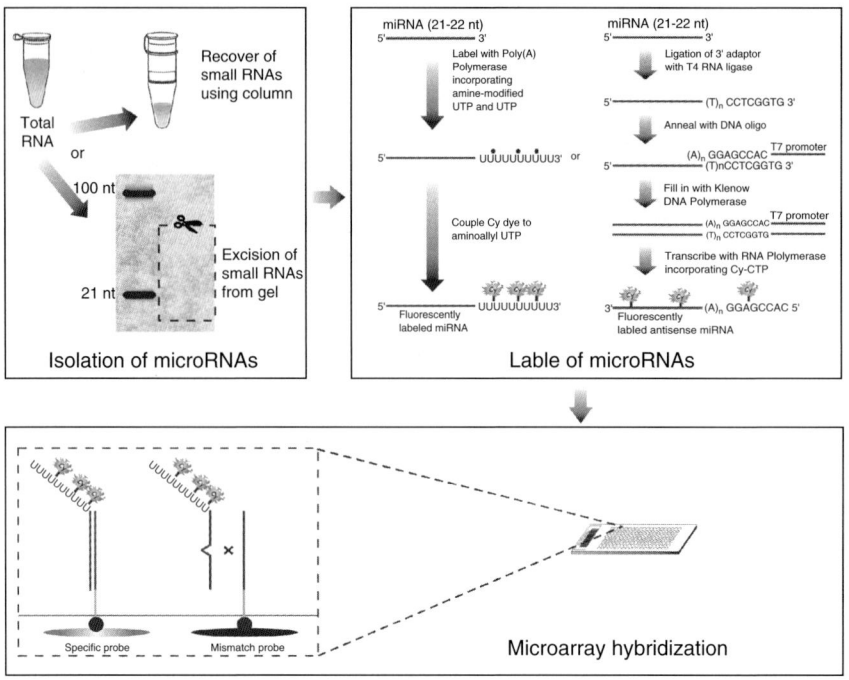

Figure 15. Detection of microRNAs using microarrays. Size fractions of total RNA containing microRNAs can be isolated by excision from poly-acrylamide gel or by size-fractioning using commercial column-purification kits (*left upper panel*). Right upper panel shows two types of labeling. On the left, Poly(A) polymerase and a mixture of unmodified and amine-modified nucleotides are used to append a poly-nucleotide tail to the 3'-end of each miRNA. The amine-modified miRNAs are then cleaned up and coupled to NHS-ester modified Cy5 or Cy3 dyes. On the right, mature microRNAs are coupled to 3'-adapter containing a T7 RNA polymerase promoter. Labeled microRNAs are produced by *in vitro* transcription using T7 RNA Polymerase and fluorescent dyes. The lower panel shows the hybridization of labeled microRNAs to oligonucleotide probes. The specificity of these probes is guaranteed by no hybridization of microRNAs to mismatch probes.

regulation of endogenous genes, although the extent to which this DNA modification functions to regulate the genome is largely unknown. Aberrant DNA methylation may cause silencing of tumor suppressor genes and promote chromosomal instability in human cancers. Thus, accurate determination of cytosine methylation status in CpG dinucleotides placed in promoter regions of cancer-related genes may provide diagnostic and prognostic value for human neoplasias.

Initial studies relied on array platforms generated by PCR amplification and interrogating a limited number (~3000) of promoter CpG islands in the form of PCR amplified DNA fragments [100, 101]. These arrays were hybridized to labeled probes generated from tumor and normal cells, previously enriched in methylated CpGs by means of digestion with methylation-sensitive restriction enzymes followed by PCR amplification (Figure 16) [101, 102]. Similar strategies

Concepts on microarray design for genome

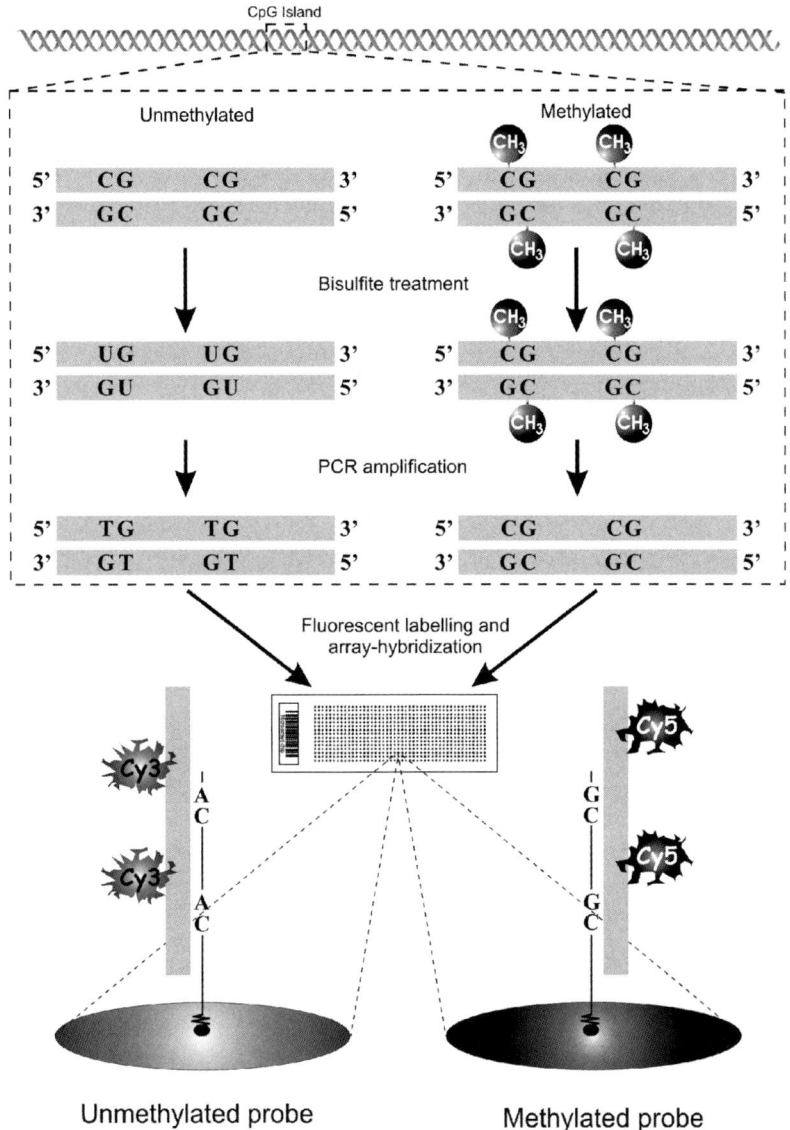

Figure 16. Restriction enzyme-based strategies for DNA methylation profiling. Methylation-sensitive restriction endonucleases and frequent cutter enzymes are used in the procedures. Adapter-specific aminoallyl-PCR's selectively enrich unmethylated (*left*) or hypermethylated (*right*) DNA fractions. DNA fragments are fluorescently labeled and hybridized to microarray which contains DNA spots representing CpG island sequences.

were developed for use of higher coverage platform such as whole-genome BAC spotted arrays [103, 104], and more recently, for the unbiased fine-mapping of methylation patterns of chromosomes 21 and 22 using tiling microarrays consisting of over 340,000 oligonucleotide probe pairs [105].

A different strategy for methylation analysis is based on the use of bisulfite oligonucleotide arrays (Figure 17). Sodium bisulfite treatment of DNA deaminates cytosine to uracil, but 5′-methylcytosine is protected. Unmethylated DNA that is treated with bisulfite contains uracil in place of cytosine and will hybridize relatively poorly to microarray oligonucleotides that contain guanines. Methylated cytosines in DNA sequences that cannot be changed by bisulfite

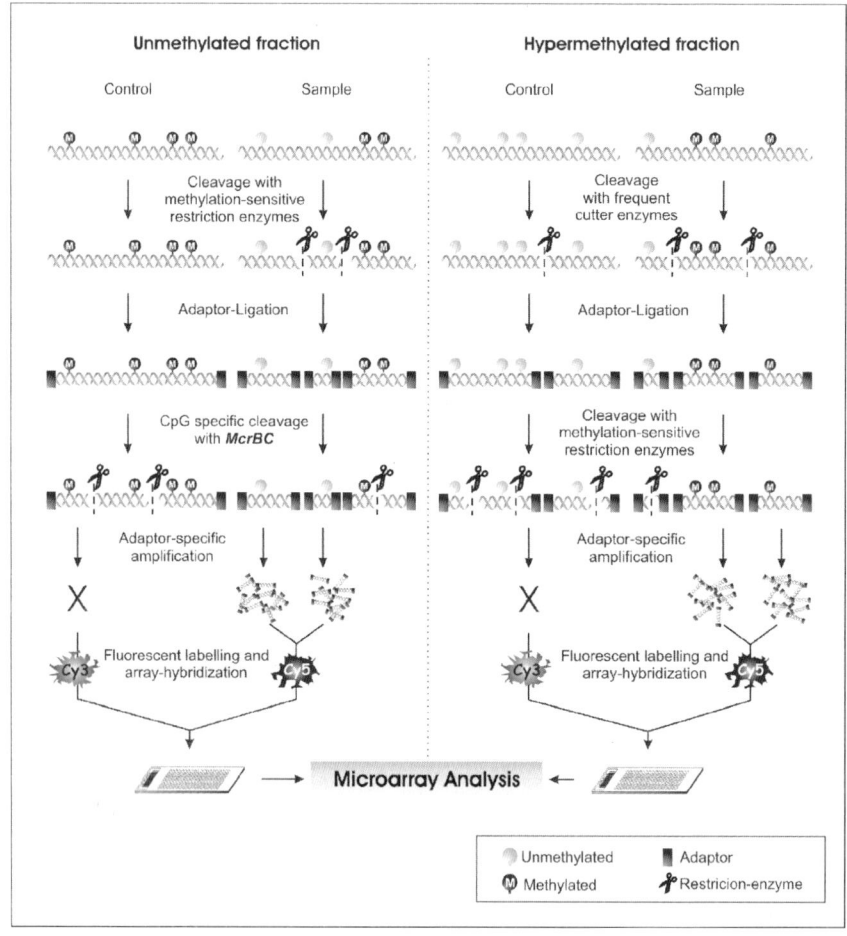

Figure 17. Bisulfite detection of methylation patterns. For each CpG island sequence, the microarray contains two oligonucleotide probes; one to detect the unmodified sequence and another to detect the bisulfite-altered sequence.

treatment will retain their ability to hybridize to the oligonucleotide arrays. Built on this principle, microarrays were designed comprising oligonucleotides corresponding to methylated and unmethylated versions of the sequence of the CpG islands of a given gene, thus allowing detection of the methylation status of multiple genes simultaneously (Figure 17). The methylated version of a probe set should differ by at least 3 nt (of 21 bp) from the unmethylated one, which will eliminate any possible cross-hybridization between methylated and unmethylated DNA of a given gene [106]. As a result, a much stronger hybridization signal should be detected for a methylated probe set if there is methylation and vice versa. Oligonucleotide arrays interrogating CpG islands (19-25 mer) may be synthesized *in situ* [107] or result from spotting of presynthesized oligonucleotides [106, 108–111]. Although informative and precise, the use of bisulfite oligoarrays may be limited in genome-wide studies due to (1) loss of probe specificity resulting from the degeneration of the code caused by the conversion of unmethylated cytosines, and (2) the difficulty to design suitable oligonucleotide probes that would exhibit similar T_m and hybridization behavior [105]. In a variation of this method, bisulfite-converted DNA from different samples can be deposited in a solid support and used to interrogate labeled synthetic probes for methylated or unmethylated versions of specific CpGs [112].

An additional approach for analysis of DNA methylation patterns involves methylated DNA immunoprecipitation (MeDIP) with an anti-methylcytosine antibody, followed by hybridization of the purified genomic fragments to whole-genome microarrays. This methodology was employed to generate methylation profiles from normal and tumor cells of all human chromosomes at 80 kb resolution and for a large set of CpG islands using a tiled whole human genome BAC array [113]. Because genome tiling arrays can be synthesized to contain relatively short (25 mer) oligonucleotides, they can potentially identify sites of DNA methylation with unparalleled precision, in some cases with single-nucleotide resolution. Recently, the MeDIP approach was used to generate the first comprehensive DNA methylation map of an entire genome (the plant *Arabidopsis thaliana*), at 35 bp resolution [114], and is also being used to investigate in the detail the methylation pattern of a region comprising 1% of the human genome using Affymetrix oligonucleotide tiling arrays [115]. As tiling arrays are universal platforms they are ideal for detecting correlations between DNA methylation and transcriptome mapping on a genome-wide scale.

3.7 ChIP-Chip

Chromatin immunoprecipitation (ChIP) coupled to hybridization onto DNA microarrays (ChIP-chip) is becoming a popular approach to investigate interactions between proteins and DNA that occur *in vivo*. In ChIP-chip experiments, cross-linked chromatin–protein complexes are extracted from a cell or tissue of interest and the DNA sheared, typically by sonication, down to relatively short (<1 kb) fragments (Figure 18). DNA fragments cross-linked to the protein of

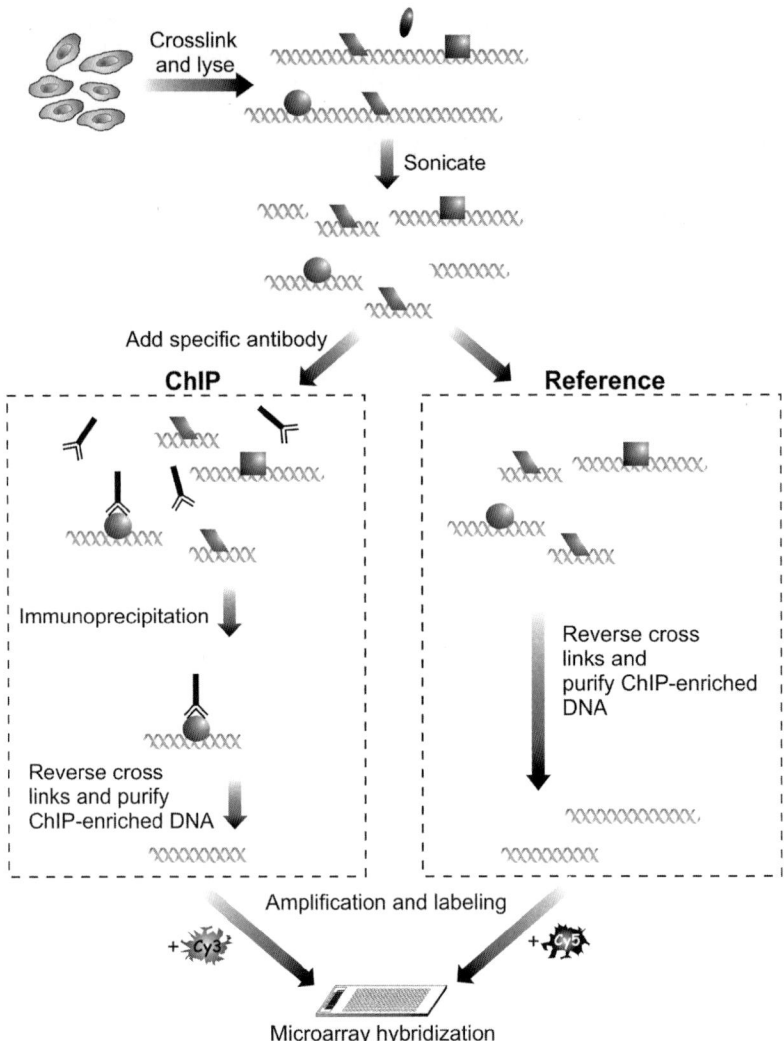

Figure 18. Schematic view of ChIP-chip procedure. Formaldehyde is used to form DNA–protein cross-links. After lyses, the extract is sonicated to shear the DNA fragments to the desired size, usually 1 kb or smaller. DNA fragments cross-linked to the protein of interest are enriched by chromatin immunoprecipitation (ChIP) using an antibody specific to that protein. Formaldehyde cross-links are then reversed and the DNA is purified. Enriched DNA is labeled with a fluorescent molecule such as Cy5 or Cy3. Genomic DNA prepared from IP input extract is generally used as a reference and similarly amplified and labeled with a different fluorescent molecule. The two probes are then combined and hybridized to the DNA microarray which contains elements that represent the entire genome.

Concepts on microarray design for genome 295

interest are enriched by immunoprecipitation with a protein-specific antibody, the formaldehyde cross-links are reversed and DNA is purified. Amplification of immunoprecipitated DNA is usually required for microarray-based detection. Amplified immunoprecipitated DNA is labeled with a fluorescent molecule and hybridized to DNA microarrays, along with a reference (usually an aliquot of gDNA used as input for immunoprecipitation reactions) labeled with a different fluorophore (Figure 18) [49].

Different array platforms have been used in ChIP-chip experiments: spotted double-strand cDNA/DNA arrays, spotted oligonucleotide arrays, and *in situ* synthesized oligonucleotide arrays. Initial studies in yeast using arrays comprised of PCR fragments spanning the whole yeast genome identified binding sites for individual transcription factors and of protein complexes related to DNA replication, recombination, and chromatin structure [49]. ChIP-chip studies in mammalian genomes have utilized different types of PCR amplicon arrays, including arrays tiling a specific genomic region of interest, mainly CpG island arrays and promoter arrays [116]. Recently, an array comprising PCR fragment probes for 1% of the nonrepetitive complement of the human genome sequence (ENCODE array) was used to identify functional promoters in an unbiased fashion [117]. Arrays comprising of tiling DNA fragments (either PCR fragments or oligonucleotides) were devised as a way to increase the resolution of ChIP-chip experiments. The advent of commercially available whole-genome high-density oligonucleotide tiling arrays (Affymetrix, Agilent, NimbleGen) represented an additional gain in resolution and also eliminated the problems associated to PCR manipulations and mechanical spotting by relying instead on *in situ* oligonucleotide synthesis. The high-coverage of the whole-genome tiling arrays has paved the way for the unbiased mapping of DNA-interacting protein factors. As an example, a study using Affymetrix arrays representing essentially all nonrepetitive sequences on human chromosomes 21 and 22 found that most binding sites for the transcriptional factors Sp1, cMyc, and p53 were located far from the transcription start sites of known protein-coding genes [118]. It should be noted that the optimal length of arrayed fragments is a balance between the cost of having many elements and the desire for increased resolution, keeping in mind that arrayed elements shorter than the average size of a sheared chromatin fragment (generally 500–1000 bp) will not increase resolution [49]. A comprehensive comparison of using PCR spotted arrays and long- and short-oligonucleotide arrays for ChIP-chip experiments is not available and therefore, the best array platform for ChIP-chip experiments is not yet established.

3.8 Genotyping

SNPs are the most abundant form of genetic variation in the human genome, with estimates of more than 10 million common SNPs [119, 120]. These single nucleotide changes in human genes may cause genetic disorders and could provide important help for explaining, e.g. disease susceptibility and cancer predisposition [121]. Therefore, it is clear that the accurate and robust detection of such SNPs plays a central role in the field of DNA diagnostics [122]. Microarray-based genotyping assays have been used in genome-wide linkage analysis of SNP markers associated to several diseases, such as prostate cancer [123], rheumatoid arthritis [124], and systemic lupus erythematosus [125].

Among the numerous methods for analyzing genomic variations, microarrays are one of the most powerful tools for high-throughput SNP genotyping. Genotyping platforms, released by Affymetrix, can interrogate up to 100,000 SNPs in parallel. An alternative technology, BeadArrays, developed by Illumina (San Diego, California) and not discussed in detail here, is particularly powerful for genotyping up to 500,000 SNPs in parallel. Essentially, the method involves (a) whole-genome amplification (WGA) to generate large amounts of amplified complex gDNA. (b) Hybridization of the WGA product to a specific and sensitive oligonucleotide probe array (50 mers). (c) An array-based allele-specific primer extension (ASPE) reaction that scores the captured SNP targets by incorporating multiple biotin-labeled dNTP nucleotides into the appropriate allelic probe, followed by a sensitive detection and signal amplification step to read the incorporated labels [120].

Microarray-based systems use different molecular strategies for distinction between SNP alleles. The robustness of these multiplexed systems is determined by the reaction principles applied for SNP allele distinction and the microarray formats used. The major reaction principles – allele-specific oligonucleotide (ASO) hybridization, ASPE, and single-base extension (SBE) [126] – can be applied using different combinations of PCR strategies, array types, SNP selection, and labeling protocols and will not be discussed in this chapter [119]. However, a brief example of each reaction principle is presented. GeneChip assays (Affymetrix) use the difference in thermal stability between a perfectly matched and mismatched ASO probe and its DNA target to distinguish between the SNP alleles [127]. Systems that use ASPE (Figure 19) or SBE [128] reaction principles are enzyme-assisted and provide a highly specific SNP genotyping. This is due to the high accuracy of nucleotide incorporation by DNA polymerase or the high specificity of DNA ligase in joining two adjacent and perfectly matched DNA strands. In SBE–TAG systems [128, 129], a hybrid oligonucleotide primer containing a generic sequence tag followed by a locus-specific sequence is hybridized adjacent to the SNP and extended with fluorescent dideoxynucleotides. Multiple SBE reactions are performed in solution with each SBE primer marked by a different unique sequence tag. The multiplex reaction is analyzed after hybridization to a generic tag array, which is generated by spotting the reverse complements of the sequence tags onto a glass microscope slide. Genotyping method using ASPE systems [120, 130] is briefly described in the legend to Figure 19.

Concepts on microarray design for genome

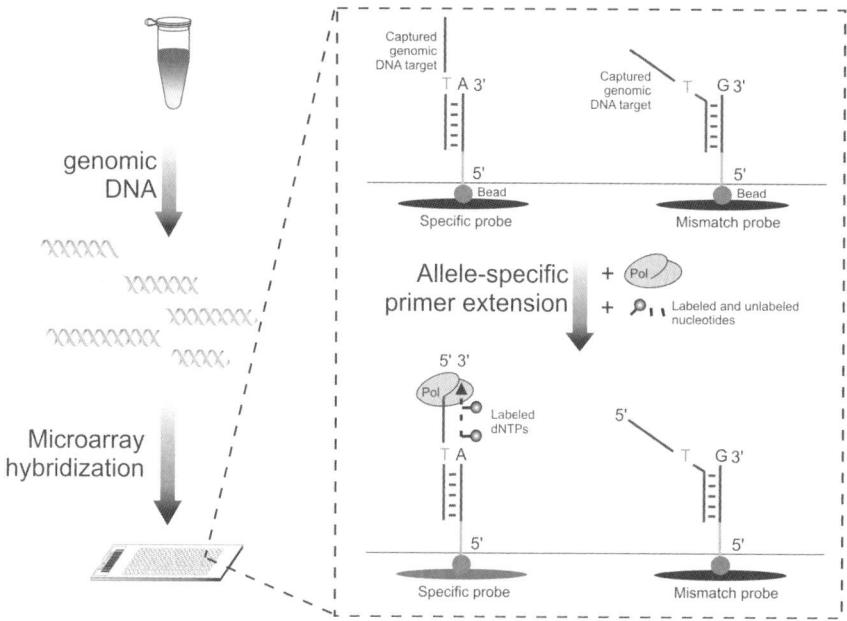

Figure 19. Genotyping on DNA microarrays. Amplification of genomic DNA (gDNA) generates fragments that are hybridized to specific and sensitive oligonucleotide probes on microarray. An allele-specific primer extension (ASPE) reaction scores the captured SNP targets by incorporating multiple biotin-labeled dNTP nucleotides into the appropriate allelic probe. For a given single nucleotide polymorphism (SNP) on a give strand, two or more different allele-specific oligonucleotide probes are designed to capture different SNPs, since polymerase extension occurs preferentially from matched 3'-termini, enabling appropriate scoring of the SNP [120].

3.9 Intronic Transcription

A detailed analysis of gene content and structure arising from the whole human genome sequencing revealed that introns comprise on average 95% of the protein-coding genes and about 30% of the human genome [131, 132]. Experimental analysis using genome tiling arrays of chromosomes 21 and 22 has permitted an unbiased probing of transcribed regions in the genome and revealed 5.3 kb of novel transcribed sequences within or overlapping intronic regions of well characterized genes, of which 2.7 kb (51%) are antisense to protein-coding genes [51]. In addition, tiling arrays of the whole human genome have permitted to extend these analyses and detected expressed messages in liver mapping to 1529 and 1566 novel transcriptionally active intronic regions, respectively arising from either the antisense or the sense strands of the corresponding gene [13]. Microarray-based evidence of ubiquitous transcriptional activity streaming from intronic genomic segments was also reported for *Drosophila melanogaster* [133]. In that work, it was observed that 41% of probes representing the full complement of intronic and intergenic regions of the *D. melanogaster* genome are transcriptionally active [133]. The expressed intronic and intergenic sequences are more likely to be evolutionarily conserved than nonexpressed ones, and about 15% of them appear to be developmentally regulated [133]. Ubiquitous intronic transcription has been confirmed by other experimental approaches, including mapping 3'-ends of transcripts with SAGE [134], 5'-ends using cap analysis of gene expression (CAGE) [135], MPSS [136], and high-throughput full-length cDNA cloning and sequencing [137]. It has become apparent that introns, as well as intergenic regions constitute major sources of non-protein-coding RNAs [138], and experiments with genomic tiling arrays or RACE have shown that intronic RNAs are long (400–2000 nt) transcripts [40, 139, 140].

To investigate the expression level of intronic messages in human tissues, Reis et al. [40] selected for microarray experiments a subset of ~1000 totally intronic EST clusters identified by informatics analysis, along with an additional 2000 clusters from exonic segments of known genes. A representative cDNA clone from each selected cluster was used for construction of spotted cDNA microarrays enriched in intronic transcripts. Hybridization of these intronic microarrays with 27 prostate tumor samples and corresponding adjacent normal tissue revealed that in prostate, the fraction of expressed messages arising from exonic or intronic transcripts were similar [40]. Moreover, the expression levels of 23 intronic noncoding transcripts are significantly correlated (p-value < 0.001) to the degree of prostate tumor differentiation (Figure 20). Intronic *RASSF1*, the most correlated gene, is expressed both as sense and antisense messages as shown by strand-specific reverse transcriptase assay. In addition to antisense *RASSF1*, a number of antisense intronic messages significantly correlated to the degree of prostate tumor differentiation were shown by RACE–PCR and strand-specific Northern blotting to be long (0.6–1.1 kb) and unspliced [40].

Reis et al. [40] provide the first report implicating a large set of natural intronic antisense transcripts in a human disease. It is not yet a settled issue if in tumors the expression of intronic antisense RNAs is a true mechanism of regulation or just reflects errors in promoter recognition/transcription initiation. Nevertheless, discovery of intronic antisense messages correlated to the degree of malignancy in prostate tumors [40] has an impact on the molecular diagnosis of cancer in general, arguing for inclusion of noncoding intronic transcripts into the arsenal of tools used for molecular diagnostics, so far almost exclusively populated by microarrays that contain only exonic protein-coding transcripts.

An intronic array may be a practical and effective compromise between a biased array that only probes the protein-coding messages, and whole human genome tiling arrays. Intronic arrays may facilitate a comparative analysis of intronic and protein-coding exonic transcription under a different number of physiological and pathological conditions. This approach should advance current knowledge about the diverse biological roles of these noncoding RNAs, which are likely involved in the control of gene expression [141].

Concepts on microarray design for genome

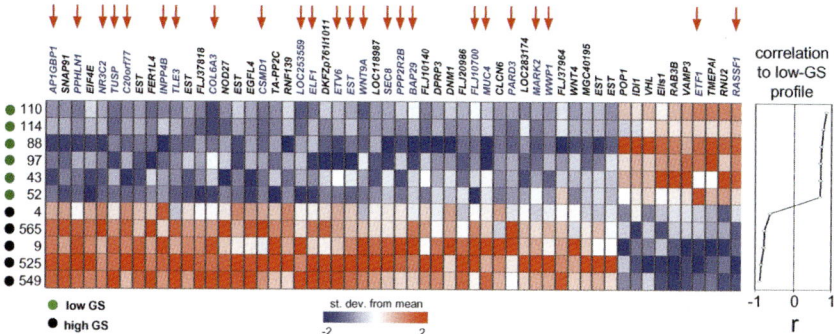

Figure 20. Intronic transcripts correlated to the degree of prostate tumor malignancy. The expression profile of 27 prostate patient tumor samples with different degrees of tumor malignancy (Gleason score, GS) was analyzed [40]. Pearson correlation and bootstrap resampling were used to identify messages significantly correlated to GS among the low-GS (scores 5 and 6) and high-GS (scores 9 and 10) samples. Left panel shows the expression matrix of 56 genes significantly correlated to the low-/high-GS class distinction (*p*-value < 0.001). Genes (columns) are ordered by their Pearson correlation and patients (rows) by their correlation (r) to the low-GS profile, which is shown on the right panel. Red arrows point to intronic gene fragments, i.e. messages transcribed from the intronic regions of the protein-coding genes whose names are indicated. Expression level of each gene is represented by the number of standard deviations above (red) or below (blue) the average value for that gene across all samples.

We believe that an array that samples the intronic noncoding regions of the genome from which there is previous evidence of transcription, along with the corresponding protein-coding regions, will help to identify the gene families, biological processes or functional gene categories of greatest relevance to intronic gene expression under diverse physiological conditions. High-density tiling arrays of selected chromosome regions containing these relevant genes should permit further detailed studies of intronic expression patterns. The information gathered from such a combined approach should help accelerate the acquisition of information about the emerging diverse roles of intronic messages.

4. CONCLUSIONS

Industry-quality high-density oligoarrays have paved the way for large-scale analyses aimed at a comprehensive evaluation of the functional genome of eukaryotes. This will be an important tool for the detailed characterization of yet unappreciated aspects of biology such as epigenomic regulation of chromatin, chromosomal stability, and gene transcription during development as well as in diseased states. Also, the analysis of SNPs in a multiplexed, genome-wide scale using genotyping arrays opens unprecedented opportunities for association studies and linkage disequilibrium analyses in man. Whole human genome tiling arrays are already commercially available, however multiple arrays are needed to cover all chromosomes at high resolution and this platform is still very expensive, making it not accessible to most researchers. Moreover, the enormous amount of data generated in these experiments can not be efficiently processed and interpreted. For transcriptome studies, a more practical and cost-effective alternative is the design of gene-oriented tiling oligoarrays interrogating genomic sequences that span transcriptionally active exonic and intronic regions relative to known gene *loci*. Active regions can be determined by compiling the existing experimental evidence already available at transcriptome databases (mRNA, EST, MPSS, SAGE, etc.). This approach would narrow the genomic space that is interrogated by such gene-oriented tiling arrays, focusing on the most relevant regions for the study of gene expression.

REFERENCES

1. Schena M, Shalon D, Davis RW, Brown PO (1995) Quantitative monitoring of gene expression patterns with a complementary DNA microarray. Science 270:467–470
2. MacBeath G, Schreiber SL (2000) Printing proteins as microarrays for high-throughput function determination. Science 289:1760–1763
3. Fang Y, Frutos AG, Lahiri J (2002) Membrane protein microarrays. J Am Chem Soc 124:2394–2395
4. Min DH, Mrksich M (2004) Peptide arrays: towards routine implementation. Curr Opin Chem Biol 8:554–558
5. Shin I, Cho JW, Boo DW (2004) Carbohydrate arrays for functional studies of carbohydrates. Comb Chem High Throughput Screen 7:565–574
6. MacBeath G, Koehler AN, Schreiber SL (1999) Printing small molecules as microarrays and detecting protein-ligand interactions en masse. J Am Chem Soc 121:7967–7968
7. Kononen J, Bubendorf L, Kallioniemi A, Barlund M, Schraml P, Leighton S, Torhorst J, Mihatsch MJ, Sauter G, Kallioniemi OP (1998) Tissue microarrays for high-throughput molecular profiling of tumor specimens. Nat Med 4:844–847
8. Ziauddin J, Sabatini DM (2001) Microarrays of cells expressing defined cDNAs. Nature 411:107–110
9. Lipshutz RJ, Fodor SP, Gingeras TR, Lockhart DJ (1999) High density synthetic oligonucleotide arrays. Nat Genet 21:20–24
10. Duggan DJ, Bittner M, Chen YD, Meltzer P, Trent JM (1999) Expression profiling using cDNA microarrays. Nat Genet 21:10–14
11. Wick I, Hardiman G (2005) Biochip platforms as functional genomics tools for drug discovery. Curr Opin Drug Discov Devel 8:347–354
12. Sievertzon M, Nilsson P, Lundeberg J (2006) Improving reliability and performance of DNA microarrays. Expert Rev Mol Diagn 6:481–492
13. Bertone P, Stolc V, Royce TE, Rozowsky JS, Urban AE, Zhu X, Rinn JL, Tongprasit W, Samanta M, Weissman S, et al. (2004) Global identification of human transcribed sequences with genome tiling arrays. Science 306:2242–2246
14. Cheng J, Kapranov P, Drenkow J, Dike S, Brubaker S, Patel S, Long J, Stern D, Tammana H, Helt G, et al. (2005) Transcriptional maps of 10 human chromosomes at 5-nucleotide resolution. Science 308:1149–1154
15. Hessner MJ, Meyer L, Tackes J, Muheisen S, Wang XJ (2004) Immobilized probe and glass surface chemistry as variables in microarray fabrication. BMC Genomics 5:53
16. Taylor S, Smith S, Windle B, Guiseppi-Elie A (2003) Impact of surface chemistry and blocking strategies on DNA microarrays. Nucleic Acids Res 31(16):e87

17. Bertone P, Trifonov V, Rozowsky JS, Schubert F, Emanuelsson O, Karro J, Kao MY, Snyder M, Gerstein M (2006) Design optimization methods for genomic DNA tiling arrays. Genome Res 16:271–281
18. Pinkel D, Segraves R, Sudar D, Clark S, Poole I, Kowbel D, Collins C, Kuo WL, Chen C, Zhai Y, et al. (1998) High resolution analysis of DNA copy number variation using comparative genomic hybridization to microarrays. Nat Genet 20:207–211
19. Telenius H, Carter NP, Bebb CE, Nordenskjold M, Ponder BAJ, Tunnacliffe A (1992) Degenerate oligonucleotide-primed PCR – general amplification of target DNA by a single degenerate primer. Genomics 13:718–725
20. Watson SK, deLeeuw RJ, Ishkanian AS, Malloff CA, Lam WL (2004) Methods for high throughput validation of amplified fragment pools of BAC DNA for constructing high resolution CGH arrays. BMC Genomics 5(6):1–8
21. Lockwood WW, Chari R, Chi B, Lam WL (2006) Recent advances in array comparative genomic hybridization technologies and their applications in human genetics. Eur J Hum Genet 14:139–148
22. Shalon D, Smith SJ, Brown PO (1996) A DNA microarray system for analyzing complex DNA samples using two-color fluorescent probe hybridization. Genome Res 6:639–645
23. Stoughton RB (2005) Applications of DNA microarrays in biology. Annu Rev Biochem 74:53–82
24. Dorris DR, Nguyen A, Gieser L, Lockner R, Lublinsky A, Patterson M, Touma E, Sendera TJ, Elghanian R, Mazumder A (2003) Oligodeoxyribonucleotide probe accessibility on a three-dimensional DNA microarray surface and the effect of hybridization time on the accuracy of expression ratios. BMC Biotechnol 3:6
25. Stafford P, Sendera T, Kaysser-Kranich T, Palaniappan C (2003) High-quality microarray data using CodeLink Bioarray Platform. Life Sci News 13:1–5
26. Barczak A, Rodriguez MW, Hanspers K, Koth LL, Tai YC, Bolstad BM, Speed TP, Erle DJ (2003) Spotted long oligonucleotide arrays for human gene expression analysis. Genome Res 13:1775–1785
27. Tan PK, Downey TJ, Spitznagel EL, Xu P, Fu D, Dimitrov DS, Lempicki RA, Raaka BM, Cam MC (2003) Evaluation of gene expression measurements from commercial microarray platforms. Nucleic Acids Res 31:5676–5684
28. Yauk CL, Berndt ML, Williams A, Douglas GR (2004) Comprehensive comparison of six microarray technologies. Nucleic Acids Res 32(15):e124
29. de Reynies A, Geromin D, Cayuela JM, Petel F, Dessen P, Sigaux F, Rickman DS (2006) Comparison of the latest commercial short and long oligonucleotide microarray technologies. BMC Genomics 7:51
30. Larkin JE, Frank BC, Gavras H, Sultana R, Quackenbush J (2005) Independence and reproducibility across microarray platforms. Nat Methods 2:337–343
31. Irizarry RA, Warren D, Spencer F, Kim IF, Biswal S, Frank BC, Gabrielson E, Garcia JGN, Geoghegan J, Germino G, et al. (2005) Multiple-laboratory comparison of microarray platforms. Nat Methods 2:345–349
32. Petersen D, Chandramouli GVR, Geoghegan J, Hilburn J, Paarlberg J, Kim CH, Munroe D, Gangi L, Han J, Puri R, et al. (2005) Three microarray platforms: an analysis of their concordance in profiling gene expression. BMC Genomics 6:63
33. Knight J (2001) When the chips are down. Nature 410:860–861
34. Singh-Gasson S, Green RD, Yue YJ, Nelson C, Blattner F, Sussman MR, Cerrina F (1999) Maskless fabrication of light-directed oligonucleotide microarrays using a digital micromirror array. Nat Biotechnol 17:974–978
35. Fodor SPA, Read JL, Pirrung MC, Stryer L, Lu AT, Solas D (1991) Light-directed, spatially addressable parallel chemical synthesis. Science 251:767–773
36. Hughes TR, Mao M, Jones AR, Burchard J, Marton MJ, Shannon KW, Lefkowitz SM, Ziman M, Schelter JM, Meyer MR, et al. (2001) Expression profiling using microarrays fabricated by an ink-jet oligonucleotide synthesizer. Nat Biotechnol 19:342–347

37. Kane MD, Jatkoe TA, Stumpf CR, Lu J, Thomas JD, Madore SJ (2000) Assessment of the sensitivity and specificity of oligonucleotide (50 mer) microarrays. Nucleic Acids Res 28:4552–4557
38. Hardiman G (2004): Microarray platforms – comparisons and contrasts. Pharmacogenomics 5:487–502
39. Su AI, Wiltshire T, Batalov S, Lapp H, Ching KA, Block D, Zhang J, Soden R, Hayakawa M, Kreiman G, et al. (2004) A gene atlas of the mouse and human protein-encoding transcriptomes. Pro Natl Acad Sci USA 101:6062–6067
40. Reis EM, Nakaya HI, Louro R, Canavez FC, Flatschart AV, Almeida GT, Egidio CM, Paquola AC, Machado AA, Festa F, et al. (2004) Antisense intronic non-coding RNA levels correlate to the degree of tumor differentiation in prostate cancer. Oncogene 23:6684–6692
41. Shingara J, Keiger K, Shelton J, Laosinchai-Wolf W, Powers P, Conrad R, Brown D, Labourier E (2005) An optimized isolation and labeling platform for accurate microRNA expression profiling. RNA – A Publication of the RNA Society 11:1461–1470
42. Johnson JM, Castle J, Garrett-Engele P, Kan Z, Loerch PM, Armour CD, Santos R, Schadt EE, Stoughton R, Shoemaker DD (2003) Genome-wide survey of human alternative pre-mRNA splicing with exon junction microarrays. Science 302:2141–2144
43. Conrad DF, Andrews TD, Carter NP, Hurles ME, Pritchard JK (2006) A high-resolution survey of deletion polymorphism in the human genome. Nat Genet 38:75–81
44. Fraga ME, Esteller M (2002) DNA methylation: a profile of methods and applications. Biotechniques 33:632–649
45. Schubeler D, Scalzo D, Kooperberg C, van Steensel B, Delrow J, Groudine M (2002) Genome-wide DNA replication profile for *Drosophila melanogaster*: a link between transcription and replication timing. Nat Genet 32:438–442
46. van Steensel B, Henikoff S (2003) Epigenomic profiling using microarrays. Biotechniques 35:346–350, 352–354, 356–357
47. Ren B, Cam H, Takahashi Y, Volkert T, Terragni J, Young RA, Dynlacht BD (2002) E2F integrates cell cycle progression with DNA repair, replication, and G(2)/M checkpoints. Genes Dev 16:245–256
48. Kapranov P, Cawley SE, Drenkow J, Bekiranov S, Strausberg RL, Fodor SP, Gingeras TR (2002) Large-scale transcriptional activity in chromosomes 21 and 22. Science 296:916–919
49. Buck MJ, Lieb JD (2004) ChIP-chip: considerations for the design, analysis, and application of genome-wide chromatin immunoprecipitation experiments. Genomics 83:349–360
50. Lippman Z, Gendrel AV, Black M, Vaughn MW, Dedhia N, McCombie WR, Lavine K, Mittal V, May B, Kasschau KD, et al. (2004) Role of transposable elements in heterochromatin and epigenetic control. Nature 430:471–476
51. Kampa D, Cheng J, Kapranov P, Yamanaka M, Brubaker S, Cawley S, Drenkow J, Piccolboni A, Bekiranov S, Helt G, et al. (2004) Novel RNAs Identified From an In-Depth Analysis of the Transcriptome of Human Chromosomes 21 and 22. Genome Res 14:331–342
52. Ishkanian AS, Malloff CA, Watson SK, deLeeuw RJ, Chi B, Coe BP, Snijders A, Albertson DG, Pinkel D, Marra MA, et al. (2004) A tiling resolution DNA microarray with complete coverage of the human genome. Nat Genet 36:299–303
53. Royce TE, Rozowsky JS, Bertone P, Samanta M, Stolc V, Weissman S, Snyder M, Gerstein M (2005) Issues in the analysis of oligonucleotide tiling microarrays for transcript mapping. Trends Genet 21:466–475
54. Adams MD, Soares MB, Kerlavage AR, Fields C, Venter JC (1993) Rapid cDNA sequencing (expressed sequence tags) from a directionally cloned human infant brain cDNA library. Nat Genet 4:373–380
55. Velculescu VE, Zhang L, Vogelstein B, Kinzler KW (1995) Serial analysis of gene expression. Science 270:484–487
56. Brenner S, Johnson M, Bridgham J, Golda G, Lloyd DH, Johnson D, Luo S, McCurdy S, Foy M, Ewan M, et al. (2000) Gene expression analysis by massively parallel signature sequencing (MPSS) on microbead arrays. Nat Biotechnol 18:630–634

57. Vilo J, Kivinen K (2001) Regulatory sequence analysis: application to the interpretation of gene expression. Eur Neuropsychopharmacology 11:399–411
58. van Someren EP, Wessels LFA, Backer E, Reinders MJT (2002) Genetic network modeling. Pharmacogenomics 3:507–525
59. Dudley AM, Aach J, Steffen MA, Church GM (2002) Measuring absolute expression with microarrays with a calibrated reference sample and an extended signal intensity range. Proc Natl Acad Sci USA 99:7554–7559
60. Carter MG, Sharov AA, VanBuren V, Dudekula DB, Carmack CE, Nelson C, Ko MSH (2005) Transcript copy number estimation using a mouse whole-genome oligonucleotide microarray. Genome Biol 6(7):R61
61. Anderson L, Seilhamer J (1997) A comparison of selected mRNA and protein abundances in human liver. Electrophoresis 18:533–537
62. Vrana KE, Freeman WM, Aschner M (2003) Use of microarray technologies in toxicology research. Neurotoxicology 24:321–332
63. Greshock J, Naylor TL, Margolin A, Diskin S, Cleaver SH, Futreal PA, deJong PJ, Zhao SY, Liebman M, Weber BL (2004) 1-Mb resolution array-based comparative genomic hybridization using a BAC clone set optimized for cancer gene analysis. Genome Res 14:179–187
64. Mantripragada KK, Buckley PG, de Stahl TD, Dumanski JP (2004) Genomic microarrays in the spotlight. Trends Genet 20:87–94
65. Ylstra B, van den IJssel P, Carvalho B, Brakenhoff RH, Meijer GA (2006) BAC to the future! or oligonucleotides: a perspective for micro array comparative genomic hybridization (array CGH). Nucleic Acids Res 34:445–450
66. Davies JJ, Wilson IM, Lam WL (2005) Array CGH technologies and their applications to cancer genomes. Chromosome Res 13:423–423
67. Black DL (2003) Mechanisms of alternative pre-messenger RNA splicing. Annu Rev Biochem 72:291–336
68. Thanaraj TA, Clark F, Muilu J (2003) Conservation of human alternative splice events in mouse. Nucleic Acids Res 31:2544–2552
69. Modrek B, Lee CJ (2003) Alternative splicing in the human, mouse and rat genomes is associated with an increased frequency of exon creation and/or loss. Nat Genet 34:177–180
70. Pan Q, Shai O, Misquitta C, Zhang W, Saltzman AL, Mohammad N, Babak T, Siu H, Hughes TR, Morris QD, et al. (2004) Revealing global regulatory features of mammalian alternative splicing using a quantitative microarray platform. Mol Cell 16:929–941
71. Le K, Mitsouras K, Roy M, Wang Q, Xu Q, Nelson SF, Lee C (2004) Detecting tissue-specific regulation of alternative splicing as a qualitative change in microarray data. Nucleic Acids Res 32:e180
72. Blanchette M, Green RE, Brenner SE, Rio DC (2005) Global analysis of positive and negative pre-mRNA splicing regulators in Drosophila. Genes Dev 19:1306–1314
73. Ule J, Darnell RB (2006) RNA binding proteins and the regulation of neuronal synaptic plasticity. Curr Opin Neurobiol 16:102–110
74. Sugnet CW, Srinivasan K, Clark TA, O'Brien G, Cline MS, Wang H, Williams A, Kulp D, Blume JE, Haussler D, Ares M (2006) Unusual intron conservation near tissue-regulated exons found by splicing microarrays. PLoS Comput Biol 2:22–35
75. Frey BJ, Mohammad N, Morris QD, Zhang W, Robinson MD, Mnaimneh S, Chang R, Pan Q, Sat E, Rossant J, et al. (2005) Genome-wide analysis of mouse transcripts using exon microarrays and factor graphs. Nat Genet 37:991–996
76. Clark TA, Sugnet CW, Ares M (2002) Genomewide analysis of mRNA processing in yeast using splicing-specific microarrays. Science 296:907–910
77. Li CX, Kato M, Shiue L, Shively JE, Ares M, Lin RJ (2006) Cell type and culture condition-dependent alternative splicing in human breast cancer cells revealed by splicing-sensitive microarrays. Cancer Res 66:1990–1999
78. Castle J, Garrett-Engele P, Armour CD, Duenwald SJ, Loerch PM, Meyer MR, Schadt EE, Stoughton R, Parrish ML, Shoemaker DD, Johnson JM (2003): Optimization of oligonucleotide

arrays and RNA amplification protocols for analysis of transcript structure and alternative splicing. Genome Biol 4:R66
79. Yeakley JM, Fan JB, Doucet D, Luo L, Wickham E, Ye Z, Chee MS, Fu XD (2002) Profiling alternative splicing on fiber-optic arrays. Nat Biotechnol 20:353–358
80. Li HR, Wang-Rodriguez J, Nair TM, Yeakley JM, Kwon YS, Bibikova M, Zheng C, Zhou LX, Zhang K, Downs T, et al. (2006) Two-dimensional transcriptome profiling: Identification of messenger RNA isoform signatures in prostate cancer from archived paraffin-embedded cancer specimens. Cancer Res 66:4079–4088
81. Zhang CL, Li HR, Fan JB, Wang-Rodriguez J, Downs T, Fu XD, Zhang MQ (2006) Profiling alternatively spliced mRNA isoforms for prostate cancer classification. BMC Bioinformatics 7:202
82. Fehlbaum P, Guihal C, Bracco L, Cochet O (2002) A microarray configuration to quantify expression levels and relative abundance of splice variants. Nucleic Acids Res 33(5):e47
83. Srinivasan K, Shiue L, Hayes JD, Centers R, Fitzwater S, Loewen R, Edmondson LR, Bryant J, Smith M, Rommelfanger C, et al. (2005) Detection and measurement of alternative splicing using splicing-sensitive microarrays. Methods 37:345–359
84. Mockler TC, Chan S, Sundaresan A, Chen HM, Jacobsen SE, Ecker JR (2005) Applications of DNA tiling arrays for whole-genome analysis. Genomics 85:655–655
85. Vanhee-Brossollet C, Vaquero C (1998) Do natural antisense transcripts make sense in eukaryotes? Gene 211:1–9
86. Lavorgna G, Dahary D, Lehner B, Sorek R, Sanderson CM, Casari G (2004) In search of antisense. Trends Biochem Sci 29:88–94
87. Lagos-Quintana M, Rauhut R, Meyer J, Borkhardt A, Tuschl T (2003) New microRNAs from mouse and human. RNA – A Publication of the RNA Society 9:175–179
88. Bartel DP (2004) MicroRNAs: Genomics, biogenesis, mechanism, and function. Cell 116:281–297
89. Michael MZ, O'Connor SM, Pellekaan NGV, Young GP, James RJ (2003) Reduced accumulation of specific microRNAs in colorectal neoplasia. Mol Cancer Res 1:882–891
90. Calin GA, Dumitru CD, Shimizu M, Bichi R, Zupo S, Noch E, Aldler H, Rattan S, Keating M, Rai K, et al. (2002) Frequent deletions and down-regulation of micro-RNA genes miR15 and miR16 at 13q14 in chronic lymphocytic leukemia. Proc Natl Acad Sci USA 99:15524–15529
91. Eis PS, Tam W, Sun LP, Chadburn A, Li ZD, Gomez MF, Lund E, Dahlberg JE (2005) Accumulation of miR-155 and BIC RNA in human B cell lymphomas. Proc Natl Acad Sci USA 102:3627–3632
92. Johnson SM, Grosshans H, Shingara J, Byrom M, Jarvis R, Cheng A, Labourier E, Reinert KL, Brown D, Slack FJ (2005) RAS is regulated by the let-7 MicroRNA family. Cell 120:635–647
93. Liu CG, Calin GA, Meloon B, Gamliel N, Sevignani C, Ferracin M, Dumitru CD, Shimizu M, Zupo S, Dono M, et al. (2004) An oligonucleotide microchip for genome-wide microRNA profiling in human and mouse tissues. Proc Natl Acad Sci USA 101:9740–9744
94. Babak T, Zhang W, Morris Q, Blencowe BJ, Hughes TR (2004) Probing microRNAs with microarrays: tissue specificity and functional inference. RNA – A Publication of the RNA Society 10:1813–1819
95. Barad O, Meiri E, Avniel A, Aharonov R, Barzilai A, Bentwich I, Einav U, Glad S, Hurban P, Karov Y, et al. (2004) MicroRNA expression detected by oligonucleotide microarrays: system establishment and expression profiling in human tissues. Genome Res 14:2486–2494
96. Thomson JM, Parker J, Perou CM, Hammond SM (2004) A custom microarray platform for analysis of microRNA gene expression. Nat Methods 1:47–53
97. Sun Y, Koo S, White N, Peralta E, Esau C, Dean NM, Perera RJ (2004) Development of a micro-array to detect human and mouse microRNAs and characterization of expression in human organs. Nucleic Acids Res 32:e188

98. Miska EA, Alvarez-Saavedra E, Townsend M, Yoshii A, Sestan N, Rakic P, Constantine-Paton M, Horvitz HR (2004) Microarray analysis of microRNA expression in the developing mammalian brain. Genome Biol 5:R68
99. Calin GA, Liu CG, Sevignani C, Ferracin M, Felli N, Dumitru CD, Shimizu M, Cimmino A, Zupo S, Dono M, et al. (2004) MicroRNA profiling reveals distinct signatures in B cell chronic lymphocytic leukemias. Proc Natl Acad Sci USA 101:11755–11760
100. Cross SH, Charlton JA, Nan XS, Bird AP (1994) Purification of Cpg islands using a methylated DNA-binding column. Nat Genet 6:236–244
101. Huang THM, Perry MR, Laux DE (1999) Methylation profiling of CpG islands in human breast cancer cells. Hum Mol Genet 8:459–470
102. Hatada I, Kato A, Morita S, Obata Y, Nagaoka K, Sakurada A, Sato M, Horii A, Tsujimoto A, Matsubara K (2002) A microarray-based method for detecting methylated loci. J Hum Genet 47:448–451
103. Misawa A, Inoue J, Sugino Y, Hosoi H, Sugimoto T, Hosoda F, Ohki N, Imoto I, Inazawa J (2005) Methylation-associated silencing of the nuclear receptor 1I2 gene in advanced-type neuroblastomas, identified by bacterial artificial chromosome array-based methylated CpG island amplification. Cancer Res 65:10233–10242
104. Ching TT, Maunakea AK, Jun P, Hong CB, Zardo G, Pinkel D, Albertson DG, Fridlyand J, Mao JH, Shchors K, et al. (2005) Epigenome analyses using BAC microarrays identify evolutionary conservation of tissue-specific methylation of SHANK3. Nat Genet 37:645–651
105. Schumacher A, Kapranov P, Kaminsky Z, Flanagan J, Assadzadeh A, Yau P, Virtanen C, Winegarden N, Cheng J, Gingeras T, Petronis A (2006) Microarray-based DNA methylation profiling: technology and applications. Nucleic Acids Res 34:528–542
106. Yu YP, Paranjpe S, Nelson J, Finkelstein S, Ren B, Kokkinakis D, Michalopoulos G, Luo JH (2005) High throughput screening of methylation status of genes in prostate cancer using an oligonucleotide methylation array. Carcinogenesis 26:471–479
107. Balog RP, de Souza YEP, Tang HM, DeMasellis GM, Gao B, Avila A, Gaban DJ, Mittelman D, Minna JD, Luebke KJ, Garner HR (2002) Parallel assessment of CpG methylation by two-color hybridization with oligonucleotide arrays. Anal Biochem 309:301–310
108. Shi HD, Maier S, Nimmrich I, Yan PS, Caldwell CW, Olek A, Huang THM (2003) Oligonucleotide-based microarray for DNA methylation analysis: principles and applications. J Cell Biochem 88:138–143
109. Gitan RS, Shi HD, Chen CM, Yan PS, Huang THM (2002) Methylation-specific oligonucleotide microarray: a new potential for high-throughput methylation analysis. Genome Res 12:158–164
110. Adorjan P, Distler J, Lipscher E, Model F, Muller J, Pelet C, Braun A, Florl AR, Gutig D, Grabs G, et al. (2002) Tumour class prediction and discovery by microarray-based DNA methylation analysis. Nucleic Acids Res 30:e21
111. Yan PS, Shi HD, Rahmatpanah F, Hsiau THC, Hsiau AHA, Len YW, Liu JC, Huang THM (2003) Differential distribution of DNA methylation within the RASSF1A CpG island in breast cancer. Cancer Res 63:6178–6186
112. Wang Y, Zheng W, Luo J, Zhang D, Zuhong L (2006) *In situ* bisulfite modification of membrane-immobilized DNA for multiple methylation analysis. Anal Biochem 359(2):183–188
113. Weber M, Davies JJ, Wittig D, Oakeley EJ, Haase M, Lam WL, Schubeler D (2005) Chromosome-wide and promoter-specific analyses identify sites of differential DNA methylation in normal and transformed human cells. Nat Genet 37:853–862
114. Zhang XY, Yazaki J, Sundaresan A, Cokus S, Chan SWL, Chen HM, Henderson IR, Shinn P, Pellegrini M, Jacobsen SE, Ecker JR (2006) Genome-wide high-resolution mapping and functional analysis of DNA methylation in *Arabidopsis*. Cell 126:1189–1201
115. Hayashi H, Nagae G, Tsutsumi S, Kaneshiro K, Kozaki T, Kaneda A, Sugisaki H, Aburatani H (2006) High-resolution mapping of DNA methylation in human genome using oligonucleotide tiling array. Hum Genet 120(5):701–711

116. Bulyk ML (2006) DNA microarray technologies for measuring protein-DNA interactions. Curr Opin Biotechnol 17:422–430
117. Kim TH, Barrera LO, Qu C, Van Calcar S, Trinklein ND, Cooper SJ, Luna RM, Glass CK, Rosenfeld MG, Myers RM, Ren B (2005) Direct isolation and identification of promoters in the human genome. Genome Res 15:830–839
118. Cawley S, Bekiranov S, Ng HH, Kapranov P, Sekinger EA, Kampa D, Piccolboni A, Sementchenko V, Cheng J, Williams AJ, et al. (2004) Unbiased mapping of transcription factor binding sites along human chromosomes 21 and 22 points to widespread regulation of noncoding RNAs. Cell 116:499–509
119. Syvanen AC (2005) Toward genome-wide SNP genotyping. Nat Genet 37:S5–S10
120. Gunderson KL, Steemers FJ, Lee G, Mendoza LG, Chee MS (2005) A genome-wide scalable SNP genotyping assay using microarray technology. Nat Genet 37:549–554
121. Huber M, Losert D, Hiller R, Harwanegg C, Mueller MW, Schmidt WM (2001) Detection of single base alterations in genomic DNA by solid phase polymerase chain reaction on oligonucleotide microarrays. Anal Biochem 299:24–30
122. Huber M, Mundlein A, Dornstauder E, Schneeberger C, Tempfer CB, Mueller MW, Schmidt WM (2002) Accessing single nucleotide polymorphisms in genomic DNA by direct multiplex polymerase chain reaction amplification on oligonucleotide microarrays. Anal Biochem 303:25–33
123. Schaid DJ, Guenther JC, Christensen GB, Hebbring S, Rosenow C, Hilker CA, McDonnell SK, Cunningham JM, Slager SL, Blute ML, Thibodeau SN (2004) Comparison of microsatellites versus single-nucleotide polymorphisms in a genome linkage screen for prostate cancer-susceptibility loci. Am J Hum Genet 75:948–965
124. John S, Shephard N, Liu GY, Zeggini E, Cao MQ, Chen WW, Vasavda N, Mills T, Barton A, Hinks A, et al. (2004) Whole-genome scan, in a complex disease, using 11,245 single-nucleotide polymorphisms: comparison with microsatellites. Am J Hum Genet 75:54–64
125. Sigurdsson S, Nordmark G, Goring HHH, Lindroos K, Wiman AC, Sturfelt G, Jonsen A, Rantapaa-Dahlqvist S, Moller B, Kere J, et al. (2005) Polymorphisms in the tyrosine kinase 2 and interferon regulatory factor 5 genes are associated with systemic lupus erythematosus. Am J Hum Genet 76:528–537
126. Syvanen AC (2001) Accessing genetic variation: genotyping single nucleotide polymorphisms. Nat Rev Genet 2:930–942
127. Matsuzaki H, Dong SL, Loi H, Di XJ, Liu GY, Hubbell E, Law J, Berntsen T, Chadha M, Hui H, et al. (2004) Genotyping over 100,000 SNPs on a pair of oligonucleotide arrays. Nat Methods 1:109–111
128. Hirschhorn JN, Sklar P, Lindblad-Toh K, Lim YM, Ruiz-Gutierrez M, Bolk S, Langhorst B, Schaffner S, Winchester E, Lander ES (2000) SBE-TAGS: an array-based method for efficient single-nucleotide polymorphism genotyping. Proc Natl Acad Sci USA 97:12164–12169
129. Fan JB, Chen XQ, Halushka MK, Berno A, Huang XH, Ryder T, Lipshutz RJ, Lockhart DJ, Chakravarti A (2000) Parallel genotyping of human SNPs using generic high-density oligonucleotide tag arrays. Genome Res 10:853–860
130. Fan JB, Oliphant A, Shen R, Kermani BG, Garcia F, Gunderson KL, Hansen M, Steemers F, Butler SL, Deloukas P, et al. (2003) Highly parallel SNP genotyping. Cold Spring Harb Symp Quant Biol 68:69–78
131. Lander ES, Linton LM, Birren B, Nusbaum C, Zody MC, Baldwin J, Devon K, Dewar K, Doyle M, FitzHugh W, et al. (2001) Initial sequencing and analysis of the human genome. Nature 409:860–921
132. Venter JC, Adams MD, Myers EW, Li PW, Mural RJ, Sutton GG, Smith HO, Yandell M, Evans CA, Holt RA, et al. (2001) The sequence of the human genome. Science 291:1304–1351
133. Stolc V, Gauhar Z, Mason C, Halasz G, van Batenburg MF, Rifkin SA, Hua S, Herreman T, Tongprasit W, Barbano PE, et al. (2004) A gene expression map for the euchromatic genome of *Drosophila melanogaster*. Science 306:655–660

134. Saha S, Sparks AB, Rago C, Akmaev V, Wang CJ, Vogelstein B, Kinzler KW, Velculescu VE (2002) Using the transcriptome to annotate the genome. Nat Biotechnol 20:508–512
135. Carninci P, Kasukawa T, Katayama S, Gough J, Frith MC, Maeda N, Oyama R, Ravasi T, Lenhard B, Wells C, et al. (2005) The transcriptional landscape of the mammalian genome. Science 309:1559–1563
136. Jongeneel CV, Delorenzi M, Iseli C, Zhou D, Haudenschild CD, Khrebtukova I, Kuznetsov D, Stevenson BJ, Strausberg RL, Simpson AJ, Vasicek TJ (2005) An atlas of human gene expression from massively parallel signature sequencing (MPSS). Genome Res 15:1007–1014
137. Okazaki Y, Furuno M, Kasukawa T, Adachi J, Bono H, Kondo S, Nikaido I, Osato N, Saito R, Suzuki H, et al. (2002) Analysis of the mouse transcriptome based on functional annotation of 60,770 full-length cDNAs. Nature 420:563–573
138. Mattick JS, Makunin IV (2006) Non-coding RNA. Hum Mol Genet 15(1):R17–R29
139. Rinn JL, Euskirchen G, Bertone P, Martone R, Luscombe NM, Hartman S, Harrison PM, Nelson FK, Miller P, Gerstein M, et al. (2003) The transcriptional activity of human Chromosome 22. Genes Dev 17:529–540
140. Kapranov P, Drenkow J, Cheng J, Long J, Helt G, Dike S, Gingeras TR (2005) Examples of the complex architecture of the human transcriptome revealed by RACE and high-density tiling arrays. Genome Res 15:987–997
141. Reis EM, Louro R, Nakaya HI, Verjovski-Almeida S (2005) As antisense RNA gets intronic. Omics 9:2–12

INDEX

IIS enzyme 19, 125, 130, 139, 143
ACB-PCR *see* allele-specific competitive blocker-polymerase chain reaction
accuracy 23, 85, 89, 92, 143, 150, 155, 213, 217, 223, 274, 278, 296
acetic anhydride 167, 169, 174, 176, 178, 180
adapter sequence 10, 14, 16, 17, 19, 43, 53, 54, 57, 61, 74, 76, 103, 106, 107, 113-130, 132, 133, 148, 149, 191, 201, 202, 230, 247
Affymetrix 265, 271-274, 276, 284, 293, 295, 296
AFLP, *see* amplification fragment length polymorphism
Agilent 265, 268, 271-273, 276, 277, 284, 295
alkaline acetic chloride 167, 169, 174, 176, 178
alkaline potassium permanganate 167, 169, 174, 176, 178, 181
allele-specific competitive blocker-polymerase chain reaction 211, 212, 230
allele-specific PCR 211, 212, 217, 229
alternative splicing 265, 266, 279, 284-288
amplification fragment length polymorphism 125, 126, 142, 146
annealing temperature 29, 32, 36, 44, 209
antisense transcription 265, 268, 269, 288
arbitrary primed PCR 125
artificial nucleases 211, 218, 227, 228
aziridinylbenzoquinone 167, 168, 171

background 8, 12-16, 18, 21, 32, 35, 41, 53, 55-57, 60-62, 64, 70, 71, 74, 75, 77, 79, 80, 82, 92, 101, 105, 132, 133, 135, 140, 141, 146, 150, 152, 153, 156, 157, 183, 191, 193, 195, 196, 199, 201, 217, 218, 222, 223, 230, 235, 242, 245, 247-251, 258, 274
biotinylation 97, 101, 106, 107, 142, 170
biotinylation of initial RNA 101
bisulfite oligonucleotide array 266, 292

$C_o tA$ fraction 203, 241, 252, 259, 261
cap switch 53, 56, 199, 201, 204, 205
carboxylation of pyrimidine bases 167, 169
catalytic molecular beacon 85, 90, 92
cDNA library construction 1, 2, 24, 29, 61, 113, 138
cDNA microarray 265, 269, 270, 281, 298
cDNA normalization 13, 17, 33, 98-102, 104, 105, 107, 109, 120
cDNA representation 97, 99, 146
cDNA subtraction 1, 4, 7, 53, 55, 79, 80
CEL I 211, 219, 222, 256
CGH, *see* comparative genomic hybridization
chemical cleavage 23, 24, 215-217, 219
chemical modification of mispaired nucleotides 211, 215, 248
chimera 150, 187, 190, 228, 241, 242, 252, 254
chimeric clone 187, 190, 251, 253
chimeric duplex 187, 193, 196

chromatin immunoprecipitation 266, 279, 293, 294, 295
chromosome walking 29, 30, 40, 42
CHS, *see* covalently hybridized subtraction
class II restriction endonuclease 125
class IIS restriction endonuclease 19, 125, 130, 134
clone coverage 97, 100
cloning of deleted sequences 125, 126, 150, 196
cloning selection 187-189, 191
CODE, *see* cloning of deleted sequences
coincidence cloning 1-3, 20-22, 24, 29, 33, 187-193, 195, 197, 199, 202, 242, 244, 254
common sequences 8, 20, 168, 170, 171, 177, 178, 187, 189
comparative genomic hybridization 265-276, 279, 282-284
competitive genomic hybridization 1, 3
competitor DNA 187, 191, 203, 224, 247, 248, 253, 258, 259, 261
covalent modification 167-169, 171, 174-178, 180, 184
covalently hybridized subtraction 2, 18, 167, 168, 171-177, 179-184
covalently modified subtracter 167-169, 174, 176-178, 180, 181, 183
CpG island 265, 279, 280, 290-293, 295

DARFA, *see* differential analysis of restriction fragments amplification
differential display 4, 17, 18, 24, 29, 33, 34, 36-38, 40, 42, 126, 127, 140, 141
deamination of purines
DEASH, *see* DNA enrichment by allele-specific hybridization
differential analysis of restriction fragments amplification 125, 126, 142, 144
differential display 4, 17, 24, 28, 29, 33, 34, 36-38, 40, 42, 126, 127, 140, 141
differential DNA 22, 167, 169, 170, 177, 178, 212
differential transcripts 1
differential screening 42, 53-56, 61, 64, 67, 70, 71, 74, 76, 80, 82, 126, 137, 138, 140, 141

differential sequence 1, 5, 9, 10, 12, 20, 24, 57, 168, 171, 177, 178, 188, 243-245
differential subtraction chain 125-127, 146, 148-150
differentially expressed genes 1, 2, 4, 30, 36, 40, 53, 55, 56, 59-61, 74, 99, 127, 139, 147, 153, 167, 168, 177, 179, 180, 182, 183
differentially expressed mRNA 60, 61, 97, 98
differentially methylated 1, 25
differentially regulated genes 53, 54
DNA enrichment by allele-specific hybridization 125, 126, 154-156
DNA microarray 3, 87, 89, 92, 265-270, 277, 281, 289, 293-295, 297, 298
DNA probe 85, 86, 90, 92, 93, 229, 232, 245, 268, 269, 273, 274, 278, 280
DNAzyme 85, 90-92, 94, 228
driver 1, 5-10, 12, 13, 15, 16, 18-20, 39, 55-62, 64-68, 70, 76, 77, 80, 82, 105, 106, 125-131, 134-139, 142, 144-154, 156, 244-246, 257, 258
DSC, *see* differential subtraction chain
DSN, *see* duplex specific nuclease
DSNP, *see* duplex-specific nuclease preference
duplex specific nuclease 97, 98, 101, 109-116, 119-121, 211, 212, 225
duplex-specific nuclease preference 211, 212, 225, 227

EDS, *see* enzymatic degrading subtraction
efficiency of SSH 54, 59, 60
EMC, *see* enzymatic mismatch cleavage
endonuclease V 211, 221, 222, 250
enrichment 2, 7, 9, 10, 13, 18, 19, 39, 53-55, 57, 59-61, 125-129, 133-135, 139, 140, 144-150, 153-155, 169, 174, 178, 187, 191, 193, 230, 245-247
enrichment factor 187, 191, 193
enrichment value 13, 125, 134, 245, 247
enzymatic degrading subtraction 125, 126, 146, 147
enzymatic mismatch cleavage 211, 212, 221, 222
enzymatic removal 97, 101

Index

enzyme-amplified electronic transduction 211, 234, 235
epigenomic microarray 265, 266, 279
equalization 13, 54, 63
EST 1-3, 14, 17, 18, 104, 266, 277, 280, 281, 284, 285, 298, 300
evolutionary conserved sequence 2, 20, 21, 25, 29, 187, 195, 197, 202, 204, 254, 262
expression profiling 156, 265, 280, 282

false positive 12, 13, 16, 39, 53, 55, 70, 80, 82, 153, 178, 274, 278
first strand cDNA 6, 41, 45, 65, 66, 97, 104-106, 109, 112, 113
fluorescence resonance energy transfer 85, 88, 94, 225
fluorophore 85, 87, 89-92, 217, 228, 295
frequent cutter 10, 20, 21, 97, 98, 101, 107, 109, 191, 195, 197, 199, 203, 246, 248, 250, 256, 259, 260, 291
frequent transcript 1, 3
FRET, *see* fluorescence resonance energy transfer
full-length cDNA 62, 97-99, 104, 105, 107, 109, 112, 199, 205, 278

gene atlas 265, 282
gene expression 2-4, 9, 17-19, 33, 35, 70, 92, 125-127, 140, 142, 143, 146, 156, 167-169, 179, 181, 188, 197, 266, 271, 274, 278-282, 284, 288, 289, 298-300
gene-oriented array 266, 277-279
genetic differences 125, 127, 142, 156
genetic regulation 97, 98
genome complexity 1, 246, 284
genome size 1, 90, 190, 245
genomic microarray 265, 270, 279
genomic polymorphism 1, 22
genomic repeat 19, 22, 187, 191, 197, 199, 204, 241, 242, 247, 248, 254, 258
genomic repeat expression monitor 22, 187, 197, 199-202, 204-206
genomic signature tag 125, 142
genomic subtraction 1, 4, 184, 246
glycosylase mediated polymorphism detection 211, 231

GREM, *see* genomic repeat expression monitor
GST, *see* genomic signature tag

heat denaturation 53, 56, 59, 222, 232
high complexity 53, 55, 289
highly abundant cDNA 53, 56
hybridization kinetics 3, 56, 101, 109, 148, 241, 242, 244
hybridization probe 76, 81, 85, 86, 88, 139, 222
hybridization rate 5, 93, 101, 246
hybridization specificity 20, 24, 125, 241, 242, 248
hybridization temperature 234, 241, 248, 252
hybridization time 9, 53, 55, 79, 190, 191, 245, 247
hydroxyapatite column 5, 97, 102-104, 191, 247
hydroxylamine 211, 215, 249

immobilization 97, 101, 104
immobilized 3, 12, 61, 85, 87, 89, 92, 93, 100, 104, 105, 109, 189, 193, 230, 231, 269
incomplete cDNA 97, 99
intronic transcription 297-299
inverted repeats 14, 19, 31, 85, 87, 151, 194, 252, 256
in-situ synthesis 265, 268, 271, 289, 295
in vitro cloning 29, 30, 33, 41, 42, 80

LCS, *see* linker capture subtraction
level of enrichment 55, 59
LigAmp 211, 230
ligation mediated enrichment 125, 146, 153, 162, 163
linear probe 85, 86, 92, 93
linkage disequilibrium 299
linker capture subtraction 125, 126, 146-148

mathematical model 9-12, 53, 57, 59
MDR, *see* mispaired DNA rejection
melting point 241, 248
methylated DNA immunoprecipitation 292, 293, 305

methylation 21, 22, 197, 209, 266, 279, 280, 288-293
methylation site 22, 187, 197
microRNAs 266, 289, 290
microarray 1, 3, 17, 24, 36, 86, 89, 92, 94, 126-128, 155-157, 265-273, 275-281, 283-298
microarray chip 3, 265-269, 271, 272, 277, 278, 280, 281
microarray probe 265, 277, 278, 281, 286, 287
miRNAs, see microRNAs
mirror orientation selection 2, 16, 18, 53-57, 59-61, 63, 70, 71, 73, 74, 77, 80-82, 153
mirror oriented selection, see mirror orientation selection
mismatch 24, 85, 89, 92, 150, 203, 211, 212, 214, 215, 219, 221-224, 229-231, 242, 249, 252-255, 261, 266, 274, 290
mismatch discrimination 85, 89, 92, 242
mismatch-sensitive nuclease 196, 202, 203, 221, 223, 241, 251-254, 261
mismatched target 85, 86, 277
mispaired base 85, 250
mispaired DNA rejection 21, 150, 249-251, 253-255
mispaired nucleotides 1, 23, 24, 211, 214, 215, 217, 218, 230, 231, 244, 249, 255
modified purine 167
modified pyrimidine 167, 174
molecular beacon 85-94
MOS, see mirror orientation selection
mRNA study 97, 99
multiplex 42, 85, 89, 226, 296
multiplex PCR 24, 29, 30, 34, 42, 43
Mung bean nuclease 10, 19, 12, 15, 147-150, 203, 224, 241, 252-255, 261
mutant allele 1, 234
mutant strand 1, 24, 215, 233
mutation 1, 22-25, 92, 93, 154, 211-235, 249, 254, 255
mutation detection 23, 24, 93, 211, 213-217, 220, 222-228, 231, 232, 234, 249, 255
MutS 211, 230, 231

nanoparticle 85, 232, 233
negative subtraction chain 125, 126, 146, 150
NGSCC, see non-methylated genomic sites coincidence cloning
NimbleGen 265, 268, 271, 272, 275, 284, 295
noncoding transcripts 266, 288, 298
non-methylated genomic sites coincidence cloning 2, 21, 22, 187, 197, 199, 208, 209
nonrepetitive DNA 187, 190, 247
normalization 1-3, 13, 14, 16, 17, 24, 29, 33, 39, 53-56, 59, 60, 97-113, 116, 119-122, 151
normalized cDNA libraries 29, 39, 97, 98, 100-102, 104, 105, 107, 111
Northern blot 1, 38, 54, 59, 61, 70, 81, 100, 122, 183, 289, 298
NSC, see negative subtraction chain
nuclease-based mutation scanning 212, 218, 225
nucleic acids hybridization 1-28, 86, 215, 241-264

oligoarray 265, 271, 284, 293, 299, 300
oligonucleotide probe 87, 265, 267, 271, 277, 278, 284, 285, 287, 289, 290, 292, 293, 296, 297
ordered differential display 24, 29, 33, 37
osmium tetroxide 211, 215, 249

pan handle 29, 31, 57, 85-88, 193, 194
PCR-only-based approaches 187, 188, 190, 193
PCR selection effect 10, 19, 241, 243
PCR suppression 14-17, 20, 24, 29-31, 34, 37, 42-44, 51, 79, 86, 101, 107-109, 111, 187, 188, 190, 195, 202, 205, 250, 251, 255
PEER, see primer extension enrichment reaction
perfectly matched hybrids 241, 242, 249
PERT, see phenol emulsion reassociation te
phenol 5, 9, 46, 66, 72, 77, 125, 154, 170, 203, 208, 245, 258-261
phenol emulsion reassociation technique 2, 9, 146, 241, 245

Index

plasmid vector 5, 10, 62, 74, 97, 193
phagemid vector 97, 102, 103
physical isolation of imperfectly matched DNA 24, 211, 215, 233
physical separation 36, 39, 53, 55, 57, 97, 101, 102, 105, 107, 109, 145, 187-189, 193
poly(A)+ RNA 5, 35, 36, 55, 56, 61, 62, 65, 77, 82, 97, 99, 102, 103, 105-107, 152
polymorphism recovery 1, 3, 231
poorly transcribed 12, 97, 100
potassium permanganate 167, 169, 174, 176, 178, 181, 211, 215, 249
pre-mRNA 265, 278, 284, 286
preparation of full-size cDNA 29, 34, 40
primary structure 12, 30, 125, 127
primer annealing site 12, 30, 125, 127
primer concentration 29, 32, 132
primer extension enrichment reaction 2, 18, 19, 125, 126, 128-135, 156
promoter-active repeats 22, 25, 187, 199
protocol 9, 30, 42-45, 53-55, 62-65, 70, 71, 74, 82, 97, 98, 102, 104, 105, 107, 112, 113, 126, 128, 131, 133-136, 138, 150, 153, 155, 156, 168, 172, 176, 180, 188, 191, 202, 205, 207, 209, 217, 219, 221, 222, 229, 231, 242, 247, 255, 270, 273, 278, 281, 296
PS, see also PCR suppression 29, 31, 34, 36, 38, 41-43, 45, 48, 188, 193, 195

QCM 211, 212, 231, 234
quencher 85, 87, 89-92, 227, 228

RACE 1, 2, 24, 29, 40, 41, 54, 56, 62, 65, 266, 298
random clones 54, 61
rapid amplification of cDNA ends 1, 2, 24, 29, 40, 41, 54, 62, 266
rare transcript 1, 4, 9, 13, 16, 17, 39, 59, 100, 101, 103, 106, 108, 110, 141, 193
real-time 85, 87, 89, 92, 93, 229, 230
reannealing 16, 53, 208, 222, 245
removal of the adapter sequences 54, 61

repetitive element 22, 38, 82, 187, 188, 190, 196, 197, 199-202, 204, 206, 241, 244, 247, 253, 289
repetitive sequence 9, 36, 156, 187, 191, 243, 247, 295
representational differences analysis 1, 2, 10, 126, 127, 143, 145, 169, 246
resolvase-like endonucleases 211, 218
restriction endonuclease 5, 10, 13, 15, 19, 20, 32, 40, 44, 45, 53, 59, 103, 109, 125, 130, 136, 14-150, 154, 177, 189, 191, 193, 197, 199, 201, 203, 205, 218, 224-226, 228, 245-247, 259, 260, 291
restriction fragment length polymorphisms recovery 1, 2, 12, 126, 142
RDA, see representational differences analysis
RIDGES 22, 187, 197
RNase cleavage of mismatched nucleotides 211, 224

SABE, see serial analysis of binding elements
SABRE 125-127, 146, 153, 154
SAGE, see serial analysis of gene expression
Sanger sequencing 1, 23, 213
saturating hybridization 97, 100
SBE, see single-base extension
scorpion 85, 89, 90
search for promoter sites 29, 30, 40
second strand cDNA 45, 65, 66, 97, 102, 104, 106, 109
secondary structure 85, 86, 101, 109, 111, 218, 278
selective amplification 29, 35, 36, 56, 98, 101, 107, 108, 126, 127, 153, 154, 171, 177, 181, 199-201, 205-207, 255, 256
selective PCR suppression 20, 29, 30, 43, 44, 107, 190
selectively primed adaptive driver RDA 125, 126, 146, 147
sensitivity 17, 36, 85, 89, 92, 94, 128, 156, 157, 171, 177, 216, 217, 224, 229, 234, 268, 273, 276, 278, 283, 289

sequencing-by-hybridization 125, 128
serial analysis of binding elements 125, 126, 142
serial analysis of gene expression 2, 18, 22, 125, 126, 142, 143, 188, 197, 266, 280, 298, 300
single-base extension 211, 225, 226, 266, 296
single nucleotide polymorphism 1-3, 37, 89, 125-127, 135, 211, 212, 266, 271, 297
single-strand conformational polymorphism 211, 212, 234
single-stranded DNA specific nucleases 218, 224
size of cDNA fragments 54, 62
small amount of biological material 29, 35
SNP, see single nucleotide polymorphism
sodium cyanide/sulfuric acid mixture 167, 169, 174, 178
spot hybridization 125, 134
spotted DNA microarray 265, 266, 268, 270
SSCP, see single-strand conformational polymorphism
SSH, see suppression subtractive hybridization
stem loop 24, 31, 32, 57, 85-87, 89, 92, 107, 193-195, 252, 256, 257
subcloning 1, 4, 12, 62, 74
subtracter 19, 167-172, 174-181, 183, 184
subtractive cloning 125, 126, 139, 140
subtractive hybridization 1-4, 8, 13, 15, 18, 33, 39, 53-56, 60, 66, 68, 70, 126, 135, 137, 139, 140, 142, 143, 146, 150, 155, 167-185, 188, 212, 241, 242, 256, 258
suppression adapter 14, 16, 20, 22, 29, 32, 34, 35, 40, 41, 43-45, 107, 108, 195, 197, 199, 201-205, 207, 208, 250, 256, 258-261
suppression of the PCR 29
suppression sequence 29, 32, 33

suppression subtractive hybridization 1, 2, 13-17, 19, 29, 39, 53-57, 59-62, 70, 71, 77, 80, 82, 111, 126-128, 135, 150-153, 155-157, 241, 242
Surveyor 150, 203, 211, 219, 221, 224, 249, 250, 252-255, 261

T4 endonuclease VII 211, 218, 221, 250
T7 endonuclease I 211, 218, 219, 221, 249
tagging RNA 5′ ends 53, 56
target sequence 23, 33, 53, 59, 90, 132, 191, 196, 213, 230, 257, 277
targeted differential display 29
targeted genomic difference analysis 241-243, 255
technical comments 53, 54, 59
tester 1, 5, 39, 55, 58, 127-131, 134, 136, 137, 139, 142, 144-154, 167-172, 174-181, 183-185
TGDA, see targeted genomic difference analysis
thermodynamic limitations 85, 86
thermodynamically stable duplexes 241, 244
TILLING 1, 241, 249
tilling-array 265
tracer 1, 5-10, 12, 13, 15, 16, 18-20, 39, 46, 55-57, 59-62, 64-70, 76, 77, 80, 82, 138, 139, 244-246, 257, 258
transcript concentration 97, 99, 100
transcript variants 265, 273, 284
transcriptome annotation 265, 266, 285
true genomic sequence representation 187, 191, 247

UDG, see uracil-DNA glycosylase
unmethylated CpG 21, 22, 187, 188, 197, 202, 207
uracil-DNA glycosylase 126, 146, 150, 151, 167, 168, 170, 233

whole-genome oligonucleotide array 265, 288
wild-type allele 1, 229, 232
wild-type strand 1, 24, 215, 225, 233